A Shot in the Dark

A History of the U.S. Army Asymmetric Warfare Group

Paul Jeffrey Cook

Series in American History

VERNON PRESS

www.vernonpress.com

In the Americas:
Vernon Press
1000 N West Street, Suite 1200
Wilmington, Delaware, 19801
United States

In the rest of the world:
Vernon Press
C/Sancti Espiritu 17,
Malaga, 29006
Spain

Series in American History

Library of Congress Control Number: 2023950112

ISBN: 978-1-64889-816-7

Cover design by Vernon Press. Image inspired by the Asymmetric Warfare Group's shoulder sleeve insignia.

Table of Contents

List of Figures

Acknowledgments

I am particularly indebted to a large number of retired and serving soldiers. These include Col. Bob Shaw, Col. Jim Mis, Col. Pat Mahaney, Lt. Col Scott Akerley, Lt. Col. Tom Beyerl, Maj. Vince Gothard, Command Sgt. Maj. Alfred G. Birch, Sgt. Maj. Eric Zipperer, Sgt. Maj. Jose Gordon, and especially Lt. Col. Mike Bowman of the Asymmetric Warfare Group. Mike began his service with the AWG as a Department of the Army civilian and is a living encyclopedia of the unit. I owe a special debt of gratitude to Gen. Dennis Reimer, Gen. Peter Schoomaker, Gen. George W. Casey, Jr., Gen. James D. Thurman, Gen. Richard Cody, Gen. Scott Wallace, Lt. Gen. Daniel P. Bolger, Lt. Gen. James J. Lovelace, Lt. Gen. Michael Oates, Maj. Gen. Christopher Hughes, Maj. Gen. James Schwitters, and Brig. Gen. Donald Schenk. Friends and colleagues Col. David Clonts, Col. Frank Hall, and Capt. Jeff Ryals provided me what I could not provide myself: the insights of officers who accepted the risks of serving at battalion/squadron or brigade level in Iraq at a time when the place was *really* dangerous. I cannot thank enough Lt. Col. Leah Reid of the U.S. Army General Management Office, Ms. Karen Lewis and retired Chief Warrant Officer and Department of the Army civilian Del Stewart of the U.S. Army Training and Doctrine Command Historian's Office, and Mr. Scott Gower at the National Defense University archives at Fort Lesley J. McNair.

I also wish to thank Dr. Greg Urwin, Dr. Jay Lockenour, and Dr. Alan McPherson for serving as my dissertation committee with a special debt to Greg who concurrently served as my Ph.D. advisor. All three made my time at Temple University rewarding. Rounding out my committee was Dr. Nick Schlosser of the Center of Military History. For Nick, this was an additional duty, and one I appreciate his taking on. Finally, I want to thank Vangie Campbell from Temple's history department. She was instrumental in helping me navigate the transition from soldier to student by solving the issues that might otherwise consumed so much time.

Glossary

Official definitions for many terms are ponderous and often generate as much confusion as they resolve. The definitions provided for military terms are simplified versions of official terms.

Additional Skill Identifier (ASI): A digraph awarded to soldiers after completing specific courses that allow the Army Human Resources Command (HRC) to assign soldiers to designated positions.

Air Assault: A type of operation in which assault troops are flown onto an objective area by helicopter and then either land on or rappel onto the objective.

Airborne: A type of unit trained to conduct parachute assaults onto an objective in enemy territory.

Aircraft Survivability Equipment (ASE): Systems carried aboard aircraft to limit the effectiveness of air defense weapons. It includes electronic countermeasures, warning systems, and infra-red jammers.

Area of Interest (AI): The location of all enemy activities that might affect the friendly force.

Area of Operations (AO): The area in which a commander has the authority and responsibility for military operations.

Army Acquisition Executive (AAE): The AAE's primary function is developing, acquiring, fielding, and sustaining equipment and services to meet Army needs.

Army Asymmetric Warfare Office (AAWO): An agency within the ArmyG-3/5/7 that both formulated Army policy and supervised the Asymmetric Warfare Group.

Army Campaign Plan (ACP): The Army's plan for realizing its overall operational and modernization objectives.

Army Combat Uniform (ACU): An Army camouflage uniform introduced in 2005 to replace the multiple camouflage uniforms the Army then used. Its defining characteristic was its pattern, termed the Universal Camouflage Pattern (UCP).

Army Pre-Positioned Stocks (APS): A fleet of tanks, infantry fighting vehicles, self-propelled artillery, engineer, and support vehicles positioned either in a foreign country or on vessels to facilitate the rapid deployment of units in a contingency. The APS model means soldiers fall in on such equipment rather than having to deploy it from the United States. This significantly reduces the time required to bring a unit to combat readiness during a deployment.

Al Qaida in Iraq (AQI): That branch of the fundamentalist Islamic terrorist organization operating in Iraq. Virtually all Sunni foreign fighters in Iraq had an affiliation with the group.

Army Regulation (AR): A regulation published by Headquarters, Department of the Army and applying to the Army as a whole.

Army of the Republic of Vietnam (ARVN): The regular Army of the Republic of Vietnam.

Army Requirements and Resourcing Board (A2RB): A weekly forum replacing the Army Strategic Planning Board (ASPB).

Army Strategic Planning Board (ASPB): A weekly forum conducted at Headquarters, Department of the Army for adjusting the Army's resource priorities.

Army Test and Evaluation Command (ATEC): The command responsible for testing and evaluating equipment the Army seeks to procure.

Assistant Secretary of the Army for Manpower and Reserve Affairs (ASA M&RA): The senior political appointee responsible for the Army's personnel and reserve policies.

Asymmetric Warfare Regiment (AWR): The original concept for what became the Asymmetric Warfare Group.

Base Realignment and Closure (BRAC): A Congressionally executed process of identifying installations either no longer needed or suitable for receiving additional organizations.

Basic Combat Training (BCT): The initial period of military training all enlisted soldiers complete upon entering the Army.

Battle Dress Uniform (BDU): The Army's standard field uniform from 1982 through 2005.

Battlefield Framework: A means of describing and visualizing operations. It includes such things as the area of operations, main and supporting efforts, and decisive, shaping and sustaining operations.

"Big 5": The Army's five major weapon systems fielded in the 1980s. These are the M1 Abrams main battle tank, the M2/M3 Bradley infantry/cavalry fighting vehicle, the AH-64 Apache attack helicopter, the UH-60 Blackhawk utility helicopter, and the Patriot air defense missile system. All these systems remain in use and provide the backbone of the Army's maneuver, aviation, and air defense forces.

***Boyevaya Mashina Pekhoty* (BMP):** Soviet-designed infantry fighting vehicle produced in several variants and used primarily by former Soviet republics and client states.

Center of Gravity (COG): The primary source of a force's power. Successfully attacking the COG should result in a force's collapse.

Combined Forces Land Component Command (CFLCC): Headquarters of U.S. Army Central and Third U.S. Army controlling ground operations in the Central Command's area of responsibility. It includes the Army, Marine Corps, and foreign units.

Combined Forces Land Component Command, Forward (CFLCC-F). The forward headquarters of U.S. Army Central controlling ground operations in Afghanistan.

Chemical, Biological, Radiological, Nuclear, Explosives (CBRNE): Abbreviation for chemical weapons, biological weapons, radiological weapons, nuclear weapons, and explosives of mass effect. CBRNE weapons may also be Improvised Explosive Devices (IEDs).

Combined Joint Special Operations Task Force (CJSOTF): A command consisting of Special Operations Forces from the United States and its foreign partners.

Coalition Provisional Authority (CPA): The civilian body exercising government authority over Iraq from the disestablishment of the Office of Reconstruction and Humanitarian Assistance on May 16, 2003 through June 28, 2004.

Comanche: Also known as the RAH-66, the Comanche program sought to provide the Army with a replacement for its OH-58 reconnaissance helicopter fleet.

Combat Arms: Those arms performing direct combat missions. In the United States Army, these are Air Defense Artillery, Armor, Aviation, Cavalry, Field Artillery, and Special Forces.

Combat Support Arms: Those arms that directly support the combat arms. In the United States Army, these are the Chemical Corps, Civil Affairs, the Corps of Engineers, Cyber, Military Intelligence Corps, Military Police Corps, Psychological Operations, and the Signal Corps.

Combat Service Support Branches: Those branches with the primary role of providing logistical or administrative support to the Army. In the United States Army, these include the Adjutant General's Corps, the Finance Corps, Logistics, the Army Medical Department (Dental, Medical, Medical Specialist, Medical Service, Nurse and Veterinary Corps), Ordnance Corps, the Transportation Corps, and the Quartermaster Corps.

Combatant Command: A joint command with a broad and continuing mission. They may be geographic (U.S. Africa Command, Central Command, European Command, Northern Command, Indo-Pacific Command, Southern

Command) or functional (Special Operations Command, Space Command, Strategic Command).

Combatting Terrorism Technical Support Office (CTTSO): A Department of Defense office conducting research and development on issues related to defeating terrorist threats.

Combined Arms Center (CAC): Located at Fort Leavenworth, Kansas, and subordinate to the Training and Doctrine Command (TRADOC), CAC's duties developing doctrine and supervising the Army's schools.

Combined Force Land Component Commander (CFLCC): The officer commanding all ground forces in a combat theater. These include the Army, Marine Corps, and coalition forces.

Combined Operations: Operations involving the military forces of the U.S. and one or more foreign countries together.

Command Sergeant Major (CSM): The ranking non-commissioned officer in an Army unit of battalion or larger size.

Commercial Off-the-Shelf Solutions (COTS): Materiel available commercially and procured outside the normal acquisition process. COTS provides the ability to reduce the time between identifying a requirement and obtaining the equipment.

Component Command: A joint or combined command responsible for controlling functions within a combatant command. The Land Component Commander has responsibility for controlling all actions within his area of operations. The Air Component Commander has similar responsibilities for all fixed-wing operations within the combatant command. The Maritime Component Commander controls all seaborne actions.

Courses of Action (COA): A concept or plan to accomplish a mission or task.

Crusader: An advanced long-range field artillery system planned as a replacement to the long-serving M109 155mm howitzer system.

Democratic Republic of Vietnam (DRV): The communist Vietnamese state created north of the 17th parallel following the French withdrawal from Indochina in 1955.

Deputy Chief of Staff for Personnel (G-1): The Army staff's principal human resources officer and staff section.

Direct Fire: Weapon's fire aimed directly at the target. This means the firer is observing the target and employing a weapon with a very flat trajectory. Rifles, tanks, and machine guns are most often employed in such a role.

Doctrine: Fundamental principles conducting operations and guiding the actions of operating forces.

Doctrine, Organization, Training, Materiel, Leader Development, and Facilities (DOTMLPF): Most often applied to the acquisition process and intended to ensure these elements are properly considered in developing new organizations and equipment.

Explosively Formed Penetrator (EFP): A type of improvised explosive device employing a shaped charge to defeat armored vehicles.

Explosive Ordnance Disposal (EOD): The term applied to the controlled destruction of munitions, including IEDs.

Fast Obscurant Grenade (FOG): A hand grenade rapidly emitting a smoke screen to protect against snipers.

Fuerzas Armadas Revolucionarias de Colombia **(FARC):** Translated as Revolutionary Armed Forces of Colombia, a Colombian communist-inspired insurgent movement heavily involved in the production and distribution of cocaine.

Full Operational Capability (FOC): The term applied when a unit or capability is fully capable of conducting all its assigned functions. Units normally achieve initial operational capability (IOC) and then undergo a period of training to integrate personnel and equipment and then demonstrate specific levels of collective proficiency.

Fragmentary Order (FRAGO): An order communicating a change from a previously issued order or plan. An operations order consists of five paragraphs and may contain multiple annexes. A FRAGO consists only of changes to those paragraphs or annexes.

Former Regime Element (FRE): Iraqi insurgents who previously served in the military, security services, or Ba'ath Party of Saddam Hussein al Tikriti.

G-3/5/7: The Headquarters Department of the Army staff section is responsible for operations, plans, and training. This office originally had responsibility for the AAWO, AWG and REF.

General Order (GO): An order published by a command applying to all its subordinate elements. General orders published by the Department of the Army/War Department apply to all Army units.

Global War on Terrorism (GWOT): The name applied to contingency operations against terrorist groups since September 11, 2001. This includes operations in Afghanistan, Iraq, the Horn of Africa, the Philippines, and various parts of Africa.

Ground Surveillance Radar (GSR): A portable radar that detects the movement of personnel or vehicles.

Hazardous Duty Incentive Pay (HDIP): Additional pay received by soldiers performing hazardous duty.

High Explosive Anti-Tank (HEAT): A type of munition employing a shaped charge to defeat armored vehicles such as tanks. The munition creates a high-velocity jet that penetrates the armor. It is one of the two most widely used forms of anti-tank ammunition.

High Mobility Multipurpose Wheeled Vehicle (HMMWV): An un-armored, all-wheel-drive vehicle that replaced the Jeep in military service in the 1980s. Commonly used as patrol and supply vehicles in Afghanistan and Iraq, they proved vulnerable to IED strikes. The Army used both MRAPs and armored versions of the HMMWV in response to the IED threat.

Honest John: An unguided rocket employed by the Army from the 1950s through the 1970s. The Honest John could carry conventional, chemical, or nuclear warheads.

Humanitarian Assistance (HA): The term applied to military efforts to provide civilian populations assistance with food and medical care.

Human Resources Command (HRC): The Army's personnel command whose responsibilities include assigning/re-assigning soldiers to units.

Improvised Explosive Device (IED): Explosive devices deployed in unconventional/non-military ways. They may be military explosives repurposed or constructed of locally produced components.

Improvised Explosive Device Defeat Task Force (IED Defeat Task Force): An Army organization created to centralize counter-IED efforts.

Indirect Fire: The most common means of firing artillery, indirect fire means the weapon fires its rounds in an arced trajectory with the round detonating or coming down on top of the target rather than through the target as in direct fire.

Individual Body Armor (IBA): Protective equipment worn by soldiers to protect the head, neck, torso, and groin. Attachments extend protection to the upper arms.

Infantry Fighting Vehicle (IFV): An armored fighting vehicle normally armed with both a cannon and anti-tank missiles and capable of carrying infantry. Examples include the American M2 Bradley and the Soviet BMP.

Information Operations (IO): Operations taken to affect adversary information and information systems while defending one's own information and information systems. Included in IO are psychological operations and propaganda.

Initial Operational Capability (IOC): The term applied to a new unit determining when it can first conduct operations in a limited form.

Institute of Heraldry (TIOH): The Army agency responsible for the design and approval of Army insignia including those associated with specific units.

***Insurrectos*:** Filipino insurgents who battled the U.S. Army during the Philippine War.

Intelligence Preparation of the Battlefield (IPB): An analytical model used by intelligence personnel and commanders to both visualize and explain the interaction between the terrain, weather, and forces.

Iraqi Interim Government (IIG): The government of Iraq exercising sovereign authority following the termination of the Coalition Provisional Authority (CPA) on June 28, 2004.

Iraqi Security Forces (ISF): The post-invasion military and police services of Iraq.

Joint Improvised Explosive Defeat Organization (JIEDDO): An agency originally created by the Army to counter Improvised Explosive Devices as the IED TF. The name change reflected control passing from the Army to the Department of Defense.

Joint Improvised Explosive Device Defeat Organization (JIEDDO), Army Asymmetric Warfare Office (AAWO), Asymmetric Warfare Group (AWG) and Rapid Equipping Force (REF) Review (JAAR): The JAAR was a recurring conference created by Gen. Richard Cody, Vice Chief of Staff, to review the activities of the three participating organizations.

Joint Operations: Operations involving two or more U.S. military services together.

Joint Special Operations Areas (JSOA): A geographic area within a combatant command's area of operations in which the Special Operations Commander maintains responsibility for operations.

Joint Staff Anti-Terrorism Directorate (J34): The staff office of the Joint Chiefs of Staff responsible for anti-terrorism planning and policy, targeting support, and training.

Killed-in-Action (KIA): A soldier killed because of hostile action.

Master Breacher: An individual specially trained to conduct the forced entry of buildings or rooms using explosives or other techniques.

Mere Gook Rule (MGR): An informal rule of the Vietnam War identifying the local population as inferior. Misconduct against civilians frequently reflected individuals applying this rule.

Military Assistance Advisory Group (MAAG): A U.S. military command established in a country to assist in the delivery of materiel and training to the host nation's military services.

Military Assistance Advisory Group-Vietnam (MAAGV): The American military mission in Vietnam. Originally created in September 1950 as Military Assistance Advisory Command-Indochina.

Military Assistance Command Vietnam (MACV): Command created in 1962 as the replacement for MAAGV.

Military Information Division (MID): The intelligence organization created in the Philippines in 1903 as a more capable replacement for the Bureau of Insurgent Records in Manila. The MID integrated information obtained from provost courts, combat units, and staff intelligence officers across the archipelago.

Mine Resistant Ambush Protected Vehicle (MRAP): A family of wheeled vehicles specifically designed to reduce the effects of mines. The Army fielded several versions of these vehicles to improve the protection provided to soldiers traveling along roads and trails where the IED threat was greatest. These largely replaced the HMMWV in Afghanistan and Iraq.

Mission Essential Task List (METL): A document created by a unit's commanding officer identifying the specific tasks in which a unit must establish and maintain proficiency. Commanders use the METL to plan unit training.

National Capital Region (NCR): The geographic area of Washington, D.C.; northern Virginia, and eastern Maryland.

Northern Alliance (NA): A group of anti-Taliban Afghan resistance groups. American forces worked with the Northern Alliance to topple Afghanistan's Taliban government in 2001.

Objective Force: First appearing in the late 1990s, the Objective Force was the envisioned Army of the year 2020 with equipment reflecting the technological promise of the Revolution in Military Affairs (RMA).

Officer Candidate School (OCS): A course that produced Second Lieutenants in most of the Army's branches. Conducted at Forts Belvoir, Benning, Eustis, Gordon, Knox, and Sill, the program provided most of the junior officers who served in World War II and Vietnam. Currently, OCS resides at Fort Moore.

Office of Reconstruction and Humanitarian Assistance (ORHA): The civilian body exercising authority over Iraq from April 21, 2003, through the creation of the Coalition Provisional Authority on May 16, 2003.

Overseas Contingency Operations (OCO): Military operations conducted in response to unforecast events. Additional funding allocated by the Department of Defense covers the costs of designated operations. These included ENDURING FREEDOM and IRAQI FREEDOM.

Operations and Maintenance-Army (OMA): A designated funding line within the Department of the Army budget. It essentially covers the cost of operating and maintaining equipment to maintain a required state of readiness.

OPERATION DESERT FOX: A four-day-long air campaign conducted against Iraq in December 1998.

OPERATION DESERT THUNDER: A show-of-force conducted against Iraq in 1998 in response to threat against coalition air activity by Saddam Hussein.

OPERATION INTRINSIC ACTION: A series of deployments of Army units from the United States to Kuwait beginning in 1992. These deployments saw units deploying from various installations and then drawing APS equipment stocks.

Operational Advisor (OA): The principal duty position within the AWG's operational squadrons. OA's advise commanders and units on both training and combat operations and also provide commanders with feedback on unit strengths, weaknesses, and ways to enhance performance.

Operational Detachment Alpha (ODA): The standard team employed by Special Forces units. An ODA consists of one commissioned officer, one warrant officer, and ten non-commissioned officers and includes intelligence, weapons, communications, engineering, and medical specialists.

Operational and Organizational Concept (O&O): The document that provided the initial blueprint for the AWG's mission, activities, and structure.

Operations Plan (OPLAN): A plan prepared by a combatant command to guide operations in a contingency. Such plans provide the basis for operations.

Operational Tempo (OPTEMPO): The pace of battlefield operations. A high OPTEMPO can magnify the effects of fires and maneuver by reducing the time available to an enemy to respond.

People's Army of Vietnam (PAVN): The Army of the Democratic Republic of Vietnam.

Pentomic: The name applied to a major Army reorganization of the 1950s that replaced large regiments with smaller battle groups and assigned nuclear capabilities to armored and infantry divisions.

Philippine Constabulary (PC): A paramilitary law enforcement and security organization created under Army supervision.

Philippine Scouts (PS): A unit of the Army consisting of Filipino soldiers and American officers.

Program Budget Decision (PBD): A means of reallocating funds after the Secretary of Defense approved the POM as part of the Department of Defense's budget review process.

Program Evaluation Group (PEG): A group of senior officers considering specific elements of the POM.

Program Executive Officer (PEO): The military officer or senior government civilian responsible for the cost, schedule, and performance of an acquisition program.

Program Objective Memorandum (POM): A service's recommendations to the Office of the Secretary of Defense on how it will spend funds to meet mission requirements. The POM essentially spends the service's budget by aligning funds with programs or requirements. It is the most powerful arrow in the Defense Department's bureaucratic quiver.

Program Manager (PM): A designated individual with responsibility for and authority to accomplish program objectives for development, production, and sustainment to meet operational needs.

Quds Force: A part of Iran's Revolutionary Guards responsible for international terrorism in pursuit of Iranian political objectives.

Rapid Equipping Force (REF): An organization created by the Army to expedite the fielding of equipment to soldiers in combat theaters.

Remotely Emplaced Sensors (REMS): Unattended ground sensors that would transmit information when triggered. Such sensors could be seismic (vibration), magnetic or acoustic. Deployed along boundaries and suspected infiltration routes, these sensors allowed a receiving station to identify when personnel or vehicles transited a given area.

Republic of Vietnam (RVN): The non-communist Vietnamese state created south of the 17th parallel in 1955 following the French withdrawal from Indochina.

Request for Forces (RFF): A request made by a military commander to the Department of Defense for specific forces to support operations. An RFF includes an explanation as to why a commander requires either a specific unit or capability.

Revolution in Military Affairs (RMA): A perceived major change in the nature of warfare brought about by the innovative application of new technologies, doctrine, and operational and organizational concepts.

Roundout Brigade: An Army National Guard or Army Reserve combat brigade intended to bring a Regular Army division to full combat strength. The roundout program allowed the Army to reduce the strength of its U.S.-based divisions and use the personnel savings to maintain a larger number of divisions than would otherwise be possible.

Salafist: The term applied to a branch of Sunni Islam that emphasizes traditional beliefs and also seeks to remove Western influence from Islamic societies.

School of Advanced Military Studies (SAMS): Conducted at Fort Leavenworth, KS, SAMS is a graduate course for field grade officers. Graduates receive a master's in theater operations and an initial assignment to division or corps staff as plan officers.

Service Component Command: The Army, Air Force, Marine Corps or Navy headquarters responsible for service-specific functions within a combatant command. These include administration and supply of units.

Separate Brigade: An armored or infantry brigade capable of fighting as a separate formation. Unlike an organic divisional brigade, the separate brigade has its own support and service units and thus is not reliant on a division's support structure for sustained operations.

Side Looking Airborne Radar (SLAR): An intelligence collection system that creates a radar map of both the terrain and moving objects perpendicular to the aircraft's flight path. SLAR allowed target detection at extended ranges.

Special Operations Aviation Regiment (SOAR): A specialized Army aviation unit that supports Special Operations Forces. The unit specializes in night-flying and operates a variety of highly modified versions of the UH-60 Blackhawk, CH-47 Chinook, and AH-6 Cayuse helicopter.

Special Operations Forces (SOF): Specially designated, organized, trained, and equipped forces manned with selected personnel. Such units employ unconventional tactics, techniques, and modes of employment. They include the Special Forces, the 160th SOAR, the 75th Ranger Regiment and both Psychological Operations and Civil Affairs units.

Soldier Load Program (SLP): A study conducted by the AWG focusing on the items and overall weight soldiers carried on operations.

Special Duty Assignment Pay (SDAP): An additional monthly pay received by soldiers occupying designated positions.

Special Mission Units (SMU): Special operations units organized for highly sensitive missions.

Special Operations/Special Operations Forces (SOF): Operations requiring unconventional forces/forces trained and organized for unconventional operations including counter-terrorism, strategic reconnaissance, or raids against strategic targets.

Special Operations Command-Central (SOCCENT): The Special Operations component of the U.S. Central Command, responsible for the employment of Special Operations Forces in the Middle East and Southwest Asia.

Special Operations Command-South (SOCSOUTH): The Special Operations component of the U.S. Southern Command, responsible for the employment of Special Operations Forces in Latin America and the Caribbean.

Special Forces Operational Detachment-Delta (SFOD-D): A highly specialized team of Special Force soldiers specializing in high-risk missions against strategic targets. Commonly referred to as "Delta," or "the Delta Force."

Starlight Scope: The AN/PVS-2 image intensification device. This device greatly improved a soldier's ability to see targets at night by amplifying available light.

Stryker: A family of lightly armored wheeled vehicles and the units that fight from them.

Suppression of Enemy Air Defense (SEAD): Actions taken to suppress air defense weapons. They include artillery fire, aerial bombardment, or the disruption of radars or communications.

Table of Distribution and Allowances (TDA): The document specifying the organization and equipment for units or elements that do not routinely deploy.

Table of Organization and Equipment (TO&E): The document specifying the organization and equipment for deployable units.

Tactical Air Control Party (TACP): An Air Force team attached to ground forces for the purpose of controlling close air support by communicating targets and tactical information to attacking aircraft.

Tactics, Techniques and Procedures (TTP): Methods for accomplishing a given mission or task.

Tasking Order (TASKORD): A formal order from the Department of the Army directly assigning a specific action to an organization.

Time-Phased Force Deployment List (TPFDL): Produced by the Joint Operations Planning and Execution System (JOPES), the TPFDL is a document describing the details of a unit's deployment. It includes dates, ports of embarkation/debarkation, modes of travel, and associated details.

TRADOC Systems Managers (TSM): Individuals responsible for managing DOTMLPF elements associated with specific equipment items.

Universal Camouflage Pattern (UCP): A digital camouflage pattern consisting of shades of gray, olive and tan intended to provide effective camouflage in all environments.

U.S. Army Central Command (ARCENT): The highest-level Army headquarters in the Central Command area of operations.

U.S. Army Forces Command (FORSCOM): The Army command responsible for the operational readiness of deployable Army forces within the United States. FORSCOM provides the corps headquarters, divisions, brigades or non SOF regiments the Army assigns to a combatant command.

U.S. Army Futures Command (AFC): A major command established in 2018 focusing on modernization. With its establishment, AFC assumed some functions of the existing Training and Doctrine Command. With the AWG's deactivation, it also assumed responsibilities for portions of the AWG mission.

U.S. Army MaterielCommand (AMC): The Army command responsible for acquiring and sustaining equipment, AMC is the Army's largest logistical organization. It is also responsible for operating all Army installations.

U.S. Army Special Operations Command (ARSOC): The command to which the Army administratively assigned all Special Operations Forces. From ARSOC, SOF units operate as part of a joint special operations command.

U.S. Army Training and Doctrine Command (TRADOC): The Army command responsible for developing and delivering training and doctrine. The Army's second-largest command, it operates the Army University, all the Army's service schools and training centers.

U.S. Joint Forces Command (USJFCOM): A combatant command that also maintained responsibility for developing joint doctrine, training, and interoperability.

U.S. Volunteer: An officer or unit serving for a designated, short-term period of service during wartime. U.S. Volunteer units did not count against the congressionally authorized strength of the Regular Army.

Viet Cong (VC): The communist insurgent movement in the Republic of Vietnam. The term applies to both the movement itself and its members.

Warfighting Functions: All tasks performed by a unit fall into one of the functions. As of 2021, these are intelligence, mission command, movement and maneuver, fires, protection, and sustainment.

Warrant Officer: In the U.S. Army, a warrant officer is a type of commissioned officer performing technical duties. Army warrant officers rank between the Sergeant Major of the Army and Second Lieutenant.

Weapons of Mass Destruction (WMD): Weapons intended to cause significant numbers of casualties. They include chemical, biological, radiological, nuclear, and mass blast effect weapons.

Wounded in Action (WIA): The classification of a soldier sustaining wounds resulting from combat. For members of the U.S. armed forces, being WIA results in an award of the Purple Heart Medal.

Introduction

If you dislike change, you're going to dislike irrelevance even more.[1]

General Eric K. Shinseki, Chief of Staff, United States Army

This book, my edited doctoral dissertation, examines the history of the Army's Asymmetric Warfare Group (AWG). It chronicles the circumstances surrounding the AWG's development and describes how a small group of individuals successfully challenged both a long-held self-conceptualization (how the Army saw itself and envisioned its future) and bureaucratic structures to create a new organization. It places their efforts in historical context and explores both the dynamics of change and the actions required to realize and sustain new objectives. The AWG gave the Army capabilities previously lacking and played a major role in improving unit performance and reducing casualties in Afghanistan and Iraq while conducting stability operations between 2006 and 2020.

This is a story of change, specifically that of an army evolving in response to the requirements of a long and largely unanticipated conflict waged on two continents. On September 10, 2001, the United States Army was transforming itself following the Cold War's end, the outcome of the 1991 Persian Gulf War, the lessons learned in Balkans peace enforcement, and a perceived revolution in military affairs. That latter development lay in the power of rapidly developing information technologies capable of providing combat units with heretofore unimagined access to information. The envisioned outcome of its transformation was a smaller, more strategically deployable force leveraging cyberspace's potential to deliver overwhelming power to defeat an opponent. Built primarily around two distinct fleets of combat vehicles—the Future Combat System and the Stryker family of vehicles—and possessing a high degree of digital connectivity, this force would guarantee the Army's ability to act under any set of circumstances. Wedded to high-technology approaches, the Army's transformational thinking sought to ensure that its ability overmatched any conventional military threat anywhere in the world.[2]

All this changed the morning of September 11, 2001. When the World Trade Center's towers fell that morning, the Army's existing blueprints for change fell with them. Not fully recognized at the time, the attacks on New York City and the Pentagon imposed fundamental changes on the Army's planning, procurement, recruiting, training, organization, and operations.

To deal with the problems posed by a new and capable enemy, the Army ultimately created several new institutional structures tasked with identifying and defeating new threats. The institutional Army—that part of the Army that recruit, trains, organizes, equips, and then deploys forces—did something rare in its wartime history by creating and implementing top-down, holistic, and effective solutions to the battlefield's challenges. The focus of this work, the AWG, is historically relevant for two reasons.

First, it represents a rare example of successful change imposed from the top down on an army in combat. The Army's history shows that successful changes in wartime generally come from the bottom up. The most effective problem solvers were those most intimately familiar with them: the soldiers experiencing them. The creation of hedgerow-cutting tanks in Normandy, France, in 1944 and the use of helicopters to evacuate wounded soldiers in Korea in 1950 are examples of this historical trend. The record credits the former to Sergeant Curtis G. Cullen of the 102nd Cavalry Reconnaissance Squadron and Lieutenant Steve Litton of the unit's maintenance company. The latter began with the Eighth United States Army periodically requesting support from the U.S. Air Force's 3rd Air Rescue Squadron, a unit whose primary mission was retrieving downed aircrews in enemy territory. The success of the *ad hoc* use of Air Force aircraft led to the Army ultimately creating several helicopter detachments for use in Korea and the ultimate creation of new types of medical units that continue to serve today.

These grassroots successes contrast with Vietnam-era efforts mandated from above by the Department of the Army. These included the Tunnel Explorer, Locator and Communications System (TELACS) and the XM-2 Personnel Detector Set, also known as the "people sniffer." TELACS sought to give soldiers clearing enemy tunnels the ability to communicate while the XM-2 would help them find hidden enemy soldiers. TELACS proved infeasible, while the XM-2 required specific environmental conditions (temperature, wind speed, terrain, vegetation) to detect the enemy reliably. Even when those conditions existed, the system made so much noise that it alerted the enemy to the user's presence.

While small in number, these examples illustrate a salient point: everything else being equal, the soldier in the fight has a much better understanding of what needs to happen than organizations thousands of miles away. One cannot question the desire of higher-level headquarters to help solve problems, but too often they fail to offer the right solutions at the right time.

Secondly, the AWG is historically important owing to both its focus on asymmetric warfare and its concurrently addressing current problems while identifying future threats and solutions. The United States Department of Defense defines asymmetric warfare in these terms: "In military operations the application of dissimilar strategies, tactics, capabilities, and methods to circumvent or negate an opponent's strengths while exploiting his weaknesses."[3] The AWG's basic mission was to provide comprehensive solutions to asymmetric warfare's problems. Comprehensive solutions incorporate doctrine, organizations, training, material, leader development, personnel, and facilities. The Army uses the abbreviation "DOTMLPF" to capture these elements. Title 10, United States Code, defines the Department of the Army's responsibilities. DOTMLPF represents a simple way of capturing the Title 10 responsibilities associated with introducing a new capability, the ability to complete a task or execute a course of action under specified conditions and level of performance, into the Army.[4]

DOTMLPF Explained

Fully employing a capability is impossible without sound doctrine describing its most effective employment. One example of this is the fundamental doctrine introduced with the AH-64 Apache helicopter: cross the enemy's forward line and attack a tank or motorized rifle regiment while it was still on the march. In conducting a deep attack in this manner, the helicopter unit would destroy a sizable enemy force before it could engage a defending ground force. Another example is the engagement priorities of a defending M1 tank platoon: command and control vehicles, air defense weapons, engineer assets, tanks, other armored vehicles, dismounted infantry. When combined, this doctrine would separate the parts of an attacking Soviet force to prevent it from overwhelming a defending U.S. Army division. It would also allow the most effective destruction of forces by denying the attacking force its ability to control its subordinates, engage attacking aircraft, or reduce the obstacles that would keep the attacking force in the desired engagement area—the area where the defending commander sought to kill it. The doctrine facilitated "picking apart" the enemy to destroy its ability to realize its own objectives.

Irrespective of the capability introduced, doctrine exists to provide the larger Army with a common framework for operations. When applied, doctrine ensures a common understanding of how to employ units to their fullest potential. Doctrine is most useful when it describes, rather than prescribes, action.[5]

Organizations—most significantly the specific capabilities derived from their structure—provide the means of physically expressing doctrine, training, and the outcomes of leader development on the battlefield. These have historically represented the most dynamic aspect of DOTMLPF. One way to consider organizations and their malleability is to visualize a Venn diagram. In the center of a piece of paper, one could have groups of tanks, infantry fighting vehicles, engineer equipment, fire support systems (guns, howitzers, rockets, and missiles), air defense systems, and a host of supporting items such as intelligence, signal, chemical, and logistics units (Figure 1). One could draw lines in a variety of ways to include or exclude specific elements based on the capabilities desired even as identical equipment items could exist in the different units represented by the diagram. For the career soldier, organizational change represents the doubled-edged sword that never sees the scabbard.

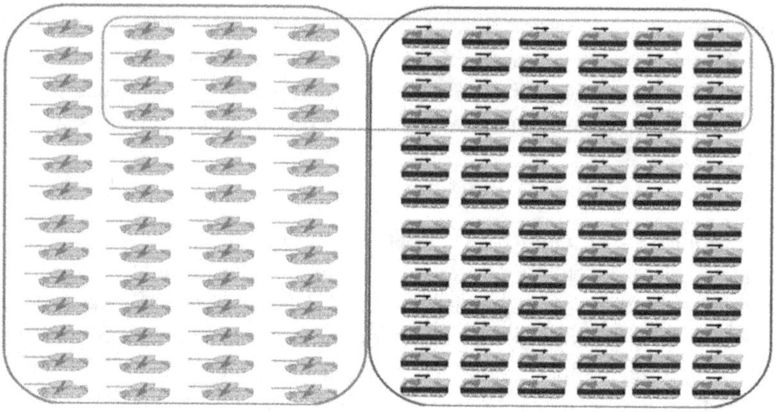

Figure 1. Changes in organization change battlefield capabilities, even as equipment remains the same

Training is arguably more important than doctrine. Considered holistically, there are two training components: technical and tactical. Technical training primarily focuses on how to operate and maintain capabilities. Most of this takes place in the Army's different service schools, where soldiers learn the basic skills associated with their specific specialty. Armor soldiers undergoing training at Fort Moore, Georgia, learn how to operate the M1 family of tanks, while some of their Ordnance counterparts learn how to repair them. Aviation officers at Fort Novosel, Alabama, learn how to fly specific aircraft under various conditions while enlisted soldiers at Fort Eustis, Virginia, learn how to repair either the aircraft itself or its various systems. When a soldier completes his or her initial training, he or she possesses the ability to perform specific tasks in an operational unit. Tactical training, on the other hand, focuses on performing tasks in a field environment. The private learns how to drive the tank at Fort Moore, but he or she learns how to really operate the tank through repeated field exercises that replicate battlefield conditions. Tactical training, in turn, provides two benefits. First, it gives a unit the ability to perform its "go-to-war" mission. Second, it informs doctrine as units develop new solutions to solve the myriad problems that crop up on the battlefield, whether real or simulated.

Materiel most often represents the physical expression of a solution to a problem. The need to defeat attacking Soviet tanks and motorized rifle regiments led to the development of the AH-64 Apache helicopter. The prospect of fighting more numerous Soviet tanks in all weather, illumination, and chemical conditions in central Europe led to the M1 Abrams tank. The need to provide tanks with necessary infantry support led to the M2 Bradley infantry fighting vehicle. In introducing new materiel, the issues of doctrine, training, leader development, and organization once again required consideration. Given this, materiel arguably is the costliest element of DOTMLPF as the Army historically has sought equipment solutions to tactical problems. Even apparently "low tech" materiel solutions carry DOTMLPF implications.

Leader Development equips officers and non-commissioned officers with the knowledge to effectively deploy and employ capabilities, which is essential to creating effective combined arms teams. As an example, leader development ensures that Armor or Infantry company grade officers (lieutenants and captains) can both employ their own weapons and integrate their efforts with the other units (artillery, engineer, intelligence, etc.) to accomplish the mission. An infantry company defending a position is much more effective if it integrates forward reconnaissance and turning, fixing, and blocking obstacles into its defensive scheme and then integrates indirect fires into its surveillance and obstacle plans. To facilitate deeper critical thinking appropriate with an assignment to positions of increased responsibility, the Army revised its formal training regulation, Army Regulation 350-1, in 2014. A major part of the revision was formalizing leader development with this realized through the "lifelong synthesis of knowledge, skills, and experiences gained through the training and education opportunities in the institutional, operational and self-development domains."[6]

Personnel raise questions of complexity perhaps even more significant than doctrine or material. For the Army, the issue is not just the "number" of people assigned to a unit or a specific piece of equipment. It is also a matter of identifying individuals with the skills and attributes required to satisfy the requirements at hand. It also presents issues of management, particularly challenging in an institution that sees a great deal of personnel turnover annually. Some of this stems from soldiers completing their terms of service and leaving the ranks, the impact of maintaining a large school system, and the need to re-assign individuals to broaden their professional and technical experience.

Finally, there are issues associated with facilities. The Army is inherently expeditionary in that it does not intend to conduct combat operations from its home stations in the United States. Despite this, the Army has significant facilities requirements. Training centers require the living and working spaces needed to train soldiers effectively. For combat arms soldiers, this includes the ranges necessary for weapons firing, which requires a lot of land. Improved weapons and fire control systems mean soldiers now expect to engage an opponent at extended distances under virtually all climatic and illumination conditions. Firing ranges must provide realistic moving or animated targets capable of engagement by both ground and air systems to enable this. Highly developed ranges, however, are not the only facilities required. Maintenance facilities including vehicle parks, hangars, repair shops, and depots as well as areas for conducting unit training also require development. Most of these same requirements exist at the posts hosting the Army's deployable units or large-scale exercises.

In developing DOTMLPF-based solutions, the AWG touched large portions of the institutional Army to include the Army's four major commands and multiple administrative agencies. Developing doctrine, training, leaders, and organizations required the AWG to interact with its parent headquarters, the Training and Doctrine Command. In providing operational advisors to deploying/deployed units, the AWG closely coordinated with the Forces Command. In developing or validating equipment solutions, the AWG engaged both the Army Futures Command and the Army Materiel Command. Properly executed, the DOTMLPF methodology allowed the AWG to provide feasible solutions to asymmetric warfare's unusual problems.

Tasked with the DOTMLPF responsibilities associated with the institutional Army and the requirement to support the operational Army, the AWG represented both a genuinely unique organization as well as an approach to military change. One can consider the AWG as the outcome of three co-existing, and at times competing, problem sets facing the Army in 2001. The first was transforming the post-Cold War Army for the expected challenges of the future. The second was the practical outcome of the ongoing terrorist attacks by groups such as Al Qaida that led to a significant change in the United States' response to this threat. This change shifted American options from isolated strikes against terrorist targets to focused operations that ultimately included not only precision weapons but both special operations and conventional forces conducting combat operations against them. In putting "boots on the ground," or "BOG," the United States dramatically escalated its response and entered a period of consistent conflict in which it remains engaged. The third problem set, closely intertwined with the second, was the enemy's exploiting asymmetric means to counter American efforts. In opting to fight on their terms, leveraging their resources to the maximum extent possible, and recognizing and exploiting American vulnerabilities, adversaries such as Al Qaida and the Afghan Taliban largely defined the War on Terror for a global audience. This had profound implications for the United States and its allies as the former sought to defeat what amounted to a global insurgency whose effects resonated far beyond the battlefields of Afghanistan or Iraq. In framing the AWG's creation, this work examines the Army's history of transformation, its experiences with asymmetric warfare, and the war in Afghanistan and Iraq. This framework facilitates an understanding of historical trends within the Army as well as the ability of external forces to drive institutional Army decision-making in unexpected ways.

Book Organization

The Asymmetric Warfare Group was ultimately the product of a unique set of circumstances, and this text's organization reflects them. Chapter 1 will explore the Army's transformation efforts from the mid-1950s through 2001. This period began with a reappraisal of the Korean War and saw significant changes in both the Army's self-conceptualization and its efforts to respond to those changes. This chapter will also illustrate the Army responding to internal and external forces with new bureaucratic processes, most notably the budget process, representing the most powerful among them. Finally, this chapter will describe the impact that a vision of the future had on the Army's ability to understand and act on new and pressing requirements absent from that vision.

Chapter 2 will consider the Army's previous experiences with asymmetric warfare. This history goes back to the Army's colonial roots and comprises the greater share of the Army's combat history. Even when focused on conventional military operations, the Army has often found itself dealing with asymmetric threats. Nineteenth-century examples of this include operations against partisan and guerrilla groups during the American Civil War and the Boxer Rebellion. Twentieth-century examples include the Philippine War and the 1916-17 expedition into Mexico, unconventional operations in Asia during the Second World War, postwar counterinsurgency efforts in Greece, counter-guerrilla operations in Korea, and the campaign against the Viet Cong during the Vietnam conflict. Following Vietnam came an extended period of involvement in Latin America driven by both political and criminal concerns with operations in El Salvador and along the Andean Ridge. The Army's role in these varied from conventional combatant to guerrilla to counterinsurgent. All these experiences imparted lessons about asymmetric warfare. The Army internalized some of these and largely ignored others, often due to focusing on a conflict's differences rather than its similarities to past campaigns.

Chapter 3 will examine the Global War on Terrorism, specifically operations in Afghanistan beginning in 2001 and Iraq in 2003 from both the strategic and tactical perspective. The strategic is critically important in considering the AWG's creation as the impact of insurgent tactics manifested themselves not only in the combat theater but also in the United States. Taliban and Al Qaida's reliance on improvised explosive devices (IEDs) and other asymmetric methods created increasingly difficult strategic challenges as the country came to question the conflicts' real costs. The tactical level is important because the AWG's efforts hinged on understanding and providing practical solutions to the problems encountered at this level.

Chapter 4 will explore the creation of the Asymmetric Warfare Group as the institutional response to the challenges posed by resilient insurgencies possessing a high degree of both tactical and technological skill. Insurgents' access to modern technology manifested itself most dramatically in the ability to create and employ the improvised explosive devices that quickly became effective casualty-producing weapons. Employing a range of sophisticated technologies and tactics became the distinguishing characteristic of the insurgent groups that the Army faced and helped shape its responses to them. The chapter will describe the impact of the insurgencies and the need for unconventional thinking and novel techniques for problem-solving to provide the Army's operational forces solutions to the threats they face.

Chapter 5 will examine the AWG's activation and its early operations. The unit's early days found its early members building a new organization with a unique culture. Even as the unit leadership undertook this task, it simultaneously dealt with the challenges of establishing a home station, manning the unit, and supporting combat operations overseas. This early period saw several critical milestones achieved. These included validating key operational and organizational concepts, establishing the AWG's selection course, and resolving administrative questions that might inhibit soldiers from seeking assignment to the organization.

Chapter 6 will chronicle a maturing AWG and ends with the Group's permanent placement under the Army's Training and Doctrine Command (TRADOC). With that administrative change, the AWG ceased operating as a unit under the auspices of the Army's Vice Chief of Staff and G3 (Operations). This new alignment benefitted the AWG by placing it within an established chain of command better facilitating its long-term growth. Its placement under TRADOC also improved its ability to communicate with the Army's various schools. This shortened "flash-to-bang" time—the time interval between the AWG observing new threats and the average soldier having the ability to understand and respond. In an environment characterized by rapidly evolving threats, such an expedited process proved valuable.

The concluding chapter will tie together the previous chapters and present some final analysis of the Army's successes (and failures) in creating, sustaining, and ultimately deactivating the AWG. The AWG's creation illustrates the interconnections between the Army's self-conceptualization, its bureaucratic processes (turning self-conceptualization into resources), and its ability to adjust these to fit changed circumstances. It will illustrate that the AWG, despite its unconventional birth and unusual mission statement, embodied a successful example of military transformation during a time of conflict. The analysis supports a key argument: that the AWG represented a largely unknown, yet highly successful unit that materially contributed to blunting the effectiveness of many insurgent tactics. Even as debates raged in and out of government about the conduct of combat operations in both Iraq and Afghanistan, the AWG demonstrated its value. Units working with the AWG experienced both fewer casualties and a concurrent improvement in their ability to operate. The dynamics of America's small unit wars began to change for the better. In Iraq in particular, writers often cite the so-called "surge" as making the critical difference and taking momentum away from the insurgents. The surge proved decisive, but without the changes enabled by the AWG, it would have been more expensive in blood and treasure.

Conclusions

A review of the Army's history shows it spends more time confronting change than battlefield opponents. While managing change is always a challenge, effecting change against the backdrop of combat makes it even more difficult. As the Army entered the Global War on Terrorism in the wake of September 11, 2001, its challenge was really the latest appearance of the old one: sustaining the transformational processes that worked while avoiding the glittering traps of those that did not. The record shows the Army did some things exceptionally well while it failed to achieve desired results in other areas. These successes (and failures) illustrate an institutional reality: to successfully adapt, the Army must maintain the ability to change its self-conceptualization and bureaucratic processes as the world in which it operates changes. Losing this ability, failing to consider itself critically, its environment, and its options means the Army will forfeit the voice that matters most—its own—as it undergoes the next inevitable transformation.

Chapter 1

A Changing Army

A great institution like the Army always is in transition.[1]

Col. Andrew J. Bacevich

Andrew Bacevich summed up things nicely: the U.S. Army always is in transition. To understand *why* this is the case it is necessary to recognize the significant forces acting on the Army and their interaction with that force and each other. These forces are both external and internal. One may group external forces into those over which the Army exerts some influence and those over which the Army has little or no control. The former includes presidential directives, the National Security Strategy, and Defense Department policy. The latter includes foreign interaction, emerging threats, and the pace and pathways of technological change. Among the internal forces are the Army's self-conceptualization, its bureaucratic processes, its institutional norms, and its culture. This chapter argues that the forces exerting the strongest influence on change within the Army are self-conceptualization and bureaucratic processes. These translate concepts into actual equipment, organizations, and capabilities. They also demonstrate the power of the other forces and determine when, how, what, and why the Army pursues change.

In considering the Asymmetric Warfare Group's (AWG) creation, a review of Army transformation from the mid-1950s through 2001 is useful. This half-century saw the U.S. Army assume five different iterations: the Pentomic, the ROAD, the Vietnam, the late Cold War, and the post-Cold War. The Pentomic Army considered atomic weapons as the centerpiece of organization and tactics as the Army sought relevancy in a period of great technical change and institutional angst. The Vietnam Army gave primacy to mobility and firepower. The late Cold-War Army emphasized high-intensity conflict in a defined locale—Central Europe—while the post-Cold War focused less on locations and more on understanding emerging capabilities. Examining all of them illustrates both transformation's various pathways and the need for the AWG. This will also illustrate the interaction of the major forces, most notably the budgeting process, upon one another as well as upon the Army.

The Pentomic Army

The Pentomic Army illustrates the power of both the budgeting process and self-conceptualization on the Army. The Korean War did not start out well for the Army, and its negotiated end following three years of bitter fighting and diplomatic wrangling left political leaders looking for new approaches to ensure national security. On January 12, 1954, Secretary of State John Foster Dulles articulated the "New Look" national security strategy. The New Look declared that the United States would "depend primarily upon a great capacity to retaliate instantly, by means and places of our own choosing. That permits a selection of military means instead of a multiplication of means. As a result, it is now possible to get, and share, more basic security at less cost."[2] This decision affirmed the centrality of the Air Force, strategic nuclear weapons, and massive retaliation and forced the Army to move on two distinct axes to reassert itself as a vital institution. The first axis changed the Army's field organization to facilitate operations on a nuclear battlefield, even if national security strategy saw the service having only a minor role, while the second carved out a role in continental air defense. Both of these fully engaged the Army in both offensive and defensive nuclear planning efforts. Perhaps just as important, they provided a means of restoring the morale of an Army struggling mightily to define its role in an era where many believed the ground soldier represented an anachronism.[3]

A major step undertaken by the Army was reorganizing its divisions to remain effective in a possible nuclear war. Within the infantry division, the existing "triangular" structure of three regiments represented a tactical liability on a nuclear battlefield. The problem's origins stemmed from three areas—communications, firepower, and mobility. The triangular infantry division depended heavily on short-range radio and wire communications, did not possess any organic weapons larger than 155mm howitzers, and could not make rapid tactical movements. The Army sought a division that eliminated those liabilities, generated more firepower, and reduced the number of administrative and support troops required to sustain it. Initial work began under the supervision of Army Chief of Staff Matthew B. Ridgway with 1954's "Atomic Field Army" or ATFA studies. ATFA wanted greater mobility and a better balance between combat and support units, and while it failed to provide the Army with a new division structure, it provided a foundation for further analysis.[4]

Starting in 1956, the triangular infantry and airborne divisions gave way to a new "Pentomic" structure, based in part on ATFA analysis. Five battle groups composed of five rifle companies and a mortar battery replaced the three infantry regiments. Armored divisions retained their existing organization of three combat commands. The MGR-1 Honest John, an unguided rocket possessing a range of 38,000 meters, provided the division's primary nuclear punch. The M28/M29 Davy Crockett recoilless rifle gave each battle group its own nuclear weapons. Reinforcing the Honest John were the 280mm howitzer and liquid-fueled MGM-5 Corporal Missile. Assigned to a corps artillery group, these weapons possessed far greater range and provided additional nuclear fires as required. The Pentomic model theoretically allowed a greater dispersion of units, thus providing them with increased protection while also allowing the offensive employment of nuclear weapons. For Army Chief of Staff Maxwell Taylor, this was the key to battlefield victory—combining soldiers with mobility and firepower. The Pentomic model meant radical changes in Army doctrine, training, leader development, and equipment. It also placed the Army in the position of trying to implement a warfighting model far beyond its technical and tactical abilities. It is perhaps more than mere historical curiosity that the Army never published an updated edition of *FM 100-5, Operations* between 1954 and 1962. The Army published three interim changes to the 1954 manual during the period emphasizing atomic considerations, but none referenced Pentomic structures. It appears the Army itself could not satisfactorily resolve the questions the Pentomic division raised.[5]

Defensive nuclear planning efforts represented the second axis and saw the Army assume a highly visible role in continental air defense. This mission represented a transformation of the Army's coast defense mission rather than an extension of the tactical anti-aircraft artillery role performed by the Coast Artillery during World War II. The coast defense mission was a historic one that consumed much of the Army's budget and manpower through 1941. Less romantically, air defense represented the payoff on investments going back to 1945 and the first tentative steps towards an effective surface-to-air missile and improvements in anti-aircraft gun systems. That mission also firmly placed the Army at the center of strategy debates over the Soviet threat. The Army might have lacked the ability to deliver a nuclear weapon inside the Soviet Union, but it would soon possess one of the most effective means of preventing the Soviets from delivering one to North America. Additionally, air defense also expressed Taylor's oft-stated position that the nation needed balanced military forces with the Army operating as part of a joint team. This expression of "jointness" was farsighted since Army doctrine would not fully embrace this concept for decades to come. Finally, it reinforced the Army's efforts to project itself as a technologically sophisticated force as it sought to change the perception of it being backward because of the extended stalemate in Korea. The effort to present a technologically sophisticated force even extended to the opening of each week's edition of the Army's public affairs television program, *The Big Picture.* Episodes opened with footage of a Nike missile launch, a nuclear artillery round detonation, and the Army's newest helicopters. Reinforcing the *Big Picture's* message were the models of Nike missiles found atop the vehicles used by Army recruiters.[6]

The 1950s presented the Army with multiple challenges, with several rooted in resolving the nagging question of its strategic role. The decade began with questions over the efficacy of ground forces at a time when many Americans placed primacy on airpower and nuclear weapons. This development sparked efforts to reinvigorate the service and develop new battlefield methods. The existing budgeting process—one that set a cap on the spending but allowed tremendous freedom to expend its allocated funds—aided the Army's efforts. A vignette from 1958 illustrates the budget process's flexibility during a period of loose bureaucratic controls. When told to cut the Army budget from $11,000,000,000 billion to $9,000,000,000, a marathon meeting resulted in one general losing $50,000,000 when he left the meeting to relieve himself. The Army lacked a recognized process for establishing and sustaining priorities and there was no significant external review. It is useful to ponder this period because the Army would again experience concerns over relevance and repeatedly strive to both prove its utility and demonstrate its capabilities—a process that received unexpected aid from foreign engagements following the Cold War's end.[7]

The Reorganization Objective Army Division (ROAD)

By the time John F. Kennedy assumed the presidency, the underlying strategic thinking behind the New Look and massive retaliation had fallen from favor as improved Soviet capabilities exposed its many fallacies. Taylor, now retired, advocated a different approach modeled on the National Military Plan he articulated while serving as Chief of Staff. Publicized in his 1960 book, *The Uncertain Trumpet,* his thinking resonated with the incoming administration. Concurrent with this was the realization within the Army that a Pentomic organization would not only fail on a nuclear battlefield; it severely limited the Army's ability to deploy forces to support other contingencies. The 1958 deployment of forces to Lebanon, OPERATION BLUEBAT, validated this concern. To restore organizational sanity to the force, the Army sought a new divisional structure. Almost simultaneously with this, the Army undertook some bold initiatives regarding its aviation component and its expanding battlefield role. These two actions would radically alter all the DOTMLPF (doctrine, organization, training, materiel, leader development, personnel, facilities) elements in providing the Army with new ways to organize and fight while concurrently creating new administrative, logistics, and training requirements. [8]

Efforts to redesign the Pentomic infantry and airborne divisions began in 1959 with the Modern Mobile Army (MOMAR) study, a short-lived effort to overcome two of the Pentomic division's most pressing problems: defeating enemy armor and insufficient mobility. MOMAR's failure to provide a suitable path for reorganization led to a second effort, known as Reorganization Objective Army Division (ROAD) starting in December 1960 under Continental Army

Command (CONARC) direction. By March 1961, CONARC completed a plan somewhat radical in its basic design: eliminating the armored division combat commands and infantry division battle groups and replacing them with a brigade headquarters company. Reorganizing the infantry and airborne divisions, at least on the surface, appeared as a giant step backward as the five battle groups gave way to three brigades that, at least superficially, strongly resembled the pre-Pentomic triangular organization. There were two significant changes, however, between the proposed organization and the older triangular structures. The brigades in the new ROAD division retained the same flexibility of the older armored division combat commands—they were command and control units only with no organic subordinate units. This structural change meant that a commander could attach or detach units as required for operations providing an exceptional level of tactical flexibility. The second change lay in the organization of the division's support command and, over time, engineer, aviation, military intelligence, signal, and air defense elements with all divisions possessing the same basic elements. The specifics of equipment and density would vary depending on the division type, but there existed a high degree of commonality. ROAD proved successful as a concept and ultimately continued as the Army's basic organizational structure through the early 2000s.[9]

As the operational Army underwent another transformation with the Pentomic division's conversion to the ROAD template, the institutional Army sought to streamline itself. While the motive for reorganizing operational forces was the realization that the structure was ultimately unworkable, the driver for institutional change came from Secretary of Defense Robert S. McNamara. McNamara embodied the activist manager in outlook and action. Bringing new management concepts and a willingness to invoke authority provided by a 1958 Department of Defense reorganization, the Secretary quickly centralized operations and placed radically new procedures within his department. In doing so, McNamara believed he could generate efficiencies in procurement, operations, and oversight. One of the most significant changes to that end lay in the Planning, Programming and Budgeting System (PPBS) and its requirement for the Program Objective Memorandum (POM). The POM lay at the heart of McNamara's efforts and effectively created a Defense Department budget ensuring the resourcing of required military capabilities as opposed to the existing system of each military department (Army, Navy, Air Force) budgeting independently. The POM took something of a long view by funding for the current year and the next four. This process eliminated unnecessary redundancies, but it also tended towards stasis as various elements of the Army sought to preserve their submissions. The POM could (and still does) change from year to year, but the process possessed a bias towards caution as opposed to bold change, particularly when it came to primary (and costly) weapons or systems procurement.[10]

For the Institutional Army, McNamara's service as Secretary of Defense meant significant changes to meet his requirements. McNamara aided the Army's efforts by commissioning a study known as "Project 80." This project is once again considered a vital component of the institutional Army's organization: the Technical Services. These branches—Army Medical Department, Chemical Corps, Corps of Engineers, Ordnance Corps, Quartermaster Corps, Signal Corps, Transportation Corps— essentially sustained the Army. Each of these had a general officer serving as the branch chief (the Surgeon General, the Chief of Engineers, etc.) who presided over the branch and its functions. These, in turn, had their roots in the War Department's original bureau system. Through the Army's long history, these services generally possessed a high degree of autonomy, and while the Army Service Forces structure checked this during World War II, postwar reorganization restored it. [11]

Officially titled "Study of the Functions, Organization, and Procedures of the Department of the Army," Project 80 resulted in a committee tasked with identifying new organizational structures within the Department of the Army. The Hoelscher Committee, named after Army Deputy Comptroller Leonard A. Hoelscher, recommended changes in both the Army and Continental Army Command (CONARC) staff in October 1961. These included altering the relationship between technical service branch chiefs and the Army's coordinating staff, creating a Force Development Command, creating a Combat Developments Command, and an Army Materiel Command. All these changes represented fundamental breaks with existing practice and, in some cases, long-held traditions, such as the submergence of most branch chiefs in other organizations the most visible. Institutional angst notwithstanding, McNamara approved the final changes recommended and the Army saw a new structure take hold on February 17, 1962. The revised structure saw most of the technical service functions transferred to the newly created Army Materiel Command with only the surgeon general of the Army and the chief of engineers retaining their positions. The newly created Combat Developments Command assumed responsibility for force organization and design without concurrent responsibility for a key DOTMLPF factor—the training required to empower the organizations created. This new structure would serve as the Army's institutional background for the next decade.[12]

The ROAD period introduced several enduring structures to the operational and institutional Army. The ROAD division, with flexible brigades and a robust base, remained the Army's primary organization through the remainder of the twentieth century. It grew over time as the Army reassessed the best way to provide intelligence, aviation, and engineer support; however, the basic structure and governing principles remained unchanged. The Army Materiel Command became the central component of the Army's massive logistics bureaucracy, a position it retains through the present. The most important structure, however, was the Program Objective Memorandum. The POM changed how the Army funded its requirements for research, procurement, equipment fielding, operating organizations and agencies, and training. The POM's need to program funding five years into the future coupled with the Defense Department's ability to change Army submissions ensured the service carefully considered its requests, particularly for its most expensive items. Requests that failed to align with Defense Department requirements or perspectives risked disapproval and thus tempered changes in the Army's vision of itself and its needs. The POM rationalized defense spending, but in so doing it also moderated change by determining what received funding—and what did not.

The Vietnam Army

As the Army moved through the Pentomic and ROAD periods, it became involved in the growing conflict in Vietnam. Beginning with advisors and ending with full combat units, the Army's roles and missions continuously evolved during its service there. Within Vietnam, the most important changes for the operational Army were the deployment and expansion of advisors and Special Forces, the proliferation of aviation units, and the creation and deployment of separate infantry and armored brigades. Operational units in Vietnam were not, however, the only units impacted by the war. The need to maintain units in Vietnam profoundly impacted the Seventh U.S. Army in Europe as it increasingly served as a bill payer to ensure adequate strength in Vietnam. Units within the United States, both Regular and Army National Guard, found themselves increasingly involved in civil disturbance duties generated by both the civil rights and anti-war movements. The demands of the war and civil disturbance operations also impacted the institutional Army and its expanding bureaucracy as it provided operating forces with the doctrine, training, and equipment these diverse missions required. The requirements of jungle fighting meant the Army needed to field a wide variety of equipment items for both its ground and air elements with each equipment item bringing DOTMLPF requirements. Changes in reconnaissance and surveillance systems, target acquisition equipment and aircraft—both add-on systems and new types—were among the most significant.[13]

Historians do not agree on a specific date for America's entry into the Vietnam War. Mark Lawrence's *The Vietnam War* makes the case for 1955. Francis J. Kelly's *Vietnam Studies: U.S. Army Special Forces* cites the year 1957, while Gabriel Kolko's *Anatomy of a War* argues for 1961. Finally, Fredrik Logevall argues for the "long 1964" in *Choosing War.* What all these scholars agree upon, however, is the fact that American advisors played a critical role in the war's early stages and that Special Forces, popularly known as "the Green Berets," filled critical advisory and combat roles. For Infantry, Armor, Field Artillery and other conventional force advisors, service in Vietnam offered responsibilities and opportunities not found in American units, particularly for company-grade officers who frequently found themselves coordinating myriad battlefield activities. For Special Forces, Vietnam presented the opportunity to apply all they had theorized concerning organizing partisan movements and counterinsurgency. Green Beret units provided vital links between a South Vietnamese militia, the Civilian Irregular Defense Group (CIDG), and the Army; performed strategic reconnaissance; conducted unconventional operations in previously denied areas; and supported conventional operations. More significantly, however, the war elevated the status of Special Forces within the larger Army. For the first time, it gave them a central role in a large-scale conflict. This represented a challenge to those wearing the green beret since much of the larger Army's leadership tended to distrust them owing to perceptions that Special Forces did not consider themselves part of the Army. While multiple reasons existed for these understandings, a major one stemmed from a 1963 Army Regulation that separated their administrative, logistical, and financial management from the larger Army. To facilitate the Army Special Forces' large-scale presence in Vietnam while sustaining capabilities in countries both adjacent and distant, the population of that elite component required rapid growth. Ultimately, the Army created six Regular Army Special Forces groups with an additional two in the Army National Guard, two in the U.S. Army Reserve, and a number of separate detachments operating outside the standard group structure. This meant Special Forces grew from approximately 2,000 soldiers in 1961 to more than 12,000 by the decade's end. Expanding Special Forces was a difficult task owing to the strict selection and training standards the Army applied and the high casualty rates associated with such duty. While the Green Berets saw reductions in their size following the war, their performance in Vietnam (and elsewhere) ensured the Army retained a robust special operations capability into the future. This capability included the ability to both organize or support insurgencies in support of American political objectives.[14]

Vietnam's demands saw the Army massively expand its aircraft fleet and with it, the organizations required to employ it. Between 1962 (the year the first Army aircraft operated in Vietnam) and 1970 the size of the Army's aircraft fleet expanded dramatically with over 10,000 aircraft serving in the country. Aircraft distribution and organization varied between units. The 1st Cavalry Division possessed four helicopter battalions to perform its airmobile mission. The 1st Aviation Brigade, operated more than 4,000 aircraft, including all the fixed-wing Army aircraft in Vietnam. The Army fully exploited its aviation capabilities during the war; however, there were several unintended consequences. The first was further erosion of the manpower pool available for ground combat units since aviation units competed with the Army's other high-technology, high-skill units for the highest-quality soldiers. The second was a massive investment in facilities, training, and maintenance at a time when President

Lyndon Baines Johnson's administration sought to limit military spending to fund other priorities. These DOTMLPF elements significantly impacted the larger Army, particularly in Europe, where units suffered the effects of fewer soldiers, fewer spare parts, and fewer dollars for training.[15]

Creating separate infantry brigades, units containing not only infantry but also fire support, reconnaissance, and support units, provided the Army with a way to economize resources. For the Regular Army, it provided an on-call capacity to perform missions not requiring full divisions by substituting units with the ability to deploy more rapidly, a highly desirable quality under the so-called Flexible Response doctrine. The Army lacked this capability during Operation BLUEBAT in 1958. For the Reserve Components, separate brigades represented a way to simultaneously improve quality while reducing costs by reducing demands for competing resources, most notably the funding required for division-level training. In the Panama Canal Zone and Alaska, the separate brigades assigned—the 196th, 171st, and 172nd—provided tailored forces capable of performing essential defense missions without making unnecessary claims on increasingly scarce money and manpower. The Army National Guard's newly created separate brigades benefitted from the reduced pool of maneuver units as it allowed the earlier introduction of new equipment. The Army Reserve saw similar benefits stemming from the deactivation of Reserve infantry divisions. The deactivation of Reserve divisions reflected a more substantial transformation affecting the Army Reserve—the virtual elimination of its combat role and conversion to a support/service support organization. Concerns over cost, equipment, and manpower efficiencies drove all these changes. The approach proved both innovative and enduring.[16]

For the Army, Vietnam represented a period of accelerated change driven by the competing demands of combat, the need to retain capable forces in Europe and Korea, and concerns over the difficulties of maintaining capable active and reserve forces. A series of highly publicized events fueled the discussion. These included the 1968 Tet Offensive in which Viet Cong attacks across Vietnam included both the U.S. military headquarters at Tan Son Nhut and the American Embassy in Saigon. The 1969 disclosure of the My Lai Massacre called both discipline and command competence into question. The results of the U.S. Army War College's 1970 *Study on Military Professionalism* sent shock waves throughout the Army as the officer corps articulated serious criticism of existing norms. Such episodes led to another crisis in confidence and once again caused the Army to call its relevance into question. Ultimately, the challenges of Vietnam exposed profound weaknesses within the institutional Army. This exposure spurred another round of headquarters' reorganization rivaling the changes made to the operational force and placed the Army on a new azimuth in the post-Vietnam era.[17]

The Late-Cold War Army

An unattributed quotation heard through the Army in the early 1980's attested that "the United States entered the Korean War with a terrible Army and came out with a good one and then entered the Vietnam War with a good one and came out with a terrible one." There is much truth in these words. From 1973 through the Cold War's end, the Army focused on exorcising the ghosts of Vietnam and creating an Army once again capable of meeting the demands awaiting it. To do this, the Army entered a long period of steady change. Some changes were dramatic while others proved more prosaic. Taken together, however, they restored the Army to its pre-Vietnam War position as a well-equipped, well-trained, well-disciplined, and well-led force. To achieve this, the Army required a new focus and an increasingly capable Soviet Army in Europe provided one.

General Creighton Abrams succeeded Gen. William Westmoreland as Army Chief of Staff in 1972 and held that office through his death in 1974. Abrams possessed an exceptionally clear understanding of the Army's most pressing problems. He served as the commanding general, Military Assistance Command Vietnam (MACV), before his elevation to the Pentagon and saw first-hand both that war's problems and their roots. Broadly speaking, Abrams had to undertake two major transformations to fix the Army. First, he had to change the Army's fundamental culture amid a period of tremendous social ferment. Second, he had to develop a generation of weapons capable of defeating a growing Soviet military that showed increasing levels of technological progress and tactical sophistication. Abrams recognized that investments in more modern weapons systems were of little use without fixing the underlying cultural problems (drug use, racism, indiscipline, low morale, and the conversion to an all-volunteer force) resulting in ineffective units.[18]

Throughout the course of his career, Abrams remained focused on the soldier. Perhaps more than any other senior officer of his day, he tended to view the lowest Army member not as a serial-numbered item but as a feeling and thinking individual. In a largely conscript Army, operating with tremendous monthly turnover and experiencing crises of confidence, morale, and even purpose, Abrams' focus on meeting his troops' expectations represented an unusual position. The prevailing sentiment in the Army was that careerists placed their personal concerns before those of the soldiers they led. Abrams noted the 1970 US Army War College *Study on Military Professionalism* and its finding that officers "perceive a significant difference between the ideal values and the actual or operative values of the officer corps."[19] The study also identified a focus on short-term objectives and the tendency of senior officers to look upward to ensure supervisor satisfaction rather than downward to meet the legitimate needs of subordinates. Comments from the study's participants included the unfair treatment of those leaders trying to maintain high standards and officers more concerned with their careers than their soldiers.

Abrams had little difficulty in turning the study's comments into observable changes. The Army's most significant institutional and cultural change in a generation—the end of conscription—occurred on his watch. The last draft call took place on December 7, 1972, with the final authority for induction expiring June 30, 1973. If Abrams could not reorient the Army's operative culture and values, soldiers would pay the price. This meant that retention rates would likely crumble and recruiting would suffer. To prevent this, Abrams focused on meeting soldier expectations for leadership and service support. "People aren't IN the Army, they ARE the Army" epitomized Abrams' position.[20] He communicated it frequently to impress upon his military audiences the need to change the institutional focus and ground it again in discipline, trust, integrity, and competency—especially in leadership. Correctly doing this satisfied soldier expectations and began the long road of once again rebuilding the Army as a competent and relevant force.[21]

If the Project 80 changes met Robert McNamara's requirements, they did not adequately meet the Army's needs. Budget reductions following the end of hostilities in Southeast Asia coupled with a realization that the Army's overall headquarters structure had grown bloated meant further changes were necessary to ensure maximum performance. The 1962 reorganization left CONARC in place. CONARC encompassed a massive organization with diverse responsibilities. These included directing most Army training programs, ensuring unit combat readiness (Regular, National Guard, and Reserve) in the United States, overseeing the operation of major installations, and serving as the Army component command for two different joint combat organizations—Strike Command and Atlantic Command. As a massive consumer of manpower and money, CONARC was an obvious subject for critical examination. Beginning in 1970, CONARC began an internal study of its organization in response to Westmoreland directing a reexamination of the institutional Army's organizational structure. Follow-on studies by the Defense Department, the Army, and CONARC considered CONARC's structure, roles, and missions with a view towards making the command more efficient. CONARC possessed two core functions—training the individual soldier and providing trained and ready units. The logical action was thus to cleave CONARC into two different commands—one focusing on training and one on readiness. This action, however, created a new problem: redundancy between the part of the CONARC training mission and the mission of the existing Combat Developments Command. The solution to this issue was also, at least superficially, a simple one—merging the Combat Developments Command into the new training development and management organization. On July 1, 1973, the Army simultaneously deactivated CONARC and activated the U.S. Army Training and Doctrine Command (TRADOC) and the U.S. Army Forces Command (FORSCOM). The task was enormous and involved over 1,000,000 soldiers, 40 installations, and hundreds of millions of dollars. The

Combat Developments Command ceased operations on December 31, 1973, with some functions moving to TRADOC headquarters and others falling under the various service schools.[22]

As Abrams set about removing the rot within the Army's human component and improving the institutional Army's structure, he laid the foundation for developing a group of weapons and equipment ultimately known as "the Big 5." The M1 Abrams tank, the M2/M3 Bradley Infantry/Cavalry Fighting Vehicle, Advanced Attack Helicopter (fielded as the AH-64 Apache attack helicopter), the Utility Tactical Transport Aircraft System (fielded as the UH-60 Blackhawk utility helicopter), and the MIM-104 Patriot surface-to-air missile, the "Big 5" came to represent the Army's warfighting future. Designed for war in Europe, these systems performed successfully in 1983's Operation URGENT FURY in Grenada (the UH-60), 1989's Operation JUST CAUSE in Panama (the AH-64 and UH-60), the 1991 Persian Gulf War (all five systems), and through the present. Despite plans to replace the ground combat vehicles, these still represent some of the most technologically advanced systems of their type. It would be long after General Abram's death before the first of these weapons entered the Army's inventory, but the criteria developed under his supervision captured both his own experience and that of recent armored conflicts and established a firm path for modernization for the next 40 years.[23]

Abrams' successor as Chief of Staff, General Frederick Weyand, favored transformational challenges residing in the human domain. If there was the proverbial elephant in Weyand's room, it was converting the Army to the All-Volunteer Force (AVF). Weyand became the first Chief of Staff whose entire tenure coincided with the All-Volunteer Force, including coming to terms with effectively recruiting it. As important as filling the ranks, however, was effective employment. Under Weyand, the Army changed its approach to training soldiers by moving to performance-oriented training. Performance-oriented training established specific tasks, conditions, and standards for a soldier to perform to demonstrate proficiency in his or her duties. In codifying task standards, the Army changed its perspective on training management. This change established the foundation for the operational Army's enduring post-Vietnam training focus: readiness through realistic training.[24]

Of equal importance to the new training model was a new overarching doctrine, revealed in the 1976 edition of *FM 100-5, Operations.* Rooted in the lessons of the 1973 Arab-Israeli War and the vision of Gen. William E. DePuy, Maj. Gen. Donn Starry, and Brig. Gen. Paul Gorman, the new doctrinal method became known as "active defense." Active defense posited rapid and violent combat by soldiers who knew both what to do and why to do it to win the first battle against the Soviet Union in central Europe. At the center of this were two elements—mechanized combined arms formations employing the tank as the centerpiece and "training under conditions approximating battle."[25] As a doctrine, active defense proved short-lived. It indelibly marked the path of future transformation, however, ensuring the ability to defeat a highly mechanized, modern force.[26]

Replacing General Weyand in 1976, Gen. Bernard Rogers saw his tour focus on organizational issues as the equipment decisions of his predecessors continued their movement towards fielding. Looking forward to the time when these systems entered service, Rogers oversaw the Division 86 project. Starting in 1978 and continuing through October 1979, Division 86 had the task of visualizing the future when full fielding of the "Big 5" was well underway and determining the optimum structure for the successful employment of these weapons. The idea behind the project was simple: ensure that the coming increases in firepower, protection, and maneuver did not exceed the average commander's ability to employ them. Division 86 considered a variety of changes to the Army's heavy division and ultimately proposed several significant changes intended to improve command and control, facilitate greater tempo, extend the division's tactical reach, and generate greater lethality. Those changes included tank platoons of four tanks each, tank and mechanized infantry battalions of four companies each, anti-armor companies in each mechanized infantry battalion, eight artillery pieces in each field artillery battery, and the creation of a combat aviation brigade for each heavy division with attack helicopters as the primary weapon.[27]

General Rogers completed his tour as the Army's Chief of Staff in June 1979. His successor, Gen. Edward C. Meyer, oversaw a second round of Division 86 studies. This effort reflected the realization that the heavy division construct so desirable in central Europe was not necessarily the right model in other parts of the world. Meyer wanted to ensure the Army could successfully meet threats outside of Europe without compromising efforts there. The challenge for Meyer was creating an effective light infantry division, an effort that would extend long beyond his tenure as Chief of Staff.[28]

Meyer's watch also found the seeds of Abrams' procurement efforts bearing fruit. The "Big 5" began entering service with the first UH-60As reaching the 101st Airborne Division (Air Assault) in August 1979, the first M1 tanks reaching the operational Army in February 1980, and the M2/M3 Bradley's appearing in 1981. All these weapons systems originated in an atmosphere of controversy owing to concerns over technology, expected performance, and increasing cost. The initial fielding saw teething problems, both expected and unexpected, reigniting long-standing debates over need, cost, and utility. Despite the early challenges, the Army achieved initial operational capability with the M1, M2, the AH-64 Apache attack helicopter and the Patriot missile system before Meyer's term ended. The Army had the tools it thought necessary to win in

Europe. Significantly, it also now possessed the tools that would dominate or influence POM submissions for decades to come.[29]

Even as fielding was underway, the Army moved forward with a bold experiment in organizational training—a National Training Center (NTC) that facilitated training with most of these systems against an uncooperative opposing force using Soviet-styled equipment and employing Soviet doctrine. At remote Fort Irwin, California, the newly Multiple Integrated Laser Engagement System, or MILES, enhanced simulated combat. MILES enabled soldiers and vehicles to inflict or become simulated casualties, which represented a quantum leap in tactical realism as soldier actions determined an engagement's outcome.[30] With battles followed by after-action reviews, or AARs, units identified strengths and weaknesses across the warfighting functions. A cadre of observer-controllers facilitated these AARs, and the Army soon realized the advantages of expanding the distribution of lessons learned. Starting in 1984, the Army began capturing and disseminating lessons learned with the NTC Lessons Learned Team at Fort Leavenworth's Combined Arms Training Activity.[31]

Meyer also oversaw a profound change in Army tactical and operational doctrine. August 1982 saw the unveiling of a new edition of *FM 100-5, Operations.* This edition discarded the active defense approach that constituted the Army's first attempt to create a post-Vietnam doctrine suitable for operations in Europe and replaced it with a fundamentally new vision: Air-Land Battle. Air-Land Battle represented a quantum leap in Army thinking as it completely revised the Army's battlefield framework and sought to exploit the full capabilities of all its new weapons. The new doctrine represented the full expression of the Army's self-conceptualization: technically sophisticated, tactically proficient, and able to decisively defeat any modern opponent.[32]

Meyer's 1979-1983 tour as Chief of Staff was the most consequential since General Abrams due to the long-term impact his changes had on the Army. The NTC remains the model for highly realistic unit training, while the systems fielded during this period constitute most of the Army's warfighting and support vehicle fleets. In the aggregate, these accomplishments gave the post-Vietnam Army revitalized tactical and technical foundations that extended well beyond Meyer's tour as Chief of Staff.[33]

Iterative technological improvements to the Army's new weapons continued under General John Wickham's service as Chief of Staff. More significant than these modifications were two new publications and two structural changes occurring between 1983 and 1986. The publications were the 1983 *Army of Excellence* study and the 1986 edition of *FM 100-5, Operations,* while the structural changes were the creation of a force of five light infantry divisions and the Battle Command Training Program (BCTP). Taken together, they reshaped the Army's approach to warfighting by creating a new vision grounded in evolving capabilities and in providing a better mix of forces to meet operational requirements. The BCTP provided a means to improve training further and develop the capacity to effectively employ a revitalized operational force built around the "Big 5."[34]

Rooted in the earlier Division 86 studies, the new Army of Excellence highlighted several unfortunate realities. These included the high cost of some Division 86 proposals, concerns over a commander's ability to control the division's subordinate brigades, delays in providing new capabilities, and lingering questions over light divisions. These questions included how many to create and how to structure them. In contrast to the problems with Division 86 was the revised edition of *Field Manual 100-5, Operations,* published in 1986. This document reflected both lessons learned and developing concepts related to conducting joint operations. These would prove vital over the next decade.[35]

In creating the BCTP, the Army markedly improved its ability to train divisions and corps in critical warfighting tasks. It achieved this by focusing on senior leaders and their staff. The BCTP provided division and corps-level command post exercises allowing commanders and staff to respond to an opposing force operating in cyberspace. When coupled with the combat-training experiences for battalions and brigades provided by the National Training Center and subsequent training centers at other locations, the BCTP gave the Army a unique training capability. Collective training—the training of units—in a realistic environment was now possible from echelons ranging from fire team or tank crew to corps commander.

There is one other event related to transformation that took place during Wickham's tenure as Chief of Staff that is also particularly noteworthy, the 1983 U.S. invasion of Grenada, better known as Operation URGENT FURY. This represented the first major American military effort since the aborted American hostage rescue in Iran (Operation EAGLE CLAW) in 1979. As forces attained their strategic objectives, senior leaders and critical observers realized URGENT FURY exposed multiple institutional weaknesses within the American defense structure. Most disturbing among them was the inability of the various services to work together effectively. Some weaknesses were essentially technical. Others included an imperfect understanding of joint planning, doctrine, and employment. These were significant issues as they led to both American casualties and delays in conducting operations. The United States Senate examined the military organizational structure in the aftermath of URGENT FURY with the resulting Locher Report recommending changes. The most enduring outcome of the invasion proved to be the Goldwater-Nichols Act, which strengthened the authority of commanders and sought to replace service-centric approaches with joint ones. To realize the desired joint approach, the act created new promotion requirements for senior officers by mandating service in joint assignments, directed the resourcing of joint headquarters, and firmly placed the chairman of the Joint Chiefs in the position of primary military advisor to the president and secretary of defense. This last provision worked a profound change in the chairman's previous role as "first among equals" and with no special authority over the uniformed service chiefs. Furthermore, Goldwater-Nichols mandated a series of changes within the Department of the Army's organization and required the Chief of Staff to have joint experience as a general officer. It also established joint service policies requiring the Army to provide both officers for joint duty assignments and ensure a promotion rate equal to that of their non-joint peers. [36]

More revealing than the law's provisions, however, was Goldwater-Nichols' impact on Army doctrine and culture. Chapter 15 of the 1982 edition of *FM 100-5, Operations* stated, "The Army will seldom fight alone (...) joint operations will be the rule rather than the exception."[37] It then offered 13 broad paragraphs that said little about operating in a joint environment with most of the discussion oriented toward delineating chains of command and the distinctions between unified commands, specified commands, and joint task forces. In contrast, the 1986 edition stressed the "imperative that Army units fight as part of a joint team" and addressed broader constructions of joint operational art, campaign planning, and joint, combined, and multi-national operations.[38] Goldwater-Nichols did not alter the fundamental trajectory of Army transformation efforts, but it grounded them within a broader joint structure and kept them linked to that structure in the succeeding years.

Gen. Carl Vuono assumed the Chief of Staff's duties in June 1987. As he assumed the duties, the elephant in his office was the Goldwater-Nichols Act and its requirements to reorganize several Army headquarters organizations and staff and concurrently reduce headquarters personnel strength by 15 percent. One of the law's provisions ultimately changed one of the institutional Army's three major commands. FORSCOM became a specified command. The organization retained its original mission of controlling all Army combat forces in the United States and providing a dedicated Army component to the U.S. Central Command. Its new missions included the joint training of designated forces, assisting civil authorities within the United States, and the land defense of the continental United States and Canada.[39]

Goldwater-Nichols also directed the formation of the U.S. Special Operations Command (USSCOCOM), an action that provided the impetus for the Army to create the Special Forces branch. Concurrent with this, the Army established the 1st Special Operations Command to serve as the Army component command to USSOCOM. Creating this command provided a single headquarters for some of the Army's most unique units as the Army consolidated Special Forces, Psychological Operations, and Civil Affairs units under the new organization.

Although Special Forces and their unique capabilities (including supporting insurgencies) garnered great attention, the Army's "corporate" focus remained on the "Big 5" and winning in Europe.[40]

The transformation underway in Army training continued under Vuono. Nineteen eighty-eight brought the publication of *Field Manual 25-100, Training the Force.* Prepared by TRADOC, *Training the Force* opened with a foreword by Vuono emphasizing the need to "develop and enforce high standards in combat readiness through tough, realistic multi-echelon combined arms training (…) to challenge and develop individuals, leaders and units."[41] Doctrine now called for commanders to develop a mission essential task list, or METL, based on the unit's wartime mission and then practice to achieve proficiency in those tasks. The manual specifically referenced *FM 100-5, Operations,* and its supporting manuals as providing common procedures and methods. Mission training plans published by various Army schools provided commanders guidance for developing unit training and assessing task proficiency. This tightly connected training to doctrine and empowered commanders to train their soldiers based on their own assessments.[42]

While all these changes reflected actions within the Army, two signal events also took place during Vuono's tour that spurred intense debate over that service's composition, roles, and doctrine. The first was the collapse of communism in Eastern Europe and the subsequent German reunification and collapse of the Soviet state. The second was the 1990 Iraqi invasion of Kuwait and Operation DESERT SHIELD/DESERT STORM to restore Kuwaiti sovereignty. With the Soviet threat eliminated and with the Army decisively demonstrating its mastery of high-intensity combat in southwest Asia, many outside the Army called for massive reductions in force structure. Concurrent with this, the institutional Army sought to determine how best to adapt the force to new strategic realities. Much had gone right in rebuilding the Army after Vietnam. The new challenge resided in sustaining what was good as the service entered a new era of uncertainty.[43]

The Post-Cold War Army

Gen. Gordon L. Sullivan replaced Vuono as Army Chief of Staff in June 1991. Upon taking office, Sullivan faced challenges not seen since the end of World War II as he sought to adapt the Army to a markedly altered strategic environment. With no power comparable to the United States on the horizon and expectations of a peace dividend facilitated by the Soviet Union's collapse, Sullivan had to guide the restructuring of an army that met the nation's needs even as the strategic environment became more complex and unstable. As he assumed his duties in June 1991, the Army was conducting humanitarian assistance operations in Iraqi Kurdistan while maintaining a presence in

Kuwait. Instability in the Middle East only represented one problem set with crises in both the United States and the Caribbean placing demands on Army resources. The 1991 Haitian coup resulted in a slowly escalating crisis culminating with an American landing on the island in September 1994 to help return the country to civilian rule. The April 1992 Los Angeles riots resulted in more than 13,500 soldiers of the Regular Army and Army National Guard establishing a joint task force with a Marine Corps infantry regiment to assist civil authorities in restoring order. In August 1992, Army units deployed to Somalia to assist in international relief efforts with this mission rapidly shifting to a combatant one. August also saw Hurricane Andrew strike Florida. This led to more than 23,000 soldiers deploying to Florida to again aid civil authorities. These five deployments found Army units performing markedly different missions, largely outside the scope of those the Army considered its primary function. They also provided visible public reminders that while the Army might desire a particular environment and associated mission, it could not create them. Finally, the five deployments demonstrated both the variety and power of external forces on the Army.[44]

Against this backdrop of deployments, Sullivan approved a new edition of *FM 100-5, Operations* and undertook several initiatives intended to better inform the Army regarding the changes it needed to make and the best way to make them. The 1993 edition of *FM 100-5* reflected the Army's 1991 DESERT STORM experience. While Operation DESERT STORM validated much of the Air-Land Battle doctrine, the Army recognized the end of the Soviet threat called for significant doctrinal revisions. DESERT STORM dramatically demonstrated the power of joint and combined operations. In the future, the Army would not only fight jointly, but with the armies of other nations. This realization led the Army to a relatively rapid review and revision of its capstone doctrine. Important changes in the 1993 edition included fuller discussions of operations other than war (OOTW), full dimension operations, joint and combined operations, and greater flexibility in establishing the battlefield framework. As with previous editions, the new *100-5* affected the institutional Army as it responded to the cascading changes in the other DOTMLPF elements.[45]

While *100-5* provided guidance to the current force, the institutional Army moved towards the future with the creation of Battle Labs, Force XXI, and the Louisiana Maneuvers. Sullivan possessed a strong sense of Army history, and he intentionally selected the name "Louisiana Maneuvers" to invoke the 1940-41 exercises in the same state that shaped the World War II organization. In a 1992 Chief of Staff message, Sullivan noted his "intellectual debt" to World War II Army Chief of Staff Gen. George C. Marshall and Army Ground Forces Commanding General, Lieut. Gen. Lesley J. McNair, who orchestrated the original 1940-41 maneuvers that enlightened Army thinking on the eve of war.

The Battle Labs, created at the various Training and Doctrine Command branch centers and schools, represented an effort to simultaneously introduce new technologies and equipment upgrades even as pressures to realize the "peace dividend" meant fewer resources for transformation. Concurrent with this, the Battle Labs (in conjunction with the Louisiana Maneuvers) also informed Force XXI. In conjunction with the Louisiana Maneuvers, Force XXI expressed the Army's understanding of a new environment, one in which the volume, accuracy, and speed of information available to a commander would significantly increase. This new environment was one in which networked elements would function in both hierarchical and peer relationships. In the words of Lieut. Gen. Paul E. Menoher, then commanding the Intelligence Center at Ft. Huachuca, Arizona, Force XXI represented "both a vision and a process: a vision of what the Army of the early twenty-first century will look like (...) and a process through which we (...) achieve this vision."[46] This corresponded with the then much-discussed revolution in military affairs, or RMA. Retired Brig. Gen. John S. Brown provided a good working definition of RMA in *Kevlar Legions: The Transformation of the U.S. Army, 1989-2005*. Brown termed it "a major change in the nature of warfare brought about by the innovative application of new technologies which, combined with dramatic changes in military doctrine and operational and organizational concepts, fundamentally alters the character and conduct of military operations."[47]

In this case, the new technology resided in the cyber domain with theorists envisioning a future of networked leaders, soldiers, and combat systems provided with a common operational picture of both themselves and the enemy. In tactical terms, this technology offered the ability to compress time while concurrently giving commanders at all levels high-quality information facilitating rapid and sound decision-making. Compressing time when the enemy did not enjoy that capability and substantially reducing decision-making risk would give the Army tremendous advantages in times of conflict. The effort to fully realize the potential of cybernetics adjusted the transformation's arc as the Army began thinking about both upgrades and replacements for some of the "Big 5" and other major equipment items. The ambitious plans included a new stealth helicopter, the RAH-66 Comanche, and the Advanced Field Artillery System known as Crusader. Both programs reflected the Army's belief that conflict with a technologically sophisticated opponent possessing sizable numbers of armored vehicles represented the greatest threat. Both programs also consumed increasing amounts of money in POM submissions between 1983 and 2004, even as POMs continued resourcing the "Big 5."[48]

Force XXI became virtually synonymous with the battlefield's full digitization. The Army's objective was full connectivity between combat, combat support, and combat service support units. Digitization profoundly affected each

DOTMLPF element as well as the Army's budget as the service embraced the so-called "revolution in military affairs" and boldly made its first large-scale strategic, operational, and tactical deployments into cyberspace. One of the Army's first major initiatives in support of this goal was fielding a "digital division." The Army assigned the 4th Infantry Division (Mechanized) the mission of digitizing its subordinate elements and demonstrating the improved command and control, intelligence, communications, fire support, maneuver, and logistical capabilities flowing from a fully networked organization employing new operating methods. Implicit in full digitization was an almost complete change in some of the most basic procedures the division employed. Most standing operating procedures, the codified ways in which units operated and managed the myriad details of a unit with over 16,000 soldiers and 811 combat vehicles and aircraft and the training associated with them, would change to exploit new capabilities.[49]

With Sullivan's June 1995 retirement, uniformed leadership of the Army passed to Gen. Dennis Reimer. The breakup of communist Yugoslavia into five (later six) sovereign states resulted in the worst ethnic violence in Europe since World War II and the deployment of American units into Bosnia Herzegovina starting in 1995. For the Army, this originally envisioned one-year deployment turned into an open-ended mission continuing through the present. For Reimer, the deployments to Bosnia and the inter-related crisis in Kosovo meant the existing Army was committed to another high-profile mission featuring most of the "Big 5" weaponry even as the Army sought the optimal path to its future.

Building on Sullivan's Force XXI construct, Reimer's tenure saw refinements in the "Army After Next" (AAN). AAN reflected, in part, the observations of two white papers—1996's "Force of Decision" and 1998's "One Team, One Fight, One Future." "Force of the Future" emphasized technological advances in changed organizations employing new doctrine while operating as part of a joint force. A paradigm shift lay at the heart of "Force of the Future" with its operative thesis arguing the need to develop a *capabilities* rather than threat-based Army. Whereas past efforts first considered likely opponents and the threats they posed, this new mental model centered on technological possibilities. In other words, the Army would develop a range of capabilities and then selectively employ them as circumstances required. This new approach made sense in a world of changing political and economic relationships between countries and the associated wide dissemination of military technologies. It considered the need to both defeat enemy capabilities and concurrently integrate those of both allies and coalition partners. Significantly, Reimer set a long-term planning horizon, in the year 2020, and the Army worked backward from that point to develop realistic characteristics, objectives, and concepts. Key among these was a force joint by design, possessing the ability to leverage emerging communications technologies fully, and logistically unencumbered. Implicit in this was building into the force a higher degree of strategic and operational agility, even as it upset long-held understandings of who, what, and how the Army would fight. "One Team, One Fight, One Future" spoke of the need to ensure that all three Army components—the Regular Army, the Army National Guard, and the Army Reserve—enjoyed full integration as the Army adapted to the new technological and operational environments. For Reimer, "One Team, One Fight, One Future" illustrated two pressing realities—the poor relationship between the Regular Army and the Army National Guard and the need to maximize the Army's return on decreasing resources. The poor relationship with the National Guard reflected the minimal role that Guard combat units played in Operation DESERT SHIELD/DESERT STORM.[50]

Regaining the Guard's trust required significant work on the institutional Army's part that included establishing "enhanced brigades," an early action initiated by Sullivan but brought to fruition by Reimer. These brigades enjoyed higher resourcing and training support priorities to ensure their availability in the event of another crisis. Simply "throwing money" at the problem, however, was insufficient. The most efficacious—and more difficult—requirement was an honest effort to reestablish bonds of trust, and this was something Reimer diligently pursued. One initiative lay in creating smaller, more deployable forces within all three components to enhance agility and reduce the need for long, post-mobilization training periods.[51]

Effective transformation required careful consideration of insights gleaned through experimentation and combat simulations. Reimer noted in a May 11, 1998, e-mail to the Army's general officers that overwhelming American capabilities would likely drive foes to seek asymmetric responses to offset existing or developing advantages. While he specifically noted nuclear, biological, or chemical responses, he also touched on transnational actors and enemies attempting to disrupt capabilities within the United States. To describe this new environment, Reimer used the acronym "VUCA": volatile, uncertain, complex, and ambiguous. This construct represented a quantum shift from Cold War thinking and the clearly defined friendly, neutral, and hostile actors that provided the Army with a reliable threat against which it could plan. Reimer observed all the technological elements associated with transforming the Army, but also maintained a keen awareness of one element technology could not fundamentally change—a high degree of personal character in the Army's leaders and soldiers. As he sought to ensure the Army capitalized on developing capabilities, he emphasized the need to remain anchored in the Army's values and traditions. At the core of his thinking was his conviction that technology was only an end to a means and his belief that "the ultimate weapon on the battlefield will always be the individual soldier."[52] While the white papers, emails, and articles are significant for providing both visions of the Army's future and concrete steps to realize them, the consequences of deploying to the troubled Balkans ultimately proved both more long-lived and influential.[53]

Reimer handed off the duties of Army Chief of Staff to General Eric R. Shinseki on June 21, 1999. Like Reimer, Shinseki was a Vietnam veteran who had seen the modern Army hit its nadir. Unlike Reimer, Shinseki had several years of personal experience with the challenges of operating in the Balkans. Prior assignments included serving as the Army's operations officer during the first transfer of authority between Army units serving in the Implementation Force in Bosnia in November 1996. He followed this by serving as Commanding General U.S. Army Europe and as Commanding General of the Stabilization Force in Bosnia-Herzegovina. He served as the Army's Vice Chief of Staff during 1999's Operation ALLIED FORCE, the air campaign against Serbia, and experienced the drama of deploying Task Force Hawk. Task Force Hawk consisted of an AH-64 battalion (24 aircraft), a multiple launch rocket system (MLRS) battalion, a mechanized task force (tanks and infantry fighting vehicles), and the support troops necessary to support them. Task Force Hawk's envisioned mission was to attack the highly mobile targets the NATO air forces struggled to destroy. Owing to a combination of changing priorities, difficulty in joint interoperability with the Air Force, diplomatic concerns, and the need to develop an operating base where there was no American presence, the Task Force ultimately failed to perform its original mission. The Army faced criticism over Task Force Hawk's failure and concurrently faced an unsettling reality: a military campaign against a foreign state succeeded without any Army units involved in the fighting. For Shinseki, this raised two fundamental questions: how did the Army remain relevant in a changing age where airpower finally appeared to be capable of delivering on the promises of Giulio Douhet, and how could the Army develop credible strategic agility? The questions harkened back to the 1950s—the Army's internal angst under the New Look national security strategy and its efforts to redefine itself. For Shinseki, the answers lay in undertaking a bold transformation of the Army, announced in 1999, and one that when complete would posture the Army for battlefield success in any environment and against any foe.[54]

Shinseki inherited both the "Big 5" and the modernization efforts initiated by Reimer and Sullivan. On this foundation, he sought to develop a new "objective force" possessing capabilities far greater than those of what he called the "legacy force." The legacy force, built around the "Big 5," represented a massive investment not only in equipment, but also in doctrine, training, leader development, organizational development, personnel, and facilities and would take years to replace. The objective force would possess seven characteristics: responsive to the full spectrum of combat, deployability, agility, versatility, lethality, survivability, and sustainability. Immediately jumping to the objective force was a technical and fiscal impossibility—the means of addressing all the DOTMLPF elements, not to mention sufficient funding, did not exist. Between "legacy" and "objective" the Army required an interim force that represented a step forward while not jeopardizing its ability to accomplish its wartime mission. It also required that this interim force inform objective force development efforts *and* engage in any contingency operations since the Army lacked the resources to sustain a large, purely experimental organization. The Army engaged in parallel efforts to develop the two new forces. At the heart of this effort were two different families of ground combat vehicles—one representing an interim solution with the other representing the final objective. The conceptual model of the legacy force, the interim force, and the objective force defined Shinseki's vision and framed virtually all presentations of his modernization and transformation efforts. These efforts resulted in two markedly different outcomes with results that shaped the Army through the present. [55]

The M1 tank and M2 infantry fighting vehicle provided the legacy force's core. While still capable, these vehicles required attention not only in resourcing but in rethinking their employment. On June 14, 2001, the Army released a new capstone doctrinal manual, *FM 3-0, Operations.* The new number, *3-0,* brought the Army in line with the Defense Department's publication system. More important than the new number, however, were two concepts: full spectrum operations and unified action. Full spectrum operations included offense, defense, stability, and support, and recognized that thc mission's unique nature dictated the proportion and relationship between these four types. The attention to stability and support recognized the experiences of the Balkans, Somalia, Haiti, the short-lived humanitarian effort in Rwanda in 1994 as well as domestic support missions through the 1990s. It also acknowledged military operations other than war that included both conflict resolution and the promotion of peace. The Army classified the latter as peacetime military engagement with this encompassing such actions as combined exercises with other states. Unified action recognized the Army's position as a member of the joint service team and the requirement to synchronize action with that of the other services, allies, or other partners. Published just two months before the attacks of September 11, this was the doctrine the Army took into Afghanistan and Iraq.[56]

The objective force was the main effort—it was where the Army invested most heavily and portrayed the Army's broadest ambitions. Conceptually it showed concepts developed under Sullivan and Reimer, especially leveraging advanced communications and computer technologies. It also displayed Shinseki's reflections on the Army's Balkans experience, especially the difficulties encountered in both deploying and operating in an area with very restricted terrain and poor ground lines of communication. For Shinseki, the objective force required a new vehicle: despite his background as an Armor officer, he believed tanks would not work in the future. He sought a new vehicle that was smaller and lighter; possessed greater mobility, lethality, and survivability; and was more sustainable. Ideally, he sought a wheeled vehicle 50-70% lighter than the M1A2 tank. Shinseki did not believe Armor's mission had disappeared, but he believed the tank had reached its limit of development and that the mobility and agility required of the mission required a new approach in vehicle development. Shinseki thus visualized a Future Combat Vehicle (FCV). The FCV, possessing a common chassis, would have various mission configurations. When procured, these vehicles would serve in the objective force as elements of the Future Combat System (FCS). Fundamentally, the Army's self-conceptualization was largely unaltered: as envisioned, FCS provided the Army with new tools to use against technologically sophisticated opponents.[57]

Sometimes described as "18+1+1," FCS consisted of 18 systems—eight different combat vehicles, two "unattended munitions," four classes of unmanned aerial vehicles (UAV), three classes of unattended ground vehicles, and a multifunction utility/logistics/equipment vehicle. Connecting these various equipment items was a high-capacity communications network, +1, and the individual soldier, the second +1. While conceptually simple, making FCS work presented the Army—and its sister services—with many technical challenges. The mission statement of a briefing the Program Manager Future Combat Vehicle presented to industry representatives interested in working on the program gives some idea of the anticipated problems. It stated, "The FCS must provide the Army with significant combat overmatch against all foreseeable enemies on the battlefield in 2010 - 2025. The system must use smaller combat formations capable of very high OPTEMPO, yet have a significantly reduced logistical support requirement."[58]

In the abstract, the enemies envisioned existed at all points on the spectrum of conflict and might avail themselves of a variety of weapons and tactics. In concrete terms, that meant FCS needed to defeat targets ranging from powerful combined arms formations using the most modern weapons and equipment to guerrillas employing unsophisticated tactics and using relatively simple weapons. Between these two poles lurked such threats as nuclear, chemical, and biological agents; long-range artillery fires; manned and unmanned aircraft; man-portable air defense missiles; organized infantry; robust reconnaissance forces; and mines. As a practical matter, there existed a pronounced bias towards the higher end of the spectrum of conflict characterized by the armored, infantry, and air defense units found in technologically sophisticated armies. To effectively perform, FCS had to simultaneously collect, communicate, process, and present intelligence information from imagery, signal, electronic, measurement and signature, and human intelligence sources. Implicit in the mission statement was the need to address these threats across vast distances with little room for error. This meant engineering new ground and air-based systems capable of operating under all weather, temperature, and terrain conditions. It also meant developing the way to securely pass all this information between the widely dispersed FCS elements. A daunting technical challenge lay in the communications requirements alone as the systems employed had to not only work amongst themselves but remain safe from enemy exploitation. Also implicit in this were monumental changes in doctrine, organizations, leader development, and training.[59]

The mission statement's second sentence described a daunting challenge: employing smaller, faster, and more easily sustained formations to defeat any enemy. Smaller formations require more capable vehicles and weapons to compensate for reduced numbers. This means developing vehicles able to engage the enemy at longer ranges while possessing the ability to survive both known and emerging threats. A faster operational tempo necessitates not only improving the ability to deliver the intelligence, operational, and logistical information required for operations, but also developing the human capacity to act on that data. A reduced logistical footprint meant reducing the requirement for all types of supply including maximizing commonality of components between the various systems constituting FCS. Taken together, the two short sentences of the mission statement presented a long list of DOTMLPF issues to solve. They also provided a massive budget item—one that would grow over time.[60]

While the Army possessed a great deal of technical expertise in a wide variety of areas, producing FCS required capabilities not resident in the Army and introduced a requirement to develop both new partnerships and systems development methods. Early on, the Army partnered with the Defense Advanced Research Projects Agency (DARPA). A February 2000 memorandum of agreement (MOA) between the Army and DARPA gave the lead for early FCS development to DARPA with the Army having a supporting role. One early outcome of this arrangement was DARPA selecting Boeing Corporation, which had teamed with Science Applications International Corporation (SAIC), as the program's lead system integrator (LSI). The LSI had responsibility for making the constellation of systems work together and by 2003, they had selected 21 different firms to work various aspects of FCS. Between signing the DARPA MOA and the LSI selecting the firms responsible for turning FCS from concept to reality, one completely unanticipated event took place that would ultimately redirect FCS efforts: the attacks of September 11, 2001. Before the attacks, Shinseki personally involved himself in driving transformation and hosted weekly meetings with principals involved in moving FCS to completion. Following the attacks, Shinseki and the Army staff had to adopt new priorities and devote their attention to the demands of wartime. Shinseki remained engaged in transformation, but the day-to-day supervision of the prior main effort, FCS, transitioned to the Director, Objective Force Task Force, Lieut. Gen. John M.

Riggs. Not fully appreciated at the time, relegating the task of shepherding FCS to "second chairs" was one of several issues that put the program on the road to ultimate cancellation.[61]

Betting heavily on the objective force, the Army made several budgeting decisions in 2002 intended to accelerate the program. These include investing more than 95 percent of its scientific and technological development funding and 20 percent of its research and development and acquisition funding to FCS critical items. This was not the only area in which the Army invested heavily, but it is noteworthy that as the fighting began in Afghanistan, the service remained committed to FCS and its ambitious capabilities suite. It was in the ambitious suite of desired capabilities that a second FCS issue appeared. Described as an all-encompassing "system of systems," FCS had become something of the proverbial elephant. In the words of retired Brig. Gen. Donald Schenk, former Program Manager for FCS, "what part you are feeling tells you what it is."[62] The Army's different constituencies all saw FCS differently—for those from the Tank-Automotive Armaments Command it was all about vehicles, for those from the Signal Corps it was all about the network, for the aviation community it was all the unmanned aerial vehicles. Shinseki might have reconciled the confusion and misunderstandings, but Riggs lacked the ability necessary to do so—a task force director, no matter how good, lacked the Chief of Staff's authority. This supervisory change would haunt the effort in years to come.[63]

While the efforts to develop the objective force encountered institutional difficulties in the wake of September 11 and revised priorities, the story was very different from the interim force that saw its position elevated. Shinseki assumed the Chief of Staff's duties in June 1999 and by December had designated the interim force's first unit as the 3rd Brigade, 2nd Infantry Division, and was evaluating 35 different light-to-medium weight armored vehicles to determine the one best suited for Army service. Validating concepts for the objective force, however, required the interim force to be more than just a medium-weight vehicle—it also called for having significant digital connectivity with other fighting vehicles as well as its commanding headquarters. The interim force could not succeed if it just possessed connectivity or if it just possessed lighter-weight vehicles—it had to possess them both in a tactically functional package capable of rapid deployment and effective combat when committed. The Army reached a milestone for the interim force in November 2000 by selecting a version of the Canadian Army's Light Armored Vehicle, Third Generation (LAV III), as the base vehicle for the interim force. In making a vehicle decision and signing a procurement contract, the Army realized another major milestone with the interim force. Two thousand and two saw a symbolic accomplishment with the formal designation of the interim armored vehicle as the Stryker. This effectively ended any references to the "interim force" within the Army: what was going to be the interim force now existed, at least in part, and was no longer a theoretical element on briefing slides. By the time Shinseki's tour as Army Chief of Staff ended, the Army had fielded one Stryker-equipped infantry brigade, was preparing it for combat, and had started the process of converting the 1st Brigade, 25th Infantry Division, as its second, and designated three others for conversion.[64]

The Stryker represented a success for Shinseki. Despite the considerable controversy that the idea generated from the beginning, Shinseki guided the program from concept to fielding in less than four years, a remarkable feat. He accomplished this despite the need to make significant survivability improvements that made both strategic deployability and battlefield use more difficult. He also overcame protests from industry and advocates, skepticism within the Army (including the procurement bureaucracy), and at times Defense Department hostility. Arguably, his efforts benefitted from the revision of priorities following September 11, 2001—it was much easier to build on a program that had more than a few concrete steps completed than one that was still mostly ethereal as was the case with FCS. Most significantly, he proved a fundamental truth about transformation—if the visionary does not create something on his watch it likely will not come to fruition as intended. The differences between Stryker procurement and the procurement cycles of the M1 tank and M2/M3 infantry/cavalry fighting vehicle which spanned over a decade are remarkable.[65]

Transformation during Shinseki's tenure went beyond re-envisioning the operational force structure. Changes took place within the Headquarters, Department of the Army and with the reserve components. The changes to the reserve component were among the most important in their history. In the first 90 days following the attacks of September 11, 2001the Army mobilized over 50,000 National Guardsmen for a variety of homeland and defense installation security tasks. Following this were two major initiatives—fielding more Civil Support Teams and programming units for overseas service. By 2003, more than 21 percent of the Army National Guard had deployed overseas with approximately 38,000 soldiers serving either in or in support of operations in Iraq with this including seven infantry battalions. The Army Reserve experienced large-scale mobilizations with troops deploying to Afghanistan and Iraq and uncovered fundamentally weak administrative processes rooted in the strategic reserve Cold War mobilization model rendered ineffective as a result of the attacks. This mobilization led to internal changes within the Army Reserve to improve both its ability to mobilize and support combat operations globally.[66]

The post-Cold War Army's transformations reflect two paths, sometimes converging and sometimes diverging. The early post-Cold War found the Army fighting a war that validated both its self-conceptualization and the investments made in the "Big 5." It subsequently found itself engaged in a range of operations widely varied in purpose, scope, and location with some challenging long-held assumptions. Juxtaposed against this background were the touted Revolution in Military Affairs and the major reordering of the global institutions that clearly defined the Army's understanding of its environment. These experiences and the changing political and technical environment allowed the service to project a clear vision for change. That image shattered on September 11, 2001, and the aftermath of that date forced the Army to redefine its ideas. It would not willingly abandon its original vision, but the realities of two major campaigns in Afghanistan and Iraq and a host of related engagements in other locations made clear previously un-envisioned changes were necessary.

Conclusions

For the United States Army internal and external forces influence change. Change, however, most reflects the Army's self-conceptualization and bureaucratic processes. The Program Objective Memorandum, dating to 1961, represented the most important of the bureaucratic processes as it translated all the Army's ideas, plans, and aspirations (essentially its self-conceptualization) into actual monetary resources. What the POM enabled; the Army possessed. Before the POM, the budgeting process essentially fulfilled the same fundamental role. That process lacked the POM's rigor, but still determined in clear terms the activities the Army could undertake. Irrespective of the rationale or importance of the desired change, the budget ultimately determined what would (or would not) happen.

What the Army sought illustrated many needs. To this day, the Joint Strategic Capabilities Plan (JSCAP), provided by the Secretary of Defense informs POM submissions by telling the Army what capabilities it will provide. Equally important, however, is the Army's self-conceptualization as this translates JSCAP requirements into actual units with specific units and equipment items.

There is a clear correlation between the DOTMLPF factors and the POM. While the most visible impacts are on equipment and facilities, the other elements both affect and reflect the POM. As the Army visualizes its future, DOTMLPF factors provide the framework for translating rough concepts into concrete plans. Those plans, however, are not able to drive any form of change unless the POM includes them and the service advocates for them. It is not enough for the Army to document a requirement on the POM, it must act as the champion if that requirement is to receive the funding desired as the Army competes with the other services for resources. The POM is the penultimate bureaucratic process as it brings life or death to doctrine, training, leader development, organizations, material, personnel, and facilities.

To understand change is to understand the role of self-conceptualization, bureaucratic process, most notably the POM, and their all-powerful influence within the Army.

Chapter 2

The Army's Historical Experience with Asymmetric Warfare

> The conventional army loses if it does not win.
> The guerrilla wins if he does not lose.[1]
>
> Henry A. Kissinger

Asymmetric Roots

From its founding on June 14, 1775, the United States Army lacked an official flag until President Dwight D. Eisenhower signed Executive Order 10670 on the service's 181st birthday in 1956. This same order directed the flag display a streamer representing each campaign in which the Army participated. By 2020, the number of streamers had reached 190. Of these, 65—over one-third—reflect actions against irregular or insurgent actors dating from the republic's earliest days to the present. The Army's oldest combat battalions carry additional streamers recognizing punitive expeditions across the North American continent. Not commemorated with streamers are additional operations inherently asymmetric in nature. These include assisting emerging governments in Greece and Korea in response to communist insurgencies in the late 1940s. They also include creating Military Assistance Advisory Groups (MAAG) in friendly countries in the 1950s and 1960s to improve their ability to counter emerging threats. The late Cold War period saw continued operations, including combat, in Latin America and the Persian Gulf. At the Cold War's end, the Army conducted short-term operations in Somalia and Haiti. More enduring efforts to both restore peace and capture war criminals began in the Balkans in the 1990s. Not all these operations involved Army units in a combat role, but all required an investment of resources. Each also provided opportunities for the Army to learn lessons that might prove valuable on future occasions. Asymmetric warfare is thus as much a part of the Army's historical DNA as conventional campaigning. Asymmetric warfare tended to involve arduous—and often frustrating—efforts conducted over extended periods with frequently inconclusive results.[2]

Some of these operations offer little insight into effective strategy or tactics. The Seminole Wars (1817-1831, 1835-1842, and 1845-1848) found an ill-prepared, ill-led, and often understrength force struggling in conditions utterly foreign to most troops. Locally decisive outcomes only (and infrequently) came about through tactics judged both immoral and illegal today: the outright destruction of Indian settlements and food supplies and ignoring the distinctions between combatants and non-combatants. The same tactical approach appeared again in campaigns against the Comanches, Southern Cheyenne, and Kiowas (1867-1875). Prevailing racial norms meant most white Americans had little regard for the indigenous population. Many senior Army leaders shared that perspective. Commanding the Military Department of the Northwest during the 1862 Dakota War, Maj. Gen. John Pope stated, "It is my purpose utterly to exterminate the Sioux."[3] In 1864, Col. John Chivington, commanding all Army units in the Colorado Territory voiced similar comments. "I have come to kill Indians, and believe it is right and honorable to use any means under God's heaven to kill Indians."[4] Unlike Pope, Chivington owed his appointment as an officer to the Army's Civil War expansion. Chivington's conduct during the March 1862 Battle of Glorieta Pass led Col. Edward R.S. Canby, then commanding all troops in New Mexico, to promote him to colonel, subject to the Territorial Governor's approval. Two years later, however, Chivington fielded a force of 800 soldiers and he possessed the will to pursue a war of extermination against Colorado's native population. In his report following an engagement that became known as the Sand Creek Massacre, Chivington claimed his soldiers killed between 500 and 600 members of the Cheyenne nation.[5]

If Pope and Chivington represented one point on the spectrum of attitudes towards the "Indian problem," the positions adopted by Brig. Gen. George Crook represented the other end. Crook held several commands across the West and justifiably acquired a reputation as a "relentless opponent of the Indian on the battlefield and a steadfast friend off it."[6] Records show Crook seeking accommodation with the natives. His most significant effort is the case of *Standing Bear v. Crook*. After the latter refused Ponca chief Standing Bear a writ of *habeas corpus* at Fort Omaha, Nebraska, Standing Bear filed suit. Crook both assisted Standing Bear in preparing his suit *and* testified on his behalf at trial. *Standing Bear v. Crook* ended with Judge Elmer Dundy finding for Standing Bear, and concurrently conferring on the Indians the legal status of American citizens. The philosophical positions of Crook, Pope, and those officers who shared their divergent opinions led to differing methods of both, combating and preserving peace once the battle ended. Throughout the Indian Wars period, tensions existed between those seeking the destruction of the natives and those seeking co-existence. These played out on the battlefield, on the reservation, and in Congress. They would do so until October 5, 1898, and the Army's last fight against an Indian tribe at Leach Lake, Minnesota.[7]

If most of the Indian Wars campaigns are of limited utility in understanding the problems of twenty-first century asymmetric warfare, others offer valid lessons. The Army's Civil War-era General Orders (GO) 100 provided a legal foundation for determining acceptable conduct by asymmetric actors. It also provided the basis for responding to violations of those norms. In the Philippines (1899-1913), establishing an American government led to conflicts that found a small Army simultaneously fighting, pacifying, and administering a diverse population. The Punitive Expedition into northern Mexico (1916-1917) offers examples of the Army rapidly introducing new technologies to aid in operations—and discovering technology's limitations. Airpower, mechanization, and advanced communications all failed to meet expectations and contributed to the expedition's ultimate failure. The Vietnam War generated multiple challenges while the American military waged a simultaneous conflict between a conventional force and an asymmetric one. These included dealing with difficult allies and disaffected populations, integrating technology, and conforming operations to legal standards. An additional consideration was the impact of constant scrutiny by a news media increasingly disinclined to accept at face value all it heard from official sources. For the Army, this historical legacy demonstrates that it experienced the fundamental challenges that would confront it in Afghanistan and Iraq beginning in 2001.[8] Concurrently, it demonstrates a recurring and sometimes difficult reality in asymmetric conflicts: the need for even more rapid and far-reaching adaptations than is the case in fights against peer/near-peer opponents.

The American Civil War and Reconstruction, 1861-1867

The Civil War conjures images of massed formations in blue or gray inflicting horrendous casualties on each other. One day of fighting at Antietam, Maryland, on September 17, 1862, resulted in 12,401 Union casualties alone. Adding Confederate casualties takes the number to over 24,000, a total exceeding the combined battle casualties of all of America's preceding conflicts. One out of every five soldiers engaged at Antietam became a casualty. The sites of large Civil War battles bristle with myriad monuments to individuals and units. These are the battles that live in popular memory. Less well known, but just as important, was a concurrent asymmetric struggle. Conducted by a variety of actors, this fight created multiple challenges for the U.S. Army. Meeting these challenges meant assigning troops to a variety of guerrilla fighting or security duties. Such assignments sapped the Army's ability to create the mass needed for decisive operations and helped prolong the war's course. Developing the means to defeat both large organized forces and multiple insurgencies resulted in the Army creating new legal and tactical doctrine.[9]

The asymmetric fight raised multiple issues for the War Department and field commanders alike. At that time, the term "insurgent" did not exist in Army parlance. Irregulars received such labels as "bushwhackers," "guerrillas," "partisans," "raiders," or "rangers." The fundamental challenges for soldiers and commanders lay in determining an individual's status and then dealing with the conduct in a way that supported national policy and was consistent with the law. Throughout the war, Union policy focused on inducing the Confederacy's return. President Abraham Lincoln believed a policy of reconciliation would heal the rupture and return the war's underlying issues to the political sphere. Such a policy thus initially served as a restraint on the war's prosecution. As for insurgents, the existing corpus of military law, the Articles of War, was largely mum. Congress approved the version in use at the war's start in 1806. Constructed in three sections, only Section 2 offered any insight on the issue. This section stated

> (T)hat in time of war, all persons (...) owing allegiance to the United States of America, who shall be found lurking as spies, in or about the fortifications or encampments of the armies of the United States (...) shall suffer death (...) by sentence of a general court martial.[10]

The actions of insurgents went far beyond simple "spying." In requiring a trial of persons "owing allegiance to the United States," the Articles inferred insurgent activity was as much a legal issue as a military one. This lack of clarity framed the fundamental issue: an individual's actual status and the actions the Army could legally take based upon it.[11]

The combination of widespread insurgencies, differing perceptions of an insurgent's legal status, and the steady erosion of military strength frustrated both soldiers and the War Department through 1863. The individual soldier's overarching concern was to end often-fruitless searches for insurgents and eliminate the physical threats they posed. The War Department worried primarily about crafting a policy that would both define the limits of legal action by irregular forces and rationalize the Army's response when confronted by such forces. The War Department thought such a policy would influence Confederate behavior. In contrast, Confederate leadership held varying ideas about guerrilla warfare's propriety and how to respond to Union actions against guerrillas. At the heart of this was the question of reprisals and concerns over an ever-escalating level of violence that could make any post-war reconciliation impossible.[12]

One officer concerned with both the guerrilla issue and its legal implications was Maj. Gen. Henry Halleck. Known to the Army as "Old Brains," Halleck gained some notoriety within the service for translating Antoine-Henri Jomini's *Vie Politique et Militaire de Napoleon* (*Political and Military life of Napoleon*) and publishing a series of lectures on military art and science in 1846. Leaving active service in 1854, he settled into legal practice in San Francisco, California, and served in the California Militia before returning to active service in 1861. After the outbreak of the Civil War, Halleck personally grappled with the problem of asymmetric warfare while commanding the Department of Missouri, an area teeming with Confederate guerrillas. Replacing Maj. Gen. John C. Fremont, an officer whose actions against guerrillas tended to inflame the problem further, Halleck enjoyed sufficient success in command to come to Lincoln's attention. Following the relief of Maj. Gen. George B. McClellan as the Commanding General of the United States Army on March 11, 1862, Halleck replaced him in July. Halleck made the development of a policy governing the treatment of asymmetric actors applicable to Union commanders in all the war's theaters one of his top priorities.[13]

On August 6, 1862, Halleck wrote Francis Lieber, a noted jurist on Columbia College's faculty, asking his views on a guerrilla's legal rights. Halleck noted the Confederate position that guerrillas should enjoy the same status and treatment as conventional combatants and the threat of reprisals if Federal authorities punished Rebel irregulars as "marauders or spies." In responding to Halleck, Lieber observed, "It is universally understood in this country (...) that a guerrilla party means an irregular band of armed men, carrying on an irregular war, not being able, according to their character as a guerrilla party, to carry on what the law terms a *regular* war."[14] Lieber followed this statement with a historical overview of European experiences, rooted mainly in the campaigns of Napoleon. In Lieber's opinion, the prevailing body of international thinking on the subject was that the "rising of the people to repel invasion entitles them to the full benefit of the law of war (...) so long as they have risen in the yet uninvaded or unconquered portions (...) of the country."[15] The discussion between the two lawyers laid the foundation for the war's most important military and enduring legal doctrine—War Department General Orders (GO) 100, dated April 24, 1863.[16]

Also known as "the Lieber Code", GO 100 attempted to address the legal questions of war in their entirety through its 157 articles. Articles 81 to 85 are particularly relevant as they focus specifically on partisans, armed enemies not belonging to the hostile army, scouts, and armed rebels. Article 81 provided legal protection to partisans, defined as "soldiers armed and wearing the uniform of their army, but belonging to a corps which acts detached from the main body for the purpose of making inroads into the territory occupied by the enemy."[17] Articles 82 through 85 effectively denied those same protections to other actors committing acts of violence or espionage against Union forces. Articles 149 through 157 focused on "Insurrection, Civil War, and Rebellion." While several articles dealt with legal definitions, Articles 156 and 157 are particularly noteworthy. The first authorized a "stricter police" of disloyal citizens, while the second read, "Armed or unarmed resistance by citizens of the United States against the lawful movements of their troops is levying war against the United States, and is therefore treason."[18] With GO 100's publication, the Army possessed the common legal doctrine necessary to support operations against both conventional and asymmetric forces.[19]

GO 100, while important, did not provide the Union Army with all the tools needed to vanquish the asymmetric threat. While guerrilla operations persisted to the war's end, the Army did undertake several initiatives to reduce its impact. These displayed the unique characteristics of a given geographic region. Along the Mississippi River, the Confederate tactic of shore-based ambushes and artillery strikes against vessels trafficking the river led to the creation of the Mississippi Marine Brigade, an initiative demonstrating the Army's ability to adapt to new circumstances. This organization, placed under the command of Brig. Gen. Alfred W. Ellet, gave the Army a force capable of both fighting aboard vessels *and* landing along the river to pursue and destroy guerrillas seeking to interdict river traffic. The directive approved by Secretary of War Edwin M. Stanton called for a brigade consisting of one infantry regiment, four cavalry squadrons, and a battery of light artillery. This force would serve aboard the vessels of the existing Ram Fleet, itself largely a creation of Charles Ellet, Jr., Alfred's father, who had convinced Stanton that the Army needed new methods to protect navigation on the Mississippi. Authorized in March 1862, the Ram Fleet's original mission was to ram Confederate naval vessels. The disappearance of such vessels from the "Father of Waters" following the Union's June 6, 1862, naval victory at Memphis forced the Confederates to rely on guerrilla attacks to impede Union shipping. That transition led to the Ram Fleet's repurposing.[20]

The issue of which service—the Army or the Navy—should exercise primacy on the Mississippi was one never fully satisfactorily resolved. The Navy saw relatively large vessels, such as those of the Ram Fleet, as rightfully belonging to them. In contrast, Stanton believed that the sort of operations conducted by the Marine Brigade should be the Army's responsibility. Ultimately, it took the intervention of President Abraham Lincoln to resolve the issue. He decided that the Marine Brigade would remain under Army control with the Navy's Rear Adm. David Porter directing its day-to-day operations. This compromise was both unsatisfactory and short-lived with the Army assuming full control of the brigade on August 24, 1863. While the brigade did not enjoy overwhelming success against guerrillas, it did exact a high price on local populations for not acting against them. Citizens in Arkansas, Louisiana, and Mississippi experienced the destruction and confiscation of property as the brigade operated ashore. In so doing, it provided the Army with an example of the harm specialized units could inflict on strategic objectives if they lacked effective supervision. This was a lesson the Army would subsequently re-learn.[21]

Away from the Mississippi, the Army employed several different techniques. In northern Missouri, Brig. Gen. John M. Schofield dispersed a combination of cavalry and infantry. Deployed at multiple locations, such forces could keep possible insurgents under observation and quickly move to any location where they might concentrate. This basic tactic saw use elsewhere and with varying degrees of effectiveness. Available Union manpower, a commander's own initiative, and the relative balance of power between forces determined success or failure. Col. Marcus L. Harrison grew frustrated with inconclusive outcomes and introduced a much more radical tactic: concentrating loyal citizens in colonies capable of defending themselves. Creating fortified locations in northwest Arkansas, Harrison's plan accomplished three critical tasks. First, it reduced Union manpower requirements by reducing the number of locations requiring security. Second, it ensured the safety of Union loyalists and their crops. A side benefit of this was participants selling their crops to the Army. Third, it identified as irregular non-participants within ten miles of a location.[22]

This model's success caused other commanders to copy it across northern Arkansas. In other locations, the Army took a different approach—small posts to protect targets it deemed vulnerable to guerrilla attack. To cover the space between these posts, a combination of infantry and cavalry patrols, armored trains, and even gunboats provided additional security and firepower. In some cases, the Army relocated local populations believed supportive of irregular actors and defoliated locations to provide better observation and fields of fire.[23]

Fortifying positions and patrols were useful reactions to the problems of asymmetric warfare. Ultimately, however, the Army needed to focus combat power against the insurgents it faced. This was not possible without accurate intelligence. Patrols could gather some of the information required, but Provost Marshals could do more. With duties including administering loyalty oaths, Provost Marshals maintained records of loyal and disloyal individuals. Such information provided a foundation for identifying suspected insurgents and their associates. When provided to units, such information made patrols and raids more effective and reduced insurgent effectiveness. [24]

The Confederacy's collapse saw the Army assuming new duties. Reconstruction meant the Army served as both governing authority and in support of civilian government as laws and circumstances changed. The First Reconstruction Act of March 2, 1867, established five military districts across the former Confederacy. The act also required "the President to assign to (...) said districts an officer (...) not below the rank of brigadier-general (...) and (...) sufficient military force to enable such officer to perform his duties and enforce his authority within the district to which he is assigned."[25] The force performing these duties, often included arresting suspected insurgents, was both small and diverse. In January 1866, the Army's strength in the South was just over 87,500 soldiers. By 1868, the number dropped to 17,765. The numbers continued to decline through 1871 when only 8,038 troops occupied the South. Initially, troop strength included Blacks serving in U.S. Colored Troops units. These units in particular enflamed Southern sensitivities. Various militia units, including units comprised of both former Union and Confederate soldiers and newly freed blacks, periodically augmented Army efforts—but not as part of a unified command. Unsurprisingly, the most challenging missions involved safeguarding blacks and southern Republicans from violence. Groups such as the Ku Klux Klan formed across the South as part of white efforts to maintain social and political control. In some areas, as much as 75 percent of the white male population claimed membership. The combination of a hostile populace and small numbers of troops available challenged the Army. Exacerbating the problem was a lack of cavalry after 1869. The demands of the Western frontier took precedence over occupation. This meant only a steadily decreasing number of infantry units were available after 1868. These units occupied strategic locations, leaving much of the South devoid of a capable Army presence. The result was a mixed performance by the Army as it sought to control the Klan and similar groups. When the Army did enjoy success in enforcing the Reconstruction Act and subsequent legislation, it often reflected the contributions of state militia. In maintaining order, an effective militia loyal to the legal authority provided the force required to move against groups like the Klan. Arkansas Governor Powell F. Clayton's organizing and deploying a large and loyal militia force played a major role in establishing public order. A guerrilla hotbed during the war, Arkansas was largely free of such activity by 1869. In North Carolina and Tennessee, militia results were less dramatic but nevertheless contributed to the Klan's decline.[26]

The era's asymmetric campaigns provided the Army with a framework for understanding such contests. The conflict established a clear legal framework for a broad range of actions in GO 100, a document still influencing international discussions on the law of war. It also gave the Army a menu of tactics and lessons applicable in future conflicts. These included the need for supervision of units conducting operations and collecting, assessing, and disseminating intelligence information. Finally, it made clear the linkages between political and military actions in such conflicts. Each shaped the other as they unfolded.[27]

The Philippines, 1899-1913

The Army's Philippine experiences sprang from two actions taken by Comm. George Dewey while commanding the U.S. Navy's Asiatic Squadron during the Spanish-American War. The first was the sudden capture of Manila Bay on May 1, 1898. The second was returning rebel leader Emilio Aguinaldo to the Philippines on May 19, 1898. Aguinaldo previously led an abortive rebellion against Spanish rule and accepted exile in Hong Kong as part of a peace settlement. In returning Aguinaldo to his homeland, Dewey transplanted a nationalist leader who believed his return meant an independent Philippines. Even as Dewey took control of

Manila Bay, the U.S. government lacked a long-term plan for the archipelago. The subsequent arrival of American soldiers in June set the stage for four separate conflicts.[28]

The first of these was a short and relatively bloodless battle against the Spanish to seize Manila, the seat of Spanish government in the islands. In response to a request from Dewey, Army troops began arriving in Manila Bay at the end of June. Almost six weeks of preparations culminated with on August 13, 1898, when the Army's 8th Corps assaulted the city. This brief battle saw American and Filipino troops working in concert to force a Spanish surrender. It also resulted in American troops working against Aguinaldo's Filipino forces by denying them access to the city. The United States lacked a definitive plan for the islands, but keeping Aguinaldo and his troops out of Manila signaled any plans would prioritize Washington's interests. This paved the way for the Army's second conflict in the islands: a fight against Aguinaldo's conventionally organized Army of Liberation (February 1899-November 1899). Aguinaldo's failure to defeat the Americans or negotiate a settlement led to the third conflict: a guerrilla campaign against Aguinaldo's *insurrectos* (November 1899-July 1902). Largely composed of former Army of Liberation soldiers, the *insurrectos* would challenge the Army through 1902 and the conflict's official end. Even after the campaign's end, however, remnants and brigands would pose problems as the Americans sought to govern the islands. Finally, a fourth conflict against the Islamic Moro tribes of the archipelago's southern islands meant the Army remained engaged in asymmetric operations through 1913. That the Moros would challenge the Army is unsurprising as they resisted Spanish rule for over 300 years. Even as the Army fought, however, it found itself responsible for civil administration. This included not only actual governance for a short period, but also conducting tasks in such diverse fields as education, public health and sanitation, law enforcement, and infrastructure development. To accomplish its many missions, the Army reviewed its experiences in the Civil War and Indian Wars while developing new approaches reflecting the unique situation in the Philippines. Ultimately, a force of American soldiers and native auxiliaries employed the complementary policies of attraction (the carrot) and chastisement (the stick)—and technological and organizational developments—to end the insurgencies.[29]

Like the Mexican War, the Spanish-American War's end found the Army establishing federal authority over a population lacking previous political ties to the United States. In 1900, the Philippines possessed an estimated population of almost 7,000,000 people scattered across 7000 islands. Demographically, the population was a mélange of ethnicities, cultures, and religious faiths. Catholicism dominated the northern islands with a considerable Muslim population in the Sulu Archipelago, Palawan, and Mindanao. Spanish authority centered in Manila on the island of Luzon, and home of most of the country's political and social elite.

Geographically, the islands displayed remarkable diversity. Some areas contained large agricultural plantations, while others remained wild. Terrain features ranged from cinder cone volcanoes to rugged mountain chains to fetid swamps. Rivers and poor roads were essential lines of communication, but their characteristics changed with the time of year as the annual monsoon brought heavy rains to some parts of the country. The islands contained more than 22,500 miles of coastline and location, time, tide, and season all affected movement along or over it. The interaction of demography and geography shaped Army operations across the islands, but almost as important were domestic American political considerations. The Spanish-American War might have been "a splendid little war," but many Americans were uncomfortable with the country becoming a colonial power at its end. Among the most vociferous were members of the Anti-Imperialist League, a group arguing against colonial expansion on both moral and economic grounds. As a result, the Army was as acutely aware of the political currents in Washington as it was in Manila through its service in the islands.[30]

The conventional campaign against Aguinaldo's Republican Army began with the failure of its marriage of convenience with the U.S. Army following Spain's surrender. Initially, both armies cooperated against the Spanish in Manila. While the armies shared a common tactical objective (defeating the Spanish garrison), their divergent strategic ones meant this relationship was short-lived. Tensions between the two forces grew after the battle for Manila. A clash between American and Filipino soldiers on Manila's outskirts on February 4, 1899, triggered general hostilities. Casualties mounted as the fighting spread with growing intensity throughout the islands.

On June 2, 1899, Aguinaldo lieutenant Pedro Paterno declared a state of war. Both sides faced a fundamental problem: weakness. For the Americans, low troop strength meant they could only conduct limited operations. For the Filipinos, the lack of expertise, modern weapons, and the internecine conflict meant they did not possess any real offensive potential. Decisive American action was not possible until newly raised regiments arrived in September. The aggressive employment of fresh American troops meant the Republican Army's destruction by November. Aguinaldo responded to these changed circumstances by proclaiming guerrilla warfare on November 13, 1899.[31]

This decision marked the beginning of third campaign: operations against the *insurrectos*. Aguinaldo's *insurrectos* represented a new type of opponent for the Army—one not seeking battlefield victory but instead mounting operations aimed at discouraging both Congress and American voters. America's 1900 election would see Republican William McKinley seeking re-election against Democratic challenger William Jennings Bryan. Bryan was the choice for the Anti-Imperialist League and others opposed to American ownership of the

Philippines. Aguinaldo sought an expensive, drawn out and inconclusive campaign inspiring one of two outcomes. The first was Congress granting independence while the second was a Bryan election followed by his granting independence. McKinley's reelection did not change Aguinaldo's strategy significantly. Until Aguinaldo declared his allegiance to the United States the real focus of *insurrecto* action lay in the United States and hopes for voter or congressional fatigue. Given the weaknesses of Aguinaldo's forces and the increasing capabilities of the Army, this strategy was rational.[32]

American policy in the Philippines sought an overarching goal: "benevolent assimilation."[33] Announced by President William McKinley on December 21, 1898, the policy stated the "earnest wish and paramount aim of the military administration to win the confidence, respect, and affection of the inhabitants of the Philippines by assuring them in every possible way that full measure of individual rights and liberties which is the heritage of free peoples."[34] Army policy aligned with McKinley's. Maj. Gen. Wesley Merritt's General Order No. 3, August 8, 1898, stated: "You have come not as despoilers and oppressors, but simply as the instruments of a strong, free government whose purposes are beneficent and which has declared itself in this war the champion of those oppressed by Spanish misrule."[35] Maj. Gen. Elwell S. Otis, Merritt's successor as commander of the 8th Army Corps and second military governor of the Philippines, directed his subordinates to "keep scrupulous faith with these people and teach them to trust us."[36] This approach posited peaceful relations between the Filipinos and the Army and confidence that most Filipinos recognized—or soon would—America's dominant position in the islands and would not seek to challenge it. Civic actions—building schools, improving communications and public health, facilitating trade, and sustaining civil administration—represented one of the policy's active elements. Engaging the political, social, and economic elite represented another. A vital component of this was the rehabilitation of former *insurrectos* who abandoned the conflict. Renouncing the insurgency and pledging loyalty to the U.S. created opportunities for security—and upward mobility. As a policy, attraction seemed to offer the pathway to McKinley's objective. As a strategy, it proved initially flawed due to early Filipino resentment toward the Americans and their role in the islands. The primary condition required for success—a willingness to trust the Army—was not initially widespread. As was the case on the western frontier, the Army offered the preferred carrot while retaining the stick. Those Filipinos unwilling to embrace attraction or maintain a pro-attraction façade while conducting or supporting guerrilla warfare experienced the chastisement policy.[37]

The Army's tactics and techniques evolved over the course of its service in the islands. These changes reflected units adapting to new requirements, technological and organizational change, and the social and physical conditions on a given island. New requirements stemmed from an ever-improving understanding of how *insurrectos* fought and the social relationships shaping guerrilla organizations. At the lowest levels, these included conducting patrols, establishing security, and identifying ambushes. It also included properly searching native villages and occupying outposts. Each of these tasks served one or more specific purposes. Patrols sought to both obtain intelligence and destroy insurgent supplies when uncovered. Patrols could consume more than several days or weeks and frequently meant troops carrying their weapons, rations, and equipment through difficult terrain. Physically taxing even without enemy contact—a frequent occurrence—these patrols provided detailed information on both the terrain and people: essential in effectively fighting a counterinsurgency.[38]

Establishing security—both on patrol and occupying an outpost—meant deploying troops to observe and report activity from all points of the compass. Identifying ambushes meant removing the *insurrectos'* ability to inflict casualties. This was often the most dangerous task as a shortage of modern firearms (and a pronounced inability to use them properly) meant that many *insurrectos* attacked with a *bolo* or large, razor-sharp knife. Such a weapon meant soldiers had to come within a few feet of an *insurrecto* without detection for a successful ambush. Active security meant soldiers broke many ambushes before *insurrectos* could strike, and this partially explains the vast disparity between U.S. and Filipino battle casualties. Units usually paid a high price when security broke down: an ambush at Lipa, Luzon, on June 10, 1901, resulted in two dead and five wounded out of fifty American soldiers engaged.[39]

Units learned how to properly search villages for weapons, food, or materials intended for *insurrecto* use. Tactics included moving small units at night to allow searches at first light, cultivating informants, and villagers to identify caches. Once occupied, villages often served as outposts that provided a base of operations for both patrols and pacification efforts. Outposts varied in size in response to local conditions, with security as the primary concern. As with patrols, a breakdown in outpost security could be catastrophic. *Insurrectos* destroyed Company C, 9th Infantry, by springing a successful surprise attack at Balangiga, Samar, on September 28, 1901. Surprised by *insurrectos* during breakfast (meal periods were the only time soldiers did not carry their loaded weapons), Company C suffered 49 dead or missing and 11 wounded out of its 76-man complement.[40]

A mix of field orders, circulars, policies, and general orders (to include some provisions of 1863's General Orders 100) guided units in their efforts. Though never officially issued in its entirety, senior leadership recognized the applicability of both the GO 100's humanitarian provisions and the punitive elements of Articles 82 through 85 that authorized severe punishments for guerrilla activity. General Order No. 8, August 22, 1898, established commissions and provost courts to try Filipinos for a variety of serious offenses, including violations of the law of war. Tragically, American commanders and soldiers did not always respect the official emphasis on benevolent assimilation and lawful conduct. The massacre at Balangiga prompted Brig. Gen. Jacob Smith to launch a brutal campaign on Samar that killed thousands. This ultimately led to a series of court-martials for some of Smith's subordinates and his own trial on a charge of conduct prejudicial to good order and discipline in 1903. Senior leaders were not the only individuals whose emotions got the better of them. Records reflect examples of abusive, harsh, or illegal conduct by small unit leaders or individual soldiers with at least 350 soldiers facing court-martial. Such conduct included the much-publicized "water cure." Soldiers used the "water cure" —forced fluid intake — to obtain information from *insurrectos* or their supporters. Frequently driving such behavior was the discovery of atrocities by *insurrectos,* which both angered and frustrated soldiers in the field.[41]

A continuous challenge confronting American troops was separating guerrillas from their means of supply. This gave rise to one of the war's most controversial tactics: concentration. Isolating the civilian population would both deny *insurrectos* food and supplies and make it easier for soldiers to identify them by their presence in prohibited areas. In Telegraphic Circular, No. 2 of December 8, 1901, Brig. Gen. J. Franklin Bell, a district commander in Luzon, directed concentrating Filipino civilians in Batangas and Laguna provinces. The circular called for individuals to relocate with any food supplies at hand. Bell also provided for both price controls and the construction of temporary shelters to ensure public health. This policy change, while effective, proved politically explosive and generated numerous outcries in the United States. Opponents noted the Spanish use of the same tactic and claims of deprivation suffered by many Filipinos. This tended to limit the tactic's application. In Batangas and Laguna, concentration lasted only 113 days before the surrender of *insurrecto* leader Miguel Malvar. The local insurgency collapsed following Malvar's capture, ending the need to continue concentration there.[42]

The fourth campaign, against the Moros, continued long after President Theodore Roosevelt declared hostilities against the Philippine republicans ended on July 4, 1902. Never well integrated into the Philippines' government under Spain, Moro resistance centered on preserving Muslim religious, cultural, and social practices. These same concerns expressed themselves under American rule. As with Aguinaldo, initial contacts with the Moro were friendly. In August 1899, Maj. Gen. Elwell Otis informed the War Department of an "agreement made with Sultan and *dattos* (chiefs) whereby sovereignty United States over entire Jolo Archipelago acknowledged."[43] Not initially understood by the Army was the independence of Moro tribes. The Moros did not recognize a single leader; thus, the agreement was with one Moro group—not all of them. By

March 1902, the Army recognized the limitations of agreements when 200 Moros attacked a reconnaissance detachment of the 15th Cavalry near Lake Lanao on the island of Mindanao. Reports of Moros murdering American soldiers and the attack on the 15th Cavalry led to efforts to negotiate with responsible tribal leaders. When these failed, the Army mounted an expedition to "punish only those given offense."[44] Such expeditions, often accompanied by high Moro casualties, would continue until 1913.[45]

If the Army often employed the proverbial stick, it was quite happy to utilize the carrot when it would achieve the objective. Captain John J. Pershing first gained renown in the Philippines as a practitioner of "waging the peace."[46] Astute enough to recognize both the fragmented nature of Moro society and the potential for peaceful accommodation, Pershing preferred negotiation when possible. Pershing recognized Moro's concerns and accommodated them when he could. When he could not, he moved slowly to build the relationships necessary to realize objectives. The result was his developing a reputation as one Moros could trust. While Pershing supervised successful campaigns, he preferred negotiated efforts over armed campaigns. When combat was necessary, Pershing's troops benefitted from intelligence provided by Moros who trusted Pershing. Sincere tribal engagement thus proved a useful tactic in governance as well as combat.[47]

The Army's presence in the Philippines was a small one: the force averaged just 40,000, with much of this unavailable for combat due to administrative or support requirements or disease. To offset the small number, the Army relied heavily on three technologies: the coal-powered vessel, the telegraph, and modern firearms. The Civil War taught the Army the benefits of both cooperating with the Navy and using riverboats to support combat forces. In the Philippines, this tactic moved from an ancillary one used by specialized units to a common one with both armed and unarmed vessels proving vital. The ability to exploit the islands' most reliable lines of communication—the rivers—gave the Army speed and agility it otherwise lacked. The Army operated ten vessels of its own, equipping four of these as fire support ships armed with a mixture of heavy weapons and cannon. Reinforcing these were Navy blockade vessels. The blockade significantly reduced inter-island communications and helped isolate *insurrectos*.[48]

The telegraph served as a force multiplier. With a wired network linking headquarters and units, commanders gained an enhanced ability to both understand the tactical situation and direct operations as they unfolded. This made them more active participants in actions increasingly undertaken by small units. To facilitate this, the Signal Corps emplaced over 10,000 miles of wire and cable to and across the major islands—and transmitted an average of 11,000 messages daily. *Insurrectos* recognized the telegraph's power and made it a point to sever wires to disrupt Army operations. As a result, a great deal of effort went into protecting and maintaining telegraph lines after installation. *Insurrectos* also employed the telegraph. The Army exploited this for intelligence as it captured *insurrecto* headquarters and the copies of telegrams sent and received.[49]

The telegraph and motor vessel helped position soldiers, but the final technological changes were most important to the combat soldier—issuing him a modern, lethal rifle and pistol. Following the Civil War, the Army adopted a new infantry weapon. The Model 1865 and 1866 was a single-shot breechloading rifle using black powder. Converted from muzzle-loading Springfield rifle muskets, the new weapons provided both a higher volume of fire and a longer range. While an improvement over earlier weapons, the "trapdoor" Springfield rifle still presented two problems for the soldier: a low rate of accurate fire and the immediate identification of his position in the smoke cloud created when discharging a round. The low rate of accurate fire reflected the challenge of reacquiring a target after reloading the weapon, a process complicated by the smoke produced by black powder. Seeking to remedy this, the Army considered almost 100 different replacements before deciding on the Model 1892 (later designated Model 1896) rifle better known as the "Krag." The Krag used a five-round magazine and smokeless powder, thus solving the most significant drawbacks of the Springfield trapdoor system. The rifle was also more accurate, and easier to maintain. The Krag's ammunition lacked the stopping power of the earlier rifle, but it possessed a much faster muzzle velocity and longer range. Initially, Regular Army units used the Krag while state Volunteer units used the trapdoor Springfields. The U.S. Volunteer units authorized in March 1899 received Krags before arriving in the Philippines. As state Volunteer units returned to the United States, the Krag became the Army's standard rifle. This both improved unit effectiveness and simplified logistics.[50]

The Army entered the Philippines with a sidearm that proved deficient: the Model 1892 .38 caliber revolver, which also used a black powder cartridge. The Model 1892 was mechanically simple, but it lacked shock action: individuals struck with the .38 round would not stop, but continue moving until suffering multiple hits. With a weapon only carrying six rounds this was a serious problem when fighting *bolo*-wielding Moros at close range. As stop-gap remedies, the Army issued 45 caliber Model 1873 revolvers from Ordnance stocks and contracted with Colt Firearms for 4,600 new .45 caliber Model 1873 revolvers. Even as it issued these older revolvers, the Army was testing semi-automatic pistols of various calibers and types. Ultimately, the Army selected the Model 1911 Colt .45 as the new standard sidearm. This weapon's cartridge generated enormous shock action, solving the problems of the earlier revolvers. It also held a seven-round magazine, was easier to reload than the revolvers it replaced, and could generate a higher volume of accurate fire. This weapon proved so successful it remained in general service through the 1991 Persian Gulf War.[51]

Three organizational decisions also played significant roles in the Philippines: creating an effective intelligence structure, creating the Philippine Scouts (PS), and organizing the Philippine Constabulary (PC). Essential in any asymmetric campaign, the creation of actionable intelligence—information-driving operations—gave units the ability to increasingly (and successfully) focus their efforts. The snippets of information building the intelligence picture came from various sources with captured documents representing a primary one. Recognizing their potential value, the 8th Corps initially established a Bureau of Insurgent Records in Manila. Division of the Philippines Special Order (S.O.) 12, dated December 13, 1900, redesignated the bureau as the Military Information Division (MID)—the first field organization of its type in the Army's history. The MID sat at the apex of a system linking titled intelligence officers, subordinate headquarters, provost courts, and combat units. Reinforcing S.O. 12, General Orders 294 of September 28, 1901, directed subordinate headquarters to collect specific intelligence information. This system of designated officers and standing requirements provided intelligence to not only support tactical operations, but also inform both the Philippines senior headquarters and officials in Washington, D.C. Creating the MID represented an essential step forward in both integrating intelligence and operational planning and assessing the effectiveness of battlefield efforts.[52]

The Philippine Scouts embodied an evolutionary reincarnation of a proven tactic from the American frontier: enlisting natives and then using their knowledge of a region. While scouts had proven their utility in the American West, there was initially significant opposition within the Army's senior leadership in the Philippines to enlisting native auxiliaries. As a result, the scouts grew incrementally—first from a small experiment with members of one tribe and then growing to encompass members from across the archipelago. Unlike their Native American counterparts, the Philippine Scouts served as conventional infantry units.[53]

The initial scout cohort consisted of Macabebes, members of a tribe hostile to Luzon's dominant Tagalog population. First organized as a company by 1st Lieut. Matthew Batson in August 1899, the "Macabebe Scouts" quickly grew to a strength of 500. Macabebes demonstrated tactical prowess from the beginning and proved their worth in the single most important action of the war, the March 1901 capture of Emilio Aguinaldo. Brig. Gen. Frederick Funston (commanding the 4th District, Department of Northern Luzon) personally led a group of Macabebes, who posed as *insurrectos*, in a march of more than 90 miles in nine days to reach Aguinaldo's camp and take him prisoner. Aguinaldo's influence had decreased due to his constant movements, which disrupted his communications, but his capture nevertheless removed the most important *insurrecto* leader. His subsequent oath of allegiance to the United States demonstrated the Army's policy of rehabilitation applied to all who submitted to American authority. This success helped validate Filipino soldiers' use and contributed to an expanded program, formally known as the Philippine Scouts.[54]

Responding to developments in the Philippines, the second session of the 56th Congress considered legislation intended to "improve the efficiency of the military establishment."[55] While much of the discussion focused on administrative details, the proposed legislation's Section 38 authorized the president "to enlist natives of those islands for service in the Army, to be organized as scouts, with such officers as he shall deem necessary for their proper control."[56] The proposal also called for recognizing Philippine Scout infantry battalions and cavalry squadrons as part of the Regular Army. Signed by the president on February 9, 1901, the law gave the Philippine Scouts permanent standing, mandated the assignment of Regular Army (not US Volunteer) officers to scout units, and made provision for Filipino enlistees to secure appointments as non-commissioned officers. The original bill also revealed congressional parsimony, ethnocentrism, or both by fixing scout pay at one-half of American soldiers of the same rank. The bill's final form ultimately gave the Secretary of War the authority to set actual pay scales for the Philippine Scouts.[57]

September 19, 1900, marked a major change in the Filipino government with William Howard Taft assuming executive power as the head of the Philippine Commission. The Commission reported to the Secretary of War and served as both a legislative and executive body. The Commission's Act 175 established the Philippine Constabulary as a para-military organization within the Philippine Civil Service. Senior leadership came from Regular Army officers. As organized, it performed two roles. First, it served as the territory's superior law enforcement agency. Second, it conducted military operations in support of the Philippine Scouts and the Army. Initially led by a cavalry officer, Maj. Henry T. Allen, the constabulary's organization and operations differed from that of the scouts in two significant ways. Constabulary members served on the islands from which recruited. Filipinos could also qualify for commissions—something then not possible in the scouts. Under Allen, the Philippine Constabulary became a capable force against *insurrectos*. In concert with the Philippine Scouts, the constabulary allowed the United States to draw down its forces in the Philippines, even as the insurrection continued in the southern islands.[58]

The Philippine War culminated in something unusual in an asymmetric conflict, a decisive victory. Success stemmed from blending sound strategy, tactical agility, and usable military intelligence with an attractive political objective that gained widespread support. Effectively linking Army tactical and civic action operations with a clearly defined political objective undercut the *insurrectos.* Providing a way for them to leave the battlefield without fear of punishment made conflict decreasingly attractive—especially for the Filipino elite (the *insurrecto* leadership) who saw better opportunities in abandoning the fight. Also significant was Army-Navy cooperation. The Navy's blockading islands to prevent *insurrectos* from receiving aid from unfriendly powers reduced *insurrectos'* effectiveness. In providing transportation and gunfire support, the Navy provided the Army with much-needed flexibility in operations. Finally, the relationship between the Army's self-conceptualization and the Philippine War warrants consideration. The Army that was successful in the Philippine War was one that largely saw itself as a constabulary force. While its most senior leaders included men who had experienced the large battles of the American Civil War, most of its officers and non-commissioned officers lacked that experience. They were soldiers who spent most of their careers in the small, isolated posts scattered across the American West. Well-versed in the fundamentals of small unit tactics—especially patrolling—they also understood the importance of diplomatic engagement. These officers had the advantage of fighting the type of war for which their years of service trained them. Self-conceptualization was thus an asset and not a liability.[59]

The Mexican Punitive Expedition, 1916-1917

From Mexico's July 1821 declaration of independence from Spain through the present, violence has been a defining characteristic of Mexico's political and social history. Between 1810 and 1940, Mexico experienced two revolutions and 35 military coups as it sought both independence and social and political stability. The Mexican Revolution of 1910-1940 saw violence spill over the American border on March 8-9, 1917, with Francisco "Pancho" Villa, a former constitutionalist general, attacking the American town of Columbus, New Mexico. This raid resulted in the deaths of American civilians and soldiers. In response, President Woodrow Wilson directed the U.S. Army to enter Mexico to capture Villa and prevent further attacks on American territory by his bands. Villa's raid was not accidental. It reflected his resentment over an earlier Wilson decision to allow Mexican forces loyal to Venustiano Carranza to transit American territory to reinforce troops battling Villa at Agua Prieta, Sonora, Mexico on November 1, 1915. This action resulted in a decisive defeat for Villa and left him seeking opportunities for revenge. Villa's actions precipitated an American campaign noteworthy for failing to achieve its stated aims while facilitating great technological change within the Army. The expedition introduced motorization, aviation, radio communications and signals intelligence to actual operations.[60]

The Mexican states of Sonora and Chihuahua contained many of the features familiar to veterans of the frontier Army that had fought the Indian Wars. These included rugged terrain, poor communications, and vast distances between small settlements. What made the Punitive Expedition an engine for change within the Army was the combination of new technologies with military application and the challenges of operating a large force on foreign soil. Domestic and international political considerations also played a significant role in shaping the expedition's course. These included American anger over Villa's raid, Mexican anger over Wilson's sending troops into Mexico, and the ongoing war in Europe. The combination of technological change and strategic failure, with the latter rooted in objectives incompatible with capabilities, make the campaign relevant in the Army's history of asymmetric warfare.[61]

Essentially, the Punitive Expedition entailed sending one infantry and two cavalry brigades over some of Mexico's most rugged terrain. Experienced leadership directed the expedition: many of the campaign's officers had served in the Philippines. This force lacked access to Mexican railroads, and it did not enjoy the support of the Mexican government in its operations. The Army needed to develop the means of both supplying this force and controlling its activities in an area roughly the size of Louisiana. The lack of Mexican support meant the Army also needed to improve its ability to conduct reconnaissance if it was to have any realistic chance of running Villa to the ground. To solve these problems, the Army turned to three immature technologies: the gasoline-powered truck, the airplane, and the radio even as most of its force remained foot or horse-bound.[62]

Dating from 1876, the internal combustion engine initially found little acceptance in either the United States or the Army. In 1901, the total American market for automobiles was 5,000 vehicles of all types—gasoline, steam-powered and electric. Between 1901 and 1910, the industry began consolidating around gasoline-powered vehicles with this process aided by Henry Ford's establishing a new assembly plant at Highland Falls, Michigan, which introduced improved methods that significantly accelerated production. Despite this, the total number of motor vehicles in the United States Army in 1907 was twelve. The numbers slowly grew with the purchase of experimental vehicles in 1911 and with trucks replacing horses in some depots by 1913.[63]

The lack of rail access and the limitations of draft animals led to the Army's Southern Department consolidating the small number of available vehicles to support the movement into Mexico. A report from the expedition commander, Brig. Gen. John J. Pershing, noted the initial assembly included vehicles in "every conceivable state of unreadiness."[64] One of the first tasks to perform was taking this conglomeration of parts and transforming it into a functioning vehicle fleet. Even after assembling the initial fleet there remained the question of tactical organization. Mexico was the Army's first attempt to employ motor vehicles operationally. There was neither doctrine nor precedent to aid in answering the question of organization. To support the campaign, the Columbus Post Quartermaster, Maj. John F. Madden, organized the first group of trucks into two truck companies, each composed of 27 one-and-a-half ton (maximum vehicle payload) trucks.[65]

The technical capabilities of these vehicles faced frequent testing under Chihuahua's harsh conditions. Despite this, the number of trucks soon increased. By April 1917, over 160 trucks operated a distribution line stretching 116 miles from Columbus, New Mexico, to the expedition's Mexican base camp at Colonia Dublán. When this number proved insufficient, the War Department ordered an additional 100 vehicles to improve the flow of supplies. Ultimately, the Army deployed 544 trucks of 13 different types from eight manufacturers. Both the War Department and Pershing recognized that the only way to sustain operations was to establish a robust supply line. Denied railroad access, motorized transport—however imperfect the equipment and organization—offered the only way to do so.[66]

Even as the Army increased the size of its vehicle fleet and consolidated it to support the Punitive Expedition, it concurrently dealt with the interrelated problem of operating the vehicles. The Army lacked the trained personnel necessary to drive and fix its trucks. If the service tried to rely solely on military personnel, the mission would fail. To solve the problem, the Army sought volunteer civilian drivers and mechanics. Given the pressures of time and location, the Army did not impose rigid recruitment standards—those who could not perform would soon find themselves identified and their services terminated. As the campaign continued, the Army established a cadre of drivers and mechanics to support operations. The Army also recognized the strengths and weaknesses of its mélange of vehicles and sought to identify the best characteristics to develop a standardized model that the service could procure and that soldiers could operate and maintain. While motor vehicles greatly extended the campaign's operational reach, maintenance remained an enduring challenge. Bad-to-non-existent roads wreaked havoc on wheels and springs. Additionally, the desert conditions meant frequent clogging of fuel lines and air filters. Finally, the difficult conditions and primitive suspensions meant drivers suffered from fatigue and incurred aches unknown to those on foot or horse.[67]

The Punitive Expedition saw the Army mount its first real efforts at integrating aircraft in support of ground forces. While the Army operated observation balloons during the Civil War, these were static observation posts. Tethered to a single location for the duration of the flight, the balloon's many disadvantages limited its utility as a weapon. The fixed-wing aircraft, however, dramatically changed all the dynamics associated with aerial observation. Freed from the tether, and possessing the ability to fly where the pilot directed, the Army's early airplanes opened new fields of activity. Army Air Service officers believed airplanes could conduct reconnaissance, observe artillery fire, and attack targets. The Punitive Expedition offered the potential to perform all these missions.[68]

In response to growing tensions with Mexico, the Army formed its first air unit, the provisional 1st Aero Squadron, on March 5, 1913, in Augusta, Georgia. Equipped with seven Wright "Pusher" aircraft, the unit represented the Army's first serious effort at turning airplanes into practical components of its force structure. The Pusher's many deficiencies, however, limited its utility as a military aircraft. With an average speed of 48 knots, and only a 240-mile range, the Pusher's only real capabilities lay in providing *something* for pilots to fly and mechanics to fix. Relocating to San Diego, California, in June, the 1st replaced its Wrights with Curtiss "Model J" aircraft in June 1914. First Lieutenants B. Q. Jones and Thomas Milling flew the Model J when conducting the Army's first combat aviation mission—a reconnaissance flight along the border to locate Pancho Villa—on April 13, 1915. The Model J, however, was another disappointing airplane. The 1st Aero Squadron's commanding officer, Capt. Benjamin Foulois, later wrote, "They looked like airplanes. We were to find that an airplane that looks like an airplane may be something else."[69] The squadron's receipt of two inferior models of aircraft foreshadowed its future. The unit soon found itself reequipped with another poorly performing aircraft, the JN-3, as it deployed to Columbus, New Mexico, in March 1916 to support the Punitive Expedition.[70]

Chihuahua's terrain offered Villa and his followers noteworthy advantages. The high peaks of the Sierra Madre and its narrow canyons provided hiding places as well as obstacles to all types of movement. This difficulty was particularly true for the 1st Aero Squadron. The JN-3's low service ceiling (nominally 6,500 feet) translated into something considerably less during operations in Mexico. Shortly after the squadron's arrival in Columbus, Pershing ordered the unit to Casas Grandes. The new base's elevation of 4,800 feet meant the aircraft could not clear any obstacle more than 1,700 feet high. Peaks exceeding that height abounded in most directions within 25 miles of the expedition's primary airfield. Pershing needed to conduct reconnaissance as far as possible in any direction if he was to have any chance of running Villa to the ground. What he received were aircraft that could not fly where they were most needed – even under optimal conditions.[71]

Aggravating this ceiling issue were problems with weather and cartography. Flying near the mountains and over the foothills when possible, pilots experienced powerful wind gusts that challenged their ability to control their aircraft. Extreme temperature changes took a toll on both pilots and machines, particularly propellers. Temperature variations caused the glue that held the layers of wood that formed a propeller to fail. Known as "propeller delamination," this made the planes dangerous to fly and added to a growing list of repair tasks.

Less than thirty days into the campaign, crashes of six of the original eight aircraft flown into Mexico left them destroyed or damaged beyond repair and the remaining two aircraft were almost unflyable. No accurate maps of the region guided flight planning or conduct even when aircraft were available. To remedy this, photographing the area for cartographic purposes was an early—and vital—mission.[72]

Service in Mexico soon proved the futility of using the JN-3 for anything but a training airplane. The rapid deterioration of the 1st Aero Squadron's capabilities compelled the War Department to quickly send more capable replacement aircraft. The Army faced a fundamental problem in trying to field replacement aircraft. If it employed the process of competitive bidding, it could get a much more capable airplane than the JN-3. The need for increased capabilities, however, meant an unwanted time delay as the Army developed performance specifications. Bidders required time to respond and, if selected, construct new aircraft. The service could bypass the competitive bidding process and purchase readily available aircraft. These, however, would most likely not meet the mission's requirements. The Signal Corps (then the Army agency responsible for aviation) opted for the latter option and purchased four Curtiss N-8 aircraft. The 1st Aero Squadron evaluated these machines and found them only slightly better than the JN-3. As a result, the squadron sent the new aircraft to San Diego's Rockwell Field, then the home of the Army's flying school, for use as trainers. Twelve Curtiss R-2 aircraft followed and these provided the expedition with the minimal air support it received. Though equipped with a more powerful engine, the R-2 remained a woefully limited airplane. Experiencing many of the same problems as the JN-3, the R-3 exemplified both the grand aspirations and profound limitations of the Army's embryonic air service. The service's members were profoundly aware of aviation developments driven by the war in Europe and believed the Army could wield an air arm as capable as those of the major combatants. Mexico, however, proved the air service's impotence. In August 1916, *New York Times* correspondent Frank C. Page wrote: "The punitive expedition has twelve of the best airplanes made in the country."[73] In contrast, Pershing took a more critical view, noting "new planes of an improved type were received, but even these were defective in many respects and were of the least service at the time they were needed the most."[74]

Radio represented the third new technology tested by the Punitive Expedition. The Army possessed extensive experience with Morse code after receiving its first telegraphs in the 1840s. While the Army relied on the hard-wired telegraph through the nineteenth century, it also employed heliographs in the American Southwest. Heliographs used mirrors to both focus and reflect the sun's rays. The heliograph permitted Morse code transmissions at distances of up to 90 miles. Lacking wires, the heliograph was far more portable than the telegraph and provided better control of tactical units in the field. Despite its considerable range, the heliograph required direct sunlight and thus could not function at night or in periods of thick cloud cover. It worked best when sending and receiving stations were on high terrain with long lines of sight, conditions occurring in only a small part of the country. Radio, in use by the Army since 1899, potentially offered the telegraph's reliability and the heliograph's portability without its many restrictions. [75]

The Army's return to Cuba in 1906 saw several radio sets deployed for high-level communications. In 1910 and 1911, changes in the Signal Corps led to its fielding five field service companies with two radio sections apiece. The new units allowed the Army to establish an operational radio network to support combat operations. The Punitive Expedition employed three major innovations seeking to capitalize on radio's greater availability: the deployment of multiple stations, the concurrent mounting of radio sets in motor vehicles, and the use of radios to monitor hostile communications.[76]

The Signal Corps' growth enabled deploying 19 "spark gap" radio sets across Chihuahua. "Spark gap" refers to how an operator transmitted a radio signal—creating an electrical spark across a physical gap between two leads to produce Morse code. Augmenting this network was a wire communications system using both telegraphs and "buzzer phones." The buzzer phone was an ingenious device—a telephone that, when conditions degraded phone performance (as was frequently the case in Mexico), doubled as a Morse code transmitter/receiver.[77]

Both radio and wire communications saw their inherent limitations displayed in Mexico. A fundamental problem was the spark gap radio itself. It simultaneously broadcasted across multiple frequencies and created interference even as it transmitted. Multiple transmissions from stations simultaneously transmitting exacerbated this problem and complicated message receipt. A variety of atmospheric conditions also affect radio waves. These include rain, lightning, and the changes within the earth's ionosphere due to sunset and sunrise. Chihuahua's strong winds also impacted radio performance by requiring frequent adjustments to antennas. The region's mountains both reflected and refracted radio waves with this effecting reception. Finally, variations in the composition of the soil itself impacted radio performance. Modern technology compensates for most problems, but in 1916 technical limitations required operators to work through them. Another basic problem was the range of most sets, roughly 25 miles. This range was satisfactory when dealing with marching infantry formations that covered short distances each day. Cavalry units, however, traveled up to 50 miles a day and thus outranged supporting communications. This meant units still had to maintain dispatch riders to maintain communications, even if they did not have to ride as far. When it had flyable airplanes and could find the cavalry, the

1st Aero Squadron could help compensate for this by dropping messages. Such a combination of events, however, rarely occurred.[78]

Vehicle-mounted radios, also known as "tractor radios," represented a technical leap forward by combining improved mobility with greater range and power—at least in theory. Such radios came in two types, a one-kilowatt set intended for division-level communications and a two-kilowatt set for Army communications. Significant differences existed between the two sets. The divisional set weighed 6,700 pounds and employed a 60-foot-high mast to support long antenna wires. Theoretically, the vehicle carrying the radio and crew could travel 25 miles an hour over "average" roads, but few such roads existed in Chihuahua. The Army set weighed 9,000 pounds and employed a mast 80 feet high. The greater weight reduced its expected rate of travel to 15 miles per hour. Pershing deployed such sets at the major headquarters of Columbus, New Mexico, and Colonia Dublán, and at various intermediate locations to keep in touch with units. While wire also linked the headquarters locations, the penchant of Mexican loyalists to cut American wire meant the radios proved useful. Despite the use of radios, 1,200 miles of supplemental wire, the redoubtable heliograph, and even the occasional carrier pigeon, communications remained an enduring problem for Pershing. Innovative efforts notwithstanding, the distances and demands exceeded capabilities.[79]

The Mexican government used the telegraph as its primary means of long-distance communications. This practice created two vulnerabilities. The first was the ability of the Army to tap into Mexican telegraph lines and monitor them. The second was the simple fact that the cutting of telegraph wires was indiscriminate, resulting in disruptions in Mexican communications in the same manner as American ones. While Villa lacked radios, the Mexican government used them to maintain communications across Chihuahua. Commercial interests also employed radio. Even with radio's challenges, listening in on the Mexican government and commercial enterprises offered opportunities. These included providing information on how the Mexican government would respond to specific American actions. Listening might also provide information on the Villa's location, thus filling the disappointing gap opened by the failures of the 1st Aero Squadron to meet expectations and the small number of cavalry units in Mexico. Such intercepts resulted in dispatching several patrols to locations where radio messages stated the Villa's location. Intercepts also provided accurate information on the location, status, and intentions of Mexican units that could interfere with the expedition. Finally, eavesdropping on government communications allowed Pershing and the Wilson administration to validate the statements made by the Mexican government publicly and in meetings with American officials. To that end, Pershing dedicated one of the tractor radios to the task of communications intelligence. Mexican officials employed a variety of codes in their traffic, but two officers, Capt. William Reed and Capt. Nicholas Campanole, supervised a codebreaking program and demonstrated radio's intelligence's potential.[80]

The Army's heavy investment in technology notwithstanding, Pershing's troops never made contact with Villa. Between March and June 1916, the Army fought nine engagements of varying size in Mexico. On March 29, elements of the 7th Cavalry engaged 200 *Villistas* at Guererro. Killing thirty *Villistas* (original reports placed the number at 60), this battle was one of the costliest of the campaign for Villa's troops. On May 5, *Villistas* again suffered heavy casualties at Ojos Azules. In one of the Army's last horse-mounted charges, the 11th Cavalry killed 42 without suffering any casualties from a force of over 330. First Lieutenant George S. Patton killed one of Villa's subordinates, Julio Cardenas, on May 14, 1916. Interestingly, the soon-to-be-famous cavalryman travelled in a Dodge touring car rather than a horse: the increasing number of vehicles allowed for their use in patrolling as well as administrative tasks. Strategically more important, however, were actions at Parrall (April 12, 1916) and Carrizal (June 21, 1916).[81]

At Parral, a detachment from the 13th Cavalry skirmished with a larger Mexican government cavalry unit in an engagement killing six Americans and over 40 Mexicans. This fight pitted conventional cavalry formations against each other as the Mexicans conducted a pursuit. At Carrizal, the 10th Cavalry clashed with government forces. Here the Army's cost was higher with nine dead and 12 wounded. Additionally, 23 cavalrymen became prisoners of the Mexican government. Carrizal became a turning point for the expedition, especially with respect to American public opinion. Increasingly, the expedition became the subject of political cartoons with this reflecting both increasing public frustration over the failure to kill or capture Villa and concerns over policy to Mexico. [82]

The Punitive Expedition was both unanticipated and conducted with one eye fixed on the war in Europe. The Army focused new technology on the challenges of Mexico and saw mixed results for its efforts. The Army did not develop new equipment for either an extended campaign or an asymmetric threat. Instead, it deployed equipment readily available (the radio and the airplane) or hastily acquired (the truck). All three of these equipment items facilitated various aspects of operations, but none focused on the tasks of actual combat. Excepting Patton's action, combat featured traditional weapons and methods. While the Punitive Expedition offers several lessons, the most significant is worth repeating: technology has never offered the sole solution to the Army's tactical challenges.

Vietnam, 1950-1974

While the Army considers March 15, 1962, the start date for the first of its 17 Vietnam campaigns, its presence went back years earlier. September 1950 saw the creation of the United States Military Assistance Advisory Group, Indochina, in Saigon. Intended to aid France in its counterinsurgency campaign against the communist-inspired Viet Minh, this unit became the Military Assistance Advisory Group, Vietnam (MAAGV), with the Republic of Vietnam's (RVN) independence in 1955. National Security Council Memorandum 5429, dated August 4, 1954, captures strategic thinking as the transition from French rule to two independent Vietnamese states played out:

> The U.S. must protect its position and restore its prestige in the Far East by a new initiative in Southeast Asia, where the situation must be stabilized as soon as possible to prevent further losses to Communism through (1) creeping expansion and subversion, or (2) overt aggression.[83]

To realize these objectives, MAAGV assisted the RVN in building a modern army capable of deterring an invasion from its communist counterpart, the Democratic Republic of Vietnam (DRV), a new communist insurgent movement, the Viet Cong (VC). From 1955 through 1962, MAAGV served as the primary American military headquarters in Vietnam, when the Military Assistance Command-Vietnam (MACV) replaced it. These were all joint organizations; however, the Army provided both the commanding generals and most of the military manpower. Ultimately, the statement of work for MACV grew far beyond that of the MAAGV. While both had a counterinsurgency mission, MACV also became the war's erstwhile supreme headquarters, coordinating American and allied activities. This coordination included considering both battlefield needs and domestic political concerns in both the United States and in ally's country. Concerns over a public response to specific actions or battlefield casualties sometimes limited a force's employment.[84]

America's involvement in Southeast Asia never achieved one of its most important objectives: marshaling the power of collective action. President Lyndon Baines Johnson hoped for significant participation under the "Many Flags" program. Great Britain played no military role in the conflict at all. France, because of its colonial legacy and failure in Indochina, remained militarily sidelined. America's North Atlantic Treaty Organization (NATO) partners ultimately contributed varying amounts of non-military aid. Australia and New Zealand provided professional units, but their numbers remained small. Faced with the possibility of seeing a reduction of American forces in Korea, the Republic of Korea sent over 34,000 soldiers and marines to Vietnam. Two Southeast Asian countries, Thailand, and the Philippines, also sent army units. By 1969, these nations had over 67,000 troops serving in Vietnam. In addition to these troops, the Army of the Republic of Vietnam (ARVN) and its auxiliaries provided an additional 1,027,000 soldiers that same year. Coordinating the activities of all these forces represented one of the war's enduring challenges. Difficult under the best of circumstances, simultaneously building the ARVN and fighting both a conventional war and a counterinsurgency magnified it.[85]

Exacerbating this challenge was a new battlefield dynamic: an international and increasingly unrestrained press corps. During the Philippine conflict, World War I, and World War II, American authorities reserved the right to censor accredited correspondents. In Korea, the rules saw some loosening, but restrictions existed to prevent reporting derogatory information about American forces. In Vietnam, however, the actions of *all* actors—American and allied—were both visible and scrutinized. The ready availability of international communications meant military censorship was unfeasible. Scenes and reports from the war influenced not only American audiences but also allied ones, which produced important political ramifications. Local allies proved a double-edged sword in Vietnam, as they would in future conflicts, with the Army widely perceived as both "owning" them and having responsibility for their performance.[86]

The Battle of Ap Bac provides a clear illustration of the Allied operations challenges and the media's ready access to both information and the means to report it. A hamlet roughly thirty miles from Saigon on the Plain of Reeds, Ap Bac was the site of a January 2, 1963, battle between roughly 150 VC and a reinforced battalion of ARVN troops equipped with the Army's newest armored personnel carrier, the M113, and supported by both Vietnamese artillery and U.S. Army helicopters. This battle was unique for two reasons. First, preparation included detailed planning. Intelligence reports for several days gave Vietnamese officers and their American advisors plenty of time to develop a realistic plan to attack Ap Bac and destroy the enemy located there. Second, taking place after a year of intensive American training, this first fight would show the VC—and the world—the ARVN's capabilities. Structurally, ARVN closely resembled its American counterpart with airborne, Ranger, and mechanized units. It was a modern Army optimized to fight the North Vietnamese regulars fielded by the People's Army of Vietnam (PAVN) rather than the VC that constituted the primary threat. Three auxiliary forces (Civil Guard, Self Defense Corps, and Civilian Irregular Defense Group) focused on defending against the Viet Cong.[87]

At Ap Bac, the ARVN outnumbered the VC almost ten-to-one. Possessing better weapons and mobility, the ARVN should have delivered a major victory. Instead, the battle exposed all the things the ARVN could not do: coordinate fire and movement, exercise decisive leadership, and fix an enemy force in a position to destroy it. Compounding the ARVN's problems was the vulnerability of its American-flown HH-21 "Shawnee" helicopters. Viet Cong gunners shot down five such aircraft, killing three Americans. Additionally, they damaged nine more aircraft and wounded six more Americans. ARVN casualties numbered more than 100. At battle's end, Ap Bac was in ARVN hands, but the Viet Cong withdrawal made ownership irrelevant.[88]

The defeat highlighted the difficulties in developing military capabilities in an army—and country—possessing cultural and social norms far different than those of their American sponsors. One of the most relevant was an army's role in society: the U.S. Army is apolitical; ARVN was a hotbed of local and national political intrigue. ARVN officers often viewed things from the political rather than the tactical perspective—and their battlefield decisions reflected this. Equally important, the Ap Bac battle demonstrated the media's ability to report events on its own terms—even if they diverged from the official position. In the words of Lt. Col. John Paul Vann, an advisor to the ARVN 7th Infantry Division at Ap Bac, the South Vietnamese troops put in "a miserable damn performance."[89] It was also a performance that the *Washington Post* and the *New York Times* reported on within 24 hours. Both papers used such language as "humiliating defeat" and "costly Viet-Nam battle angers U.S. advisers [*sic*]," contradicting the MAAGV commander, Gen. Paul Harkins, who characterized Ap Bac as a victory.[90] Throughout the war's course, similar battlefield events resulted in negative media coverage. Compounding this was the long-running MAAGV/MACV effort to present ARVN efforts as successes, even when there was none of any consequence. Ultimately, attempts to explain battlefield events would devolve into the famous "five o'clock follies," the nickname reporters bestowed on the MACV daily media briefing. That derisive label captured the Army's frustrations. Part of the problem originated with policy: when commenting on Vietnamese issues, the Army initially followed the RVN's press guidance. It also reflected the cynicism of a news media hard-pressed to believe whatever Army official said. Ap Bac provided the first example of a confrontational media that would influence public perceptions through the war's long course.[91]

A largely disaffected population provided another issue for the Army to tackle. A proud people with a long history, the ethnic Vietnamese believed they possessed a culture superior to that of their neighbors. A distinctly Vietnamese culture had existed from at least 2879 BCE, with an independent Nam Viet state founded in 207 BCE. Despite multiple conflicts, Chinese influence remained pervasive. By the time MAAGV's first members arrived, Vietnamese history included periods of independence as well as Chinese, Japanese, and French rule. With the 1955 collapse of the Nguyen dynasty, a widely unpopular Roman Catholic urban elite held power. Most of the population remained poor, rural, and divided among multiple politico-religious sects—some highly militant. The national hero was not a member of the RVN government but DRV leader Ho Chi Minh. Most Vietnamese revered Ho for forcing France from the country.

It is no surprise that many South Vietnamese supported neither the country's government nor its institutions.[92]

RVN governments tended toward a disturbing characteristic—instability. This instability reverberated through all government agencies. Too often, the decisions these governments made ran contrary to the Vietnamese citizen's understanding of justice or correctness. While the VC often demonstrated a high degree of brutality in dealing with the populace, the RVN government employed the same tactic all too often. The result was a population often suspicious of both the civilian government *and* the ARVN. The American relationship with the RVN made much of the population suspicious—and at times hostile—of the Army's efforts. Frequently, the Army's basic approach was to seek out the enemy and then employ mortars, artillery, or air support to complete his destruction. Reflecting an ingrained belief that spending money was better than spending blood, the U.S. Army tended towards area-fire weapons. It is hardly surprising that the Vietnamese who saw their villages destroyed wanted as little as possible to do with either the ARVN or the Americans. "Winning hearts and minds" were the theoretical objectives, but battlefield practice tended to undercut those efforts. That the war's early years corresponded with the zenith of Modernization theory only complicated matters. Modernization theory meant that as the Army operated across Vietnam, multiple civilian government agencies did so as well. Intent on improving every aspect of the Vietnamese peasant's life, their actions meant both further cultural dislocation and animosity towards the American presence. Most American theorists, politicians, and bureaucrats never understood the fact that many Vietnamese did not want their society changed along American lines. This animosity displayed itself in a variety of ways, but it was most dangerous when it led individuals to either join or cooperate with the VC. Ultimately, the disaffected populace provided the Army with myriad opportunities to re-learn basic truths about fighting an insurgency: legitimacy is paramount and imprudence in choosing partners makes establishing it more difficult.[93]

Techno-wizardry reappeared in Vietnam—and on a scale reminiscent of the Second World War. Korea was too close to that earlier war's end and too limited in scope to witness significant technological change. The Army of 1950-53 closely mirrored that of 1944-45. Three factors, however, largely shaped technology's arc in Vietnam. First, the American build-up corresponded to the tenure of Robert McNamara as Secretary of Defense. McNamara was an activist manager possessing an abiding faith in technology's promise. Second, Cold War competition provided the nation a massive technological production base theoretically able to deliver a broad range of sophisticated items. Third, relatively low combat troop strength in Vietnam required creative solutions as compensation. The Army deployed 81 infantry battalions and activated a

Special Forces group in Vietnam, a number far too small given its size, topography, and enemies. Aggravating the problem was a lack of validated and understood counterinsurgency doctrine. For the Army, this meant an increasing flow of devices to find and fight the Viet Cong and PAVN. [94]

Technology touched almost every facet of operations in Vietnam, but intelligence collection and analysis posed the greatest challenges. In facing two enemies, the Army faced two interconnected, but separate problems. Defeating the PAVN meant not only identifying its locations within Vietnam but also the infiltration and supply routes. Defeating the VC meant not only identifying unit locations but also individual actors who served in a variety of supporting roles. Ground surveillance systems such as the AN/PPS-5 radar (GSR) and remotely emplaced sensors (REMS) sought to extend the range of American "eyes" in both time and distance by removing the cover of darkness and allowing the identification of infiltrators at extended ranges. Aerial counterparts existed in the OV-1 Mohawk airplane with its AN/AAS-14 infra-red detection system or AN/APS-94 side-looking airborne radar (SLAR). SLAR identified moving targets in Vietnam, Laos, Cambodia, or on Vietnam's rivers at ranges of up to 100 kilometers. UH-1 helicopters deployed a variety of remote sensors that transmitted information as long as their batteries functioned. Automated data processing, still in its infancy, supported the fight against the VC with the Combined Intelligence Center-Vietnam (CIC-V) using IBM computers to maintain an ever-changing data of VC personalities, units, and locations. Targeting remained a challenge throughout the war, but focusing equipment efforts on intelligence requirements produced positive results and facilitated the effective application of scarce resources.[95]

Technology also played a role in destroying the enemy. The AN/PVS-2 image intensification device, better known as a "starlight scope," dramatically improved a soldier's ability to see in the dark. The M18 "Claymore" anti-personnel mine provided devastating firepower. Consisting of an explosive charge and 700 small ball bearings discharged across an arc of 60 degrees, soldiers could prepare the weapon for individual or group firing. Easily emplaced and recovered, Claymores often reinforced ambushes, defensive positions, or base camps. Vietnam's unique topography led to changes in how the Army equipped its field artillery. The primary cannon in most artillery units was the new M102, a 105mm howitzer with a range of 11,500 meters. At 3,300 pounds, it was much more mobile than its predecessor. Nevertheless, the terrain frequently challenged a unit's ability to position and fire the weapon. The M114A1 155mm howitzer had both greater range and weight. Heavier weight meant greater difficulty in moving the weapon. Both these weapons suffered the same basic limitation: their limited ability to rotate 360 degrees. With American tactics emphasizing the creation of fire support bases—the forward positioning of artillery in remote areas—the 360-degree firing requirement was a new one for such weapons. To improve cannon effectiveness, the Army developed several new platforms on which to place these weapons. These platforms improved weapons' tactical effectiveness.[96]

To support the 9th Infantry Division in the Mekong River delta, an area almost void of dry land, the Army developed an airborne firing platform for the M102. Transportable by heavy lift helicopter, the platform provided a stable firing platform enabling the weapon's rapid traverse. Such platforms suffered a major deficiency—they provided no protection for a cannon crew. Two officers, Maj. Dan Charlton and Capt. John Beiler of the 3rd Battalion, 34th Field Artillery, sought to solve the problem. Experiments with barges led to a design based on the Navy's P-1 pontoon. Welding multiple pontoons together created an armor-protected barge capable of carrying the cannon, its ammunition, and its crew. This system allowed the cannon's use as both an indirect and direct fire weapon.[97]

Continuing the tradition of the combat soldier solving the most pressing problems, 1st Lieut. Nathan Foster of the 8th Battalion, 6th Field Artillery, devised the platform for the M114A1. Locally fabricated, the platform allowed an understrength gun crew to rotate the weapon 360° in a matter of seconds. Foster's Battery B demonstrated the platform's utility by shifting the cannon's muzzle 33 times in one 19-hour period. During the 1st Infantry Division's Operation BIRMINGHAM, this same battery fired more than 7,200 rounds over two-and-one-half weeks and performed "innumerable shifts using the pedestal."[98] In allowing rapid and accurate fire, the platform sustained the firebase as an effective tactic. Ironically, this had the unintended consequence of contributing to the practice of tethering maneuver units to the range of supporting artillery. Sometimes referred to as "firebase psychosis," this meant artillery availability, rather than the enemy situation, dictated operations.[99]

The Army's use of some new technologies also harkened to the past. Some Civil War approaches to a counterinsurgency returned. The idea behind the Mississippi Marine Brigade experienced reincarnation as the Mekong Riverine Force, an organization pairing the Army's 2nd Brigade, 9th Infantry Division with the Navy's Task Force 117. Fighting primarily from Navy vessels, the Army also employed its own air cushion vehicles and helicopters in operations around the Mekong Delta. Defoliation, on a much larger scale than the Civil War, reemerged with both the Army and Air Force applying the herbicidal agents "Orange," "Blue," and "White" in many areas. Before the Army realized the long-term environmental impacts of such agents, analysis showed such agents served a useful purpose with a decrease in friendly killed-in-action (KIA) rates. The analysis also indicated higher enemy KIA rates in enemy-initiated engagements occurring in areas treated with these agents.[100]

In Vietnam, as was the case in the Punitive Expedition, the Army placed great stock in technology. The massive investment effort offered a much better return than it did in Mexico but still failed to meet all requirements. The Army assumed its expectations were realistic under the circumstances but failed to appreciate technology's limitations fully. It also failed to understand the relationship between technical possibilities and the impacts of policy on operations. These policies included limiting the size of the force in Vietnam and restricting ground combat. Whereas Mexico repeatedly demonstrated technology's limitations, Vietnam showed the lack of will to exploit technology's capabilities amounted to technological failure. It did not matter if technology served up the enemy if the Army could not act on the information—and, all too often, that was the case.[101]

War is always hard, dirty, brutal, and frustrating. Clausewitz spoke of fog and friction as two of war's fundamental concepts—he could have added frustration to the list. Frustration was one, but not the only, cause of the Army's difficulties in mounting operations that conformed to national values and the law of war. On March 16-17, 1968, the resentments of Company C, 1st Battalion, 20th Infantry, found release in the My Lai Massacre. Even after 50 years, My Lai resonates across American political, military, and cultural landscapes. The war's best-known atrocity is not only a tragedy for its scale, but because it was not the only example of lawless behavior by American soldiers. War always provides opportunities for such events, but counterinsurgency magnifies them. The insurgent's ability to blend into the population makes every civilian a potential combatant. Compounding the issue in Vietnam was the issue of race. Army policy was to treat the Vietnamese with respect, but not every soldier complied and some subscribed to the "mere gook rule," or MGR. This ascribed an inferior position to the Vietnamese and contributed to their mistreatment. The remedy is simple—effective leadership. As Clausewitz accurately notes, the simple is difficult in war. Combat leadership is hard—and the further one moves along the counterinsurgency spectrum the harder it gets.[102]

Throughout the war, the Army increasingly struggled with small-unit leadership—the corporals, sergeants, lieutenants, and captains who fought the battles. This tension reflected four factors. First, the "crash" nature of the initial build-up meant many units experienced either (or both) abbreviated training programs or last-minute personnel replacements who missed unit training events. This decreased leadership training opportunities and retarded the developing of competence in the tasks units would perform. The decision to deploy units and then rotate personnel through them was the second. This practice significantly eroded cohesion and made small unit leadership even more difficult, particularly when casualties resulted in even higher personnel turnover. Exacerbating this was the practice of "ticket punching." Company and battalion commanders would often serve in their positions for only a few months before moving on to a new duty position. A new officer would then assume command and get his "ticket punched" by serving as a commander in combat. Over the course of one year, a soldier could find he had three or four different company or battalion commanders. Third, a liberal policy for college deferments for college deferments juxtaposed against a declining pool of high-quality enlistees was an increase in requirements. Not only did the Army need more ground combat officers but it also needed pilots. The aviator training program averaged 3,600 graduates a year between 1962 and 1973. The only solution for the leadership problem was to accept those who met the minimum standards to ensure leadership positions remained filled. Previously, a higher number of enlistees with degrees or higher test scores meant the Army need not settle for those who met the minimum as supply exceeded demand. Officer Candidate School (OCS) provided the preponderance of company grade officers. Before 1966, many OCS candidates were long-serving NCOs with demonstrated leadership skills. The Army's rapid expansion, however, meant that this highly qualified population could not meet the demand. Increasingly, candidates came from basic combat training centers. The expansion also increased demands for competent non-commissioned officers. This ultimately led the Army to establish non-commissioned officer academies to produce sergeants and staff sergeants to help fill the gap. Dubbed "shake and bake NCOs" after a popular food product, this cohort brought motivation but lacked the seasoning of those promoted after several years of service. The war's unpopularity and course impacted soldier attitudes, particularly after 1969. This manifested itself in increasing drug abuse and assaults against authority.[103]

There was also the associated problem of more senior leadership applying a careerist mindset to the battlefield's problems. This expressed itself in leaders who focused on reporting up the chain of command rather than focusing on developing their subordinate leaders. The desire to prevent reports of poor performance or misconduct meant all too often incorrect information moved upward. Alternatively, bad news that did go up the chain often found itself stopped or modified before reaching the top. This problem was pervasive enough to lead to the Army War College's 1970 study of professionalism report (see Chapter 1). The Army had a large population of highly qualified combat leaders, but it was not enough to meet all its needs.[104]

Vietnam once again unveiled an ugly truth about counterinsurgencies: frustration can consume soldiers. Frustration and weak leadership can enable serious misconduct. Misconduct on the Army's part enflames not only the local population. Mass media magnifies the effect of misconduct and can turn domestic and international audiences against the counterinsurgency. Finally, efforts to limit knowledge of such events only make things worse by destroying credibility and legitimacy. Learned in Vietnam, this was another lesson the Army would re-learn on future battlefields in Iraq and Afghanistan.

The U.S. Army's experience in Vietnam touched on many facets of both conventional and asymmetric warfare. It also strongly echoed aspects of past conflicts with these informing some of the solutions to the war's myriad problems. Most significantly, it indelibly colored the Army's understanding of mass media—an understanding it would take into battle in Grenada, Panama, and the Persian Gulf. It also reinforced the Army's penchant for technical solutions and contributed to the enduring belief that problems always had technological solutions. The reality is that while that is often the case, "often" is not "always," and it would take another conflict—and some highly astute senior leadership—for the Army to internalize that lesson. Vietnam also reminded the Army that one of the biggest challenges in a counterinsurgency campaign is establishing legitimacy among the local population. Absent that, all accomplishments were a Potemkin—they looked good from afar but lacked any real substance. Intimately linked with Army legitimacy was the issue of local allies. Legitimacy hinged on local attitudes about a unit's conduct and attitude as well as its performance. Finally, the protracted conflict reminded the Army of the necessity of ethical and competent leaders capable of training and directing the troops. The Army's only universal truth is that soldiers are as good, or as bad, as their leaders let them be—or make them be. Tour rotation policies served to short-cut both training and leader development opportunities. Intended to provide the Army with the means to win the war, rotation instead made that virtually impossible.

Conclusions

The experiences of the Civil War, the Philippine War, the Mexican Expedition, and the Vietnam War prompt several observations. First, asymmetric opponents operate with a different set of political considerations. In all the campaigns considered, opponents never had to consider domestic political considerations in the same manner as the Army. The issue of time becomes increasingly relevant as expected outcomes do not rapidly present themselves. This was particularly true in the Punitive Expedition and Vietnam. The issue would present itself again in Afghanistan and Iraq and is one reason asymmetric campaigns are so frustrating for those involved—including a civilian public increasingly desirous of quick outcomes. "Frustration management" would have profound consequences as such events as Smith's Samar expedition or the 1st Battalion, 20th Infantry at My Lai demonstrate.

Second, the Army tends towards an over-reliance on technology to address problems best solved by putting a soldier on the ground. The problems with putting the soldier on the ground are the higher cost, the greater risk, and the increased time required to achieve results. Technology helps mitigate, but does not fully eliminate, these concerns. In the Philippines, the telegraph and coal-powered vessels allowed the timely movement of troops. Technological solutions, however, cannot plug all gaps, especially when the technology is neither mature (the airplane in the Mexican Expedition) nor supported by policy (allowing sanctuaries for the VC or PAVN in Vietnam).

Third, the best outcomes occur when the Army establishes legitimacy and credibility with the local population. Aligning all activities with clearly understood and realistic political objectives is the best way to achieve this goal. The objective of "benevolent assimilation" in the Philippines allowed commanders flexibility in their approaches, with the most significant found in the rehabilitation of former *insurrectos* and the creation of native military and security organizations. Legitimacy and credibility ultimately develop from observed actions. For the Army to achieve them, it must speak consistently with one voice and enforce high standards of individual unit conduct. This is exceedingly difficult when the Army establishes policies that undercut its ability to do this.

Fourth, creating a realistic and well-understood legal framework for operations provided commanders with a tool for regulating the actions of soldiers and civilian populations in the areas in which units operated. GO 100 proved an enduring document shaping not only operations in the American South and the Philippines but ultimately globally as it helped shape multiple conventions governing the law of land warfare through the present. Today, the Army's most recent field manual on the law of land warfare, *Field Manual 6-27, The Commander's Handbook on the Law of Land Warfare,* cites the Lieber Code as a precedent.[105]

Finally, the four conflicts, their tactical problems, and the solutions tried meant the Army had a rather deep well of historical knowledge from which to draw as it fought insurgencies in Afghanistan and Iraq. Excepting improvised explosive devices and sophisticated information operations, the Army did not face any novel problems as it found itself involved in asymmetric campaigns in Afghanistan and Iraq. A review of its campaigns pointed to a constant: the need for rapid and appropriate adaptation to the unique characteristics (most notably the physical and human terrain) faced. Its challenge, therefore, lies in re-examining its historical experiences and applying the relevant lessons to the fights at hand. This was harder than it sounds as the Army's institutional structure, tactical organization, equipment and culture largely conditioned it for the large operations it wanted rather than the counterinsurgencies it had. The result was the Army seizing on some important past lessons in Afghanistan and Iraq, and disregarding some others of equal—or even greater—importance.

Chapter 3

Afghanistan and Iraq - 1990 to 2005

History is the unfolding of miscalculations [1]
Barbara Tuchman

Backgrounds and Beginnings

This chapter examines the invasions of Afghanistan and Iraq with a focus on the Army's role in both. The early days of both conflicts found the Army exceeding its own expectations as it attained military objectives far more quickly than anticipated and at a far lower cost. Successes, however, proved short-lived. As the American and coalition presence in both countries grew, insurgent groups challenged the United States, and its partners asymmetrically. The Army suffered thousands of casualties, the majority from a weapon seldom considered before 9/11, the improvised explosive device (IED). Magnifying IED effectiveness was sophisticated information operations, that raised awareness of IEDs and their effects to a global audience. Insurgent actions would ultimately threaten strategic objectives and popular support for the wars, and would consume the attention of the Army's senior leadership as it sought remedies to the unforeseen problems insurgents presented. Solutions would slowly come, but only after the Army acknowledged it faced a war for which it was largely unprepared and invested heavily in new methods. The Asymmetric Warfare Group ultimately grew from these efforts.[2]

Despite one member of Afghanistan's ruling Taliban government forecasting an American invasion, the head mullah, Omar Mohammad, refused to believe in a full-blooded American response to his country for harboring Osama bin Laden. Saddam Hussein underestimated the determination of American President George W. Bush and British Prime Minister Tony Blair to topple Iraq's *Ba'athist* regime. In contrast, the Bush administration believed that deposing these regimes would see newly empowered people rush to embrace representative democratic government. For military planners, this meant assuming the Afghan and Iraqi people would work with their would-be liberators through a relatively brief transitional period leading to popular governments. As events unfolded, the Taliban, the Iraqis, and the Americans all saw their miscalculations translate into extended conflicts taxing them in ways unimagined before both invasions began.[3]

Conflict between Afghanistan and the United States dated to the Taliban government's earliest days. From it taking power in 1994 through September 11, 2001, Taliban decisions either irritated or angered the U.S. government. Such actions as destroying irreplaceable statues from the country's Buddhist era (305 BC-715 AD), the imposition of strict dress and behavior restrictions on women, and ultimately offering a safe haven to Osama bin Laden fueled Washington's anti-Taliban attitudes. These same actions, however, resonated with the Taliban's core constituency: Afghanistan's Pashtun population. This was hardly surprising considering how Afghanistan contrasted with the United States. Poor, agrarian, tribal, and largely imbued with the values of Islam's conservative *Deobandi* Sunni sect, the country represented a throwback to the late medieval period. Juxtaposed against this was Afghanistan's recent history: perpetual armed conflict against the Soviet Union as well as other Afghans coupled with an aggressive Pakistan using Afghanistan to advance its own political goals. These fundamental differences in outlook, values, and recent history would profoundly influence American-Afghan interactions in the months ahead.[4]

While Afghanistan experienced little of America's wrath until after 9/11, Iraq's experiences differed dramatically. From 1990 onwards, Iraq replaced Iran as the pivot around which American military policy in the Persian Gulf region revolved. The 1991 ceasefire ending the Persian Gulf War solidified the enlarged American presence in the region. This included introducing ground forces into Kurdistan for humanitarian purposes and maintaining Army pre-positioned stocks (APS) in Kuwait. Consisting of combat vehicles, fuel, ammunition, and food, APS allowed the Army to fly troops into Kuwait and draw equipment rather than ship its equipment by sea. This meant an armored division could go from a stateside garrison to a defensive position on the Iraqi border in only a few days rather than weeks or months. Following Iraqi violations of 1991 U.N. Security Council Resolution 688, the U.S. and its European allies created two "no-fly" areas (later expanded into "no ground movement areas") over the country to prevent the diminished Iraqi Air Force from attacking Iraqis seeking to undermine Saddam Hussein's regime. Despite the catastrophic losses Iraq sustained during the Gulf War and the terms of both the ceasefire and multiple United Nations resolutions, Saddam periodically challenged American resolve. These provocations led to the rapid reinforcement of Army units in the region, most notably in 1998 and 1999 during operations DESERT THUNDER, DESERT FOX, and INTRINSIC ACTION. Frequently, American operations went beyond simple "shows of force." American aircraft conducted multiple strikes on a variety of Iraqi targets, including missile sites, ground units, weapons of mass destruction, and airfields between 1996 and 2003. For the Army, these represented shaping operations—"operations that create and preserve the conditions for success of decisive operations."[5] These strikes eliminated or reduced battlefield capabilities that Iraq could not regenerate before the 2003 invasion and contributed to the invasion's success.[6]

Within the Army, the wars in Afghanistan and Iraq fused with other operations into "overseas contingency operations" (OCO) also known as the Global War on Terrorism (GWOT). While the terrorist acts of September 11, 2001, provide an emotionally powerful case for dating the GWOT, armed action precedes that date. The Taliban's decision to allow Al Qaida training camps in Afghanistan resulted in the United States conducting Operation INFINITE REACH in 1998. Intended to destroy Al Qaida camps and change Taliban behavior, the operation was largely unsuccessful. Following INFINITE REACH, the Taliban continued to commit human rights abuses and destroy priceless archaeological relics. Coupled with the magnitude of 9/11, Taliban defiance meant many countries supported new American efforts to pressure the Taliban, including using military force if necessary. Like September 11, 2001, March 19, 2003, provides a flawed marker for marking the start of hostilities against Iraq. Air strikes, diplomatic pressure, economic sanctions, and information operations against Iraq extended back to 1990: 13 years of conflict preceded the March invasion.[7]

While Afghanistan and Iraq are very different countries, they share two characteristics that also shaped battlefield events: Islam and tribalism. In Afghanistan, both government policy and popular attitudes often reflected decades of education in the fundamentalist *madresha* (religious schools) of neighboring Pakistan. In 2001, the Taliban provided a graphic example of its commitment to Islamic fundamentalism. Breaking with long-standing Afghan tradition, they required the country's small Hindu minority to wear a yellow identifying symbol. Iraq's experience with Islam differed from that of Afghanistan. In Iraq, the differences displayed that country's colonial history and subsequent modernization efforts, as well as Islamic sectarianism. Modern Iraq is an artificial construct reflecting the competition between Britain and France as the Ottoman Empire slowly collapsed at the end of World War I. The 1916 Sykes-Picot Agreement between Britain and France carved Turkish domains into several new colonial possessions or protectorates, with Iraq turned over to London. Created from the *vilayets* (provinces) of Mosul, Baghdad, and Basra, the new state contained three mutually antagonistic populations. Basra contained a largely Shi'a Muslim Arab population, while Baghdad possessed a Sunni Arab majority. Mosul, in Iraq's northern mountains, contained a Kurdish population whose proverb, "a camel is not an animal, and an Arab is not a human being" captures a common attitude.[8] Given an unwanted king from the Arabian Peninsula's Hashemite family, Iraq survived as a highly unstable monarchy until 1958 and the first of several coups. Inspired by the *Ba'ath* ideology of Syrians Michel Aflaq and Salhadin al Bitar, Iraq established the trappings of a secular state emphasizing economic and social modernization. Despite this, Iraq's deep religious roots remained a major influence—particularly in both Arab communities. For the Shi'a majority, its often, second-class status and periodic government repression caused renewed adherence to religious beliefs and also eroded the bonds necessary for a cohesive state. This situation fueled conflict with the Sunni majority as the Iraqi government collapsed following the 2003 invasion.[9]

Almost as important as Islam in both countries is the tribal character of their populations. Consisting of largely consanguineous clans with shared traditions, values, and language, the tribes pre-dated the countries themselves. For many, tribal identity mattered more than one's national identity. The histories of Afghanistan and Iraq are also histories of tribal domination, which reinforced the importance of one's tribal affiliation. Geography also played a role in both tribal development and power. In Afghanistan, the combination of inhospitable terrain, vast distances, perennial water scarcity, and low population densities meant villages largely developed independently of one another. Such conditions also precluded to developing the strong central government capable of supplanting tribal loyalty with a national one. Indeed, for the powerful tribes, fostering such a government meant weakening their own positions against the country's other tribes. Even within the tribes, subordinate clans remained acutely aware of their relative power, prestige, and influence—and the benefits (or lack thereof) those factors generated. In Iraq, years of exposure to western ideas, Soviet aid, and Ba'ath rule fooled most observers into dismissing tribal relationships as relatively unimportant. Iraq's long-standing veneer as a developing and increasingly educated and technologically sophisticated state obscured the simple fact that tribal identification remained a submerged—and powerful—force. At the heart of the tribal structure was the concept of *wasta*. *Wasta* translates roughly as "clout" or "connections" and remains a strong social force across the Middle East. *Wasta* both fueled the insurgency and played a significant role in military operations in mediating American approaches and Iraqi responses as tribes navigated the changing circumstances. When those with *wasta* supported coalition entreaties, violence declined. When they did not, violence continued—or grew in intensity.[10]

Afghanistan—The Invasion

Even as fires burned in New York City and the Pentagon after 9/11, planning for the American response was underway. By October 5, 2001, plans described the basic outlines of possible American military action. Termed "courses of action," or COAs, these described the combination of forces, their missions, and the required outcomes. Responsibility for determining and then implementing the final COA rested with the U.S. Central Command's (USCENTCOM) commanding general, Gen. Tommy Franks. Franks had a West Texas working-class background and was both an Officer Candidate School graduate and a Vietnam veteran. His post-Vietnam assignments included service on the inter-German border during the Cold War, serving as the 1st Cavalry Division's assistant division commander for maneuver during the Persian Gulf War and command of the 2nd

Infantry Division in Korea. Franks understood strategy and recognized technology's benefits. He also realized that any COA ultimately depended on the soldiers executing it. To Franks, the soldier, and not the weapon, was paramount.[11]

For the Army, the COA selected initially meant deploying significant Special Operations and light infantry forces. Early deploying units included the 5th Special Forces Group (Airborne), portions of the 160th Special Operations Aviation Regiment (SOAR), and the 1st Battalion, 87th Infantry, from the 10th Mountain Division (Light Infantry). Providing required logistics and communications support was the 528th Special Operations Support Battalion (Airborne) and the 112th Signal Battalion (Special Operations) (Airborne). Recognizing the need to engage the local population, elements of the 4th Psychological Operations Group and 96th Civil Affairs Battalion were also deployed. The 5th Special Forces and 160th SOAR also incorporated significant intelligence processing capabilities. Organized as Task Force (TF) Dagger, these units possessed the ability to conduct missions ranging from humanitarian assistance to focused raids to assisting local militias. This last mission became the campaign's initial main effort as soldiers worked with the most capable of Afghanistan's anti-Taliban groups, the Northern Alliance (NA). In early October, Task Force Dagger moved from the United States to Uzbekistan's Karshi-Khanabad Airfield, soon nicknamed "K2,"and then into Afghanistan itself.[12]

An amalgam of tribal, ethnic, and religious groups, the NA's roots extended back to Afghan resistance groups formed after the 1979 Soviet invasion . This umbrella organization's membership and objectives frequently changed in response to both international and domestic developments. International developments included Pakistan's Inter-Service Intelligence Service's supporting the Taliban and INFINITE REACH. Domestic developments included the Taliban targeting the Northern Alliance's leaders and the changes in an individual Taliban member's political or social power. By late 2001, the Northern Alliance controlled only five percent of Afghanistan's territory yet almost 50 percent of the population. As a result, it was the most viable of the anti-Taliban groups. The aftermath of September 11 meant a reappraisal of American government perspectives on the group. Conversations on supporting their anti-Taliban efforts began in mid-September and came to fruition in early October with the arrival of American assistance. This assistance ultimately proved decisive.[13]

Dubbed Operation ENDURING FREEDOM, the campaign sought the destruction of the Taliban and Al Qaida. In the words of President George W. Bush, the country was "at war with terrorism, not the Afghan people."[14] This meant military efforts simultaneously struck military objectives *and* provided humanitarian assistance to the Afghan people. American forces applied lethal force against specific objectives. To control much of this force, Army Special Forces and attached Air Force tactical air control parties operated with Northern

Alliance groups. The Americans identified targets for engagement with precision munitions while the NA performed the ground combat role normally conducted by conventional Army forces. As coalition Special Operations Forces became available, they integrated into the larger American-Northern Alliance effort. This approach both reduced logistical requirements and kept conventional units available for operations elsewhere. The theoretical timeline called for limited operations and training NA units through the winter with a spring offensive in 2002 to topple the Taliban. The plan was simple—and risky. Weather, terrain, logistics, and the attitudes and abilities of NA members could all threaten success.[15]

The first strikes of ENDURING FREEDOM took place on the night of October 7, 2001. Forty-three Air Force and Navy aircraft and 50 "Tomahawk" land missiles hit targets deemed critical to the Taliban. Coincidentally with this, two Air Force C-17 "Globemaster III" aircraft dropped more than 34,400 food and medical packets over the country. Taking place within 45 minutes of the airstrikes, these humanitarian assistance drops were the first, and recurring, steps in a much larger humanitarian assistance program. While some Globemaster III's dropped humanitarian assistance, others continued the trans-Atlantic shuttle building up TF Dagger. The arrival of the 5th Special Forces Group headquarters at K2 on October 10 and the issuing of a warning order for the 160th SOAR meant the Army could soon begin playing its role. If conspicuously absent in ENDURING FREEDOM'S first actions, Army units began assuming a prominent role by conducting a series of helicopter and parachute assaults from Uzbekistan.[16]

Army operations began with uncharacteristic failure by the 160th Special Operations Aviation Regiment. Renowned for its ability to fly under the toughest conditions, the SOAR found itself unable to prevail in the face of Afghanistan's mountain weather on its first two attempts. On October 19, and the third attempt, the regiment's aircraft succeeded in clearing the mountain barrier between Uzbekistan and Afghanistan and inserted two Special Forces Operational Detachment Alphas (ODA). In the early morning hours of October 20, ODA 595 touched down at a landing zone 50 miles south of the Taliban stronghold of Mazr-e Shariff while ODA 555 landed near Bagram. Soldiers were now on the ground in Afghanistan, and the NA had a powerful new capability: precision airstrikes. That same night saw the campaign's first parachute assault as the 3rd Battalion, 75th Infantry Regiment (Ranger), seized an airstrip near Kandahar in southern Afghanistan. In the span of six hours, Army Special Operations Forces were on the ground at points hundreds of miles apart in the country's north, east, and south. While geographically separated, the purpose linked these locations by demonstrating the Army's ability to strike almost anywhere to threaten Taliban forces. Over the next two weeks, the Army used this ability to insert additional ODAs to support Northern Alliance forces.[17]

Almost immediately, the ODAs created a disproportionate effect on the battlefield. The NA lacked an air force, but with the operational detachments able to effectively control airstrikes it essentially acquired one. American and coalition aircraft could now drop precision-guided munitions on specific targets. This greatly enhanced accuracy but also allowed engagement when the Northern Alliance could immediately exploit the strike's effects. This broke the long-standing stalemate that characterized much of the NA-Taliban fight before Special Forces teams arrived. Reinforcing airstrikes was the tactical advice provided by Special Forces soldiers as they integrated into NA units. The result was a radical change in the campaign plan as it became increasingly clear that the combination of coalition force and the Northern Alliance could destroy the Taliban government long before spring.[18]

As Special Forces soldiers roamed Afghanistan, USCENTCOM continued the flow of units and equipment to Uzbekistan. November 10 saw two major milestones reached. In Afghanistan, NA and Special Forces took the northern city of Mazr-e Shariff, long a Taliban stronghold. For USCENTCOM, the number of servicemembers supporting this joint operation reached 50,000. Of this total, less than 130 were Special Operations Forces soldiers with the Northern Alliance itself.[19]

The capture of Mazr-e Shariff demonstrated to the Afghan people the capabilities of a revitalized Northern Alliance. For the NA, it proved that the Americans were committed to its fight. Special Forces soldiers not only provided precision airstrikes but also participated in the NA's old-fashioned cavalry operations that exploited them. Over the next three weeks, other major objectives fell as the Taliban fractured, and then disintegrated. The Northern Alliance closed the Khyber Pass, a line of communication enabling external assistance to the Taliban on November 13. Kabul fell the following day. Northern Alliance forces took Kondoz, east of Mazr-e Shariff and astride one of the few roads traversing northern Afghanistan, on November 23. Finally, Northern Alliance-Special Forces pressure forced the Taliban to abandon the southern city of Kandahar on December 7. The Taliban no longer provided a government, but instead a widely dispersed group of insurgents, some Afghan but many foreign, who would continue fighting. Amazingly, the Taliban's collapse came at the cost of only five American soldiers killed. Two of these casualties occurred in a helicopter accident in Pakistan on November 11, while the remaining three were the result of an errant bomb on December 5. Unfortunately, the decision of many insurgents to continue fighting following the Taliban's fall meant the cost of operations in American casualties in Afghanistan would steadily increase.[20]

Iraq—The Invasion

For many in the Army, Iraq was the "good war" by virtue of its aligning with the Army's self-conceptualization as a force optimized to defeat modern, technologically sophisticated opponents. While considerably smaller than its 1990-1991 incarnation, Iraq's army remained a largely mechanized one equipped with over 2,200 tanks and 4,000 artillery pieces. Planning for operations in Iraq began in November 2001 and reflected one of Secretary of Defense Donald Rumsfeld's convictions—regimes facilitating terrorism were as much the enemy as terrorist groups themselves. With Iraq's history of both weapons of mass destruction (WMD) and supporting terrorism under Saddam Hussein, it was inevitable that the country would receive close attention. The presence of neo-conservatives in the Bush administration believing a post-Saddam Iraq could provide a democratic example to the Arab world only increased the pressure to act against the country. Skepticism in both the United States and other countries greeted the ideas of both continuing Iraqi WMD production and post-Saddam Iraq serving as a beacon of liberty. Such skepticism presented itself in newspaper opinions and international political debates as it became increasingly clear the United States would employ military force. Within the Department of Defense (DoD) itself, options for the war in Iraq triggered far-ranging discussions capturing differing opinions as to what form such a war would take, and the resources required to successfully conclude it.[21]

These divergent opinions reflected the tensions between academic theory and practical experience. The efficacious use of military force requires reconciling the tensions between these two. Failing to consider academic theory virtually guarantees failure while failing to consider practical experience makes failing a distinct (and expensive) possibility. When asked what a war in Iraq would require in terms of troop strength, Army Chief of Staff Gen. Eric Shinseki responded "something in the order of several hundred thousand soldiers" to satisfy post-war occupation requirements.[22] This represented a marked contrast to the proposed operational concept based on theories of rapid dominance. Known as "shock and awe," this model leveraged intelligence, rapid operations, control of the environment, and strike capabilities to rapidly overwhelm an opponent. Such an approach, in design, would result in decapitating the enemy's command and control and destroying any will to resist. Conceptually, rapid dominance originated with the perceived revolution in military affairs that followed the Persian Gulf War. When wedded to neo-conservative beliefs about how post-conflict Iraq would look, Shinseki's comments appeared out of step with DoD thinking. As such, Shinseki's comments signaled an unwanted, narrow, and service-centric view of modern war to senior defense officials, most notably Rumsfeld and Deputy Secretary of Defense Paul Wolfowitz. Rumsfeld, a retired Naval Reserve fighter pilot, and Wolfowitz, a career civilian, failed to appreciate something Shinseki knew from years of professional experience: airpower creates magnificent, temporary effects. Exploiting those effects in a meaningful way required troops and, in a country the size of California, that meant a lot of them. These fundamentally different perspectives on the war and its requirements would have significant implications for the Army, particularly after Saddam's fall from power.[23]

For General Franks, the basic issue was rapidly revising Operations Plan (OPLAN) 1003. The existing contingency plan for a war against Iraq, OPLAN 1003 was several years old and more closely resembled the techniques of 1991's Persian Gulf War than the rapid dominance model Rumsfeld favored. Notably absent were the options flowing from improved precision-guided weapons and the lessons recently learned in Afghanistan. Revisions to this plan would occur not only through the operation's start date but also through virtually the entire invasion. These changes illustrated not only tensions within the DoD but also between NATO allies and regional partners. They would also challenge the subordinate commanders responsible for making a repeatedly revised plan work.[24]

While planning efforts continued at all echelons through the actual operation itself, a firm set of contingency plans was in place by August 2002. Within these broader documents existed more tightly focused plans titled "RED," "WHITE," and "BLUE." [25] These plans showed a variety of pre-invasion conditions and operational timelines with the major differences in RED, WHITE, and BLUE residing in the number of forces immediately available. Force availability, in turn, determined how the details of the plan would unfold. Because of the need to simultaneously support operations in Afghanistan and the Balkans, and meet other requirements, the Army staff *should* have played a major role in operational planning. As it turned out, USCENTCOM's approach to planning limited the role of the service component commands. This reduced the Army staff's ability to resolve issues before they reached the level of the Secretary of Defense and contributed to much of the friction presenting itself as the operation unfolded. Timelines reflected the envisioned intervals between American actions and battlefield outcomes. The most important of these, destroying the Iraqi regime, forecast a campaign ranging from 90 to 125 days based on the option implemented. All planning forecasted a short campaign to realize very specific objectives. Tempo, the pace of operations, reflected the combination of precision strikes and rapid maneuvers by conventional forces supported by both air and Special Operations Forces. Much of the operational and tactical planning for conventional ground operations fell to the headquarters of the Army's V Corps, headquartered in Heidelberg, Germany. Commanded by Lt. Gen William S. Wallace, the corps was a lingering legacy of the Cold War. An Armor (tank) officer since graduating from West Point in 1969, Wallace previously commanded both the 11th Armored Cavalry Regiment and the 4th Infantry Division. Controlling both V Corps and the

I Marine Expeditionary Force (I MEF) was the commander of the Combined Forces Land Component Command (CFLCC), Lt Gen. David McKiernan. Like Wallace, McKiernan was also an Armor officer and had previously commanded the 1st Cavalry Division. Built around the headquarters of U.S. Army Central Command (ARCENT), CFLCC's V Corps and I MEF constituted the major American ground combat units as the invasion unfolded.[26]

By December 2002, strategic objectives had crystallized in the form of a point paper defining victory. The document articulated victory as "Saddam Hussein and his clique" no longer controlling Iraq's real estate, oil fields, diplomacy, or having the ability to communicate to his military, the public, and the world.[27] He might still be alive, but cut off from the military, political (diplomatic), economic, and information instruments of power. The issue of weapons of mass destruction remained a work in progress based on the need to scour the country for weapons and their precursors and follow up on existing and developing intelligence reports. The same document noted a vacuum in government immediately following the invasion. A military officer would initially fill this, followed by a civilian reporting to DoD. A group of Iraqis from both inside and outside the country would initially provide advice. Like Vietnam and its "Many Flags" program, the plan sought international assistance in the effort to transition the government from the military to the Iraqi people. [28]

While the Bush administration sought to pressure Saddam Hussein to voluntarily leave Iraq, it concurrently began marshaling the forces needed to execute its military options. The 3rd Infantry Division (Mechanized), stationed at Fort Stewart, Georgia, was the first major combat unit to deploy to Kuwait. Liberated during the Persian Gulf War, Kuwait was home to the Army's largest Middle East garrison and served as the primary sustainment and command and control base for any invasion. The division began invasion planning in the Spring of 2002 since it viewed Iraq as its most likely contingency. More formal planning picked up in the summer with official notification by U.S. Army Forces Command. The division's advance party left Georgia in November with the first main body units deploying on January 1, 2003. Over the next thirty days, almost 20,000 soldiers arrived in Kuwait and took possession of the combat vehicles in the APS fleet in Kuwait. The 4th Infantry Division (Mechanized) received its alert order in January and rail-loaded its equipment for shipment from Fort Carson, Colorado, to Turkey. The plan called for the division to attack from Turkey into Kurdistan. Based on discussions with the Turkish General Staff, planners believed the United States had permission to execute this plan. As divisional equipment was en route to Turkey, however, the Ankara government made a final decision closing the country to the 4th and forcing a major change in both force flow and operational planning. It was not until March that the division received its deployment order and moved to Kuwait to link up with its equipment. February and March also saw the deployment of the 101st Airborne Division (Air Assault) from Fort Campbell, Kentucky and the 82nd Airborne Division from Fort Bragg, North Carolina.[29]

For Wallace, a Defense Department decision changing V Corps force allocations created both questions and considerable additional staff work. The traditional model utilized the Time-Phased Force Deployment List (TPFDL). Moving large military units, especially armored ones, is a complex process that must resolve not only questions of what unit but also how to get the unit from its home station to where it needs to go. Conceptually, this is simple. As a practical matter, it is much more complex when one considers the mechanics of the task. Sealift, airlift, air and sea ports of embarkation, debarkation, reception, staging, onward movement, and force integration all pose problems—problems often rippling through the tasks units must perform. The TPFDL minimized these issues in programming deployments by time and transportation mode and making the information available to agencies needing it. Rather than use the TPFDL, the Defense Department moved to a Request for Forces (RFF) model. The RFF model resulted in a series of extended negotiations requiring the corps to justify a request for a capability. If the DoD denied the request, there was more negotiation. The process of request-negotiate-renegotiate simultaneously consumed time and created uncertainty. Wallace summed up the issue succinctly: "It's hard to plan for a fight when you are not clear on what tools you will have to fight with."[30] This also complicated the work of the corps' rear command post in Kuwait.[31]

Attacking on V Corps' right flank, the I Marine Expeditionary Force (I MEF) and its attached Army and British units played major roles in the invasion. Finally, American, Australian, British, Danish, and Polish Special Operations Forces moved to designated locations as part of the Combined Joint Special Operations Task Force (CJSOTF). Unit missions included searching for WMD, conducting strategic reconnaissance, and coordinating with pro-coalition Kurdish groups. With major ground and SOF units largely in position by mid-March, President Bush made one last attempt at a diplomatic solution before issuing the order to invade.[32]

On March 17, 2003, President Bush delivered a speech aimed at both the American people and a larger, global audience. He emphasized four key points. First, Saddam Hussein and his family had 48 hours to leave Iraq. Second, if this did not happen the United States and its allies would use force to end Saddam's regime. Third, military operations would be directed against the Iraqi government, and not the Iraqi people. Fourth, the Iraqi military and security services could take action to protect themselves and Iraq by complying with instructions from the American-led coalition. Saddam's decision to remain in power started the final countdown for invasion.[33]

The night of March 19 saw the war's start with the SOAR attacking Iraqi observation posts along the Kuwaiti border while U.S. Air Force F-117 stealth fighters bombed the Dora Farms complex, southwest of Baghdad in an unsuccessful attempt to kill Saddam. Full-scale operations began the next day with coalition air forces conducting over 1,700 air and missile strikes through the depths of the country. V Corps engineers spent the day clearing Iraqi obstacles to allow the 3rd Infantry Division's lead elements to start its attack at eight o'clock that night. While 3ID initiated its attack, follow-on Army forces undertook preparations to follow the division as it continued moving north towards the so-called Karbala Gap. Located between Lake Razazza to the west and the Euphrates River to the east, maneuvering through the gap allowed the corps to advance while allowing the less mobile I MEF access to Highway 1.[34]

For Wallace, the biggest challenge lay in balancing movement with sustainment. The war started with V Corps having over 700 armored vehicles and almost 500 helicopters with the numbers growing as additional forces flowed into battle. Pre-deployment exercises and rehearsals informed Wallace's thinking and validated the need to maintain as robust a logistics capability as possible. This represented a major concern as distances grew and strength, and thus requirements, increased. An additional consideration was Iraqi WMD use. The Army classifies chemical weapons into two broad categories: persistent and non-persistent. Persistent agents can contaminate areas making movement across them hazardous until decontaminated. While the Karbala Gap was nominally 19 miles wide, a combination of waterways and built-up areas made it much narrower functionally—and much more suitable as a target for persistent chemical weapons. Chemical weapons could both slow V Corps' tempo and further increase logistical requirements as units worked through the issues of decontamination and the treatment of military and civilian chemical casualties. WMD remained a concern until the corps reached Baghdad.[35]

V Corps was not the recipient of all the units now flowing into Iraq. On March 26, 963 soldiers of the Italy-based 173rd Airborne Brigade conducted an airborne assault into Drop Zone Bashur near the city of Irbil. Subordinated to the CJSOTF, the 173rd's mission was to prevent six Iraqi divisions from repositioning south of Baghdad to threaten V Corps. Following the 173rd, the 3rd Battalion, 75th Infantry (Ranger) delivered yet another parachute assault the following day against the remote H-1 airfield in far western Al Anbar Governorate. Recognizing the 173rd's vulnerability, a subsequent airlanding of M1A1 tanks and M2 fighting vehicles from the 1st Infantry Division (Mechanized) in Germany ensured the brigade had an all-weather tank killing capability. Coupled with Special Operations Forces in northern and western Iraq, these operations gave the coalition both additional airfields and control of major roads linking the country's population centers.

By month's end, Saddam was under pressure from the north, west, and south with a hostile Iran to the east.[36]

As V Corps pressed its attack, it found itself battling a variety of different units. These encounters demonstrated a largely unappreciated lesson of the Persian Gulf War: Iraq's military forces included soldiers willing to fight. Units included the well-equipped Republican Guard, the regular Iraqi Army, and the paramilitary Saddam *Fedayeen* and Ba'ath Militia. The latter two groups proved they had more courage than competence. Capt. Jeff Ryals, staff intelligence officer for the 3rd Squadron, 7th Cavalry, recalled, "It was like watching a video game. The opposition was so poorly trained they just kept advancing on us. Every instinct should have told them to seek cover and retreat, but they didn't. They kept advancing and falling, advancing and falling until none were left."[37] Officers serving with the 82nd and 101st Airborne divisions made similar observations. Iraqi willingness to fight inflicted casualties on V Corps, but failed to seriously slow it. By March 31, V Corps was in a position to break into the so-called "Red Zone" south of Baghdad. This represented the zone in which the Iraqis would likely deploy their full range of capabilities—including chemical weapons—against a force now operating over 300 miles from its supply bases. For Wallace, logistics remained a concern. Movement into the Red Zone was undesirable without at least five days of required supplies.[38] With this logistics goal realized, it was time to break through the Karbala Gap. Wallace recognized that once the corps got through the gap and across the Euphrates River, there could be no operational pause until it reached Iraq's center of gravity: Baghdad.[39]

Simultaneous attacks taking place over a distance of roughly 130 miles would both maul Iraqi forces and push American forces across the river. Corps fragmentary order (FRAGO) 149M called for operations on March 31 to force the Iraqi military to reposition and concurrently destroy their reconnaissance. As the corps' main effort, 3ID simultaneously established blocking positions on key roads and conducted deep attacks with its Combat Aviation Brigade. The division was also prepared to seize bridges across the Euphrates. Finally, the division also screened the corps' northern flank as it made the crossings of the Euphrates to prevent an Iraqi move south through the Karbala Gap. The 101st Airborne both conducted a feint towards Al Hillah and contained Iraqi forces in Al Najaf. The 82nd Airborne in the south pinned Iraqi forces at As Samawah. Taking place simultaneously, these moves forced the Iraqi military to reposition forces—an effort made in daylight and under observation by both the corps and coalition air forces. That repositioning resulted in massive destruction of Iraqi armored and mechanized units. This in turn allowed the corps to rapidly push for Baghdad. Forty-eight hours after starting the attacks of FRAGO 149M, the corps reached Baghdad International Airport (BIAP). Rumsfeld had emphasized "shock and awe." Franks focused on a joint campaign plan. Wallace employed maneuver (the simultaneous strikes) to create targets for long-range and precision fires. These fires, in turn, enabled decisive maneuver.[40]

Eleven miles separate BIAP from the west bank of the Tigris River. Adjacent to the river lay the Presidential Palace. For Wallace, the remaining question was when to move 3ID over that distance and then stay. On April 5, 3ID's 1st Battalion, 64th Armor, conducted a probe into Baghdad. The battalion was repeatedly, but ineffectively, engaged by irregular forces. After considering the situation, and the ability to both re-supply forces in the city and evacuate any wounded, Wallace decided to occupy positions in Baghdad. On April 7, the 2nd Brigade, 3ID, moved into Baghdad to stay. Seizing the strategically important Presidential Palace broke the back of Saddam's regime as it signaled to Ba'ath officials in Baghdad their leader's defeat and the need to leave the area to avoid capture by Army or Marine forces now expanding the American presence in the city. It had taken 19 days and the loss of 36 American soldiers killed in action, but the V Corps accomplished the campaign's most important objective.[41]

Even as V Corps and I MEF drove towards Baghdad, additional Army units conducted operations in other parts of Iraq as part of the effort to neutralize the WMD threat and decapitate the regime. These efforts also resulted in casualties and introduced the Army to the best weapon in the arsenal of Iraq's paramilitary or irregular forces: the improvised explosives device (IED). The 3ID experienced such attacks on its drive north, and Special Operations Forces near Haditha in Al Anbar Governorate suffered casualties to suicide attacks with IEDs. Such episodes introduced the Army to a problem with which it would increasingly struggle—and which would ultimately set off alarms from Baghdad to Washington.

Afghanistan and Iraq—Continuing Operations

The end of Taliban rule meant a shift in operational and tactical focus. The remaining Taliban and Al Qaida strongholds required attention, and there was reason to believe that terrorist mastermind Osama bin Laden was somewhere in Afghanistan. Tactical operations now focused on these missions. The Army concurrently increased its Civil Affairs efforts to aid Afghanistan in re-establishing local and national governance and assist the country's civilian population. Despite the desire to limit troops in Afghanistan, the size of the Army force steadily grew through late 2001 and early 2002. Early deploying units included portions of the 513th Military Intelligence Brigade, 11th Signal Brigade, 92nd Engineer Battalion (Heavy), and the 507th Combat Support Group. The 101st Airborne Division's 3rd Brigade, the 10th Mountain Division's 2nd Brigade, and ultimately the headquarters of the 10th Mountain Division soon joined these units.[42]

Previously alerted, the 10th Mountain Division's headquarters received its deployment order on November 25, 2001. By December 12, it was in position in

Afghanistan and assigned the mission of serving as the CFLCC forward (CFLCC-F) headquarters with Maj. Gen. Franklin L. "Buster" Hagenbeck in command. An officer with extensive light infantry experience, including combat service in Grenada, Hagenbeck orchestrated the first major operation of the war employing large numbers of conventional Army forces, Operation ANACONDA. Intended to destroy Taliban forces in the Shah-i-Khot Valley, Hagenbeck used a mix of conventional, American and coalition Special Operations, and Afghan forces. Not under Hagenbeck's control were two important elements supporting the operation: the Combined Air Component Command and Task Force 11 (Task Force 11). Subordinate to the Special Operations Command, TF 11'smission was strategic reconnaissance and assistance in targeting airstrikes.[43]

The Shah-i-Khot represents difficult terrain. Elevations range from 7,500 feet on the valley floor with surrounding peaks approaching 12,000 feet. Complicating the terrain factors was the weather: in early March, daytime temperatures hovered between 15 and 20 degrees Fahrenheit. Intelligence estimates placed Taliban strength in the Shah-i-Khot at approximately 300 individuals, supported by a variety of light weapons. Events proved this estimate of enemy strength was seriously deficient—actual Taliban numbers were over 1,000, and they possessed heavier weapons than anticipated. The basic plan called for Afghan forces, a Pushtun militia under Zia Lodin, to attack into the valley while conventional Army and coalition SOF forces occupied blocking positions on the routes out of the valley. Lodin's forces would kill or capture those who stayed to fight while the American and coalition forces would kill or capture those who tried to escape through the blocking positions. Hagenbeck envisioned a short battle against a small force with all the subordinate and supporting elements working in concert—the plan was relatively simple, even if the environment was not. [44]

It is a military maxim that no plan survives contact with the enemy, and Hagenbeck's was no exception. Hagenbeck established "D-Day" as February 28, but weather compelled a 48-hour delay. Once the attack commenced on March 2, it was only a few hours old when Lodin's forces retreated from the valley following a devastating Taliban mortar barrage. Weather and the enemy also hampered American efforts to both get into position—fog forced delays in helicopter take-offs and unexpected mortar fire inflicted growing casualties as the operation unfolded. Army forces also operated under an additional handicap: no field artillery. Attack helicopters filled some of the void created by the lack of cannons, but the small number available and the need for fuel, rearming, and essential repairs meant gaps in support. Coalition air forces also helped fill the gap, but shortfalls in initial planning created additional issues for the air component. Change was the order of the day through March 18 and the battle's end. Sixteen days of combat resulted in eight American deaths and 75 wounded, and over 800 Taliban deaths, many of them non-Afghan. Hagenbeck and others learned a lesson: the enemy's commitment to his cause was greater than originally thought. Taliban commander Maulvi Saifurrahman Mansoor stated from the Shah-i-Khot that the Taliban would "continue to wage jihad until our last breath against the Americans for the glory of Islam and for the defense of our country."[45]

Despite the challenges of the enemy, terrain, weather, and partners, ANACONDA proved successful in eliminating the Taliban from the Shah-i-Khot. Its greater significance lies not in the Taliban's defeat, but in signaling a change in Taliban tactics. After this battle, the Taliban would not present themselves as targets for large scale operations until mid-2006. Instead, the battlefield became much more distributed with smaller groups of insurgents arrayed over greater distances. ANACONDA is thus the only major battle of the Afghan war involving conventional Army forces.[46]

Less than a month after ANACONDA'S end, the Army had its first lethal IED experience in Afghanistan. An IED near Kandahar killed Sgt. 1st Class Daniel Romero, S. Sgt. Brian Craig, S. Sgt. Justin Galewski, and Sgt. Jamie Maugans. Romero was a Special Forces soldier from the Colorado Army National Guard's 5th Battalion, 19th Special Forces Group. Craig, Galewski, and Maugans all belonged to the 710th Ordnance Detachment (Explosive Ordnance Disposal). The soldiers were clearing a cache of explosives when the device detonated. Initially assessed as an accident, this was the Army's introduction to what would become its most pressing problem.[47]

April saw the first of several important changes in Army command in the country. The first was placing the CJSOTF under CFLCC-F's command. This meant not only U.S. Special Operations Forces, but their counterparts from seven other countries—a number that would grow over the next several years—changed command. Coalition contributions compensated for gaps in American force structure, necessary when operating in an area the size of the states of Arizona, Nevada, West Virginia, and Delaware combined. More importantly, this brought about greater unity of command. The experience of Operation ANACONDA and TF-11 reinforced the need for this long-standing principle of war. The second was CENTCOM's decision to request a corps headquarters to rationalize command and control of ground operations. XVIII Airborne Corps received the requirement and began deploying 365 soldiers to Afghanistan in early May. To bring the corps headquarters to operational strength, it received reinforcements from other Army units, the U.S. Marine Corps and Air Force, and several coalition countries. With the arrival of the corps' headquarters, the 10th Mountain Division shed its responsibilities as CFLCC-F. By June 1, XVIII Airborne Corps was operational as the headquarters of Combined Joint Task Force-180 (CJTF-180). CJTF-180's missions included combat operations, the training of Afghan security forces, and working with both the emerging Afghan national government and neighboring countries. To effectively do this, CJTF-180 assigned the 10th Mountain Division, and then its replacement, the headquarters of the 82nd Airborne Division, responsibility for tactical operations.[48]

During the Vietnam War, the Army stationed units in South Vietnam for years at a time and then rotated soldiers through them. In contrast, the Global War on Terrorism saw units rotating to and from the combat zone. The first unit rotations in Afghanistan took place in June 2002 with the 1st and 3rd Brigades of the 82nd Airborne Division flowing into the theater. Also rotating in were the 82nd Airborne Division's Combat Aviation Brigade and the XVIII Airborne Corps' 3rd Battalion, 229th Aviation. For Maj. Gen John Vines, the 82nd Airborne commander, this meant leveraging existing command relationships. General Hagenbeck's team was largely an *ad hoc* construction, Vines' was not. Vines also enjoyed two additional advantages over Hagenbeck when it came to tactical operations: more helicopters and better flying weather. This combination gave Vines greater flexibility in planning and conducting operations.[49]

This same period saw an increase in Civil Affairs strength. Parts of the U.S. Army Reserve's 489th Civil Affairs Battalion, home stationed in Knoxville, Tennessee, deployed in March to replace the 96th Civil Affairs Battalion. The increased demand for support called not only for deploying the rest of the battalion, but also augmenting it with soldiers from New York-based units, the 401st and 450th Civil Affairs battalions. The 489th was an experienced outfit and had only returned from a rotation to Bosnia seven months before leaving for Afghanistan. For the Civil Affairs soldiers, deployment did not consist of a series of patrols or firefights. Instead, they engaged with both local officials and ordinary Afghans as they sought to rebuild critical infrastructure and provide relief. The unit accomplished this by deploying four-soldier teams. These were perhaps the most vulnerable soldiers in the country as they operated from contracted vehicles in locations scattered across the country. The civil affairs (CA) mission was important in both practical and symbolic terms. On a practical level, CA efforts improved living and economic conditions for thousands of Afghans. Symbolically, they also differentiated the Americans from the Soviet Army. The Soviet Army's 1979 -1989 occupation did not include robust CA operations. Instead, special propaganda units would periodically dispense food and medical assistance as they tried to politically indoctrinate the populace. Working against these programs were the official policies of harsh reprisals in response to attacks and regarding all Afghans as the enemy.[50]

Creating Provincial Reconstruction Teams (PRT) further boosted both the visibility and impact of CA operations. First established in Gardez in January 2003, the PRT combined "civilian and military personnel into a single entity with the purpose of improving security, governance, and economic development."[51] This approach created a "civil-military institution able to penetrate the most unstable and insecure areas because of its military component and is able to stabilize these areas because of the capabilities brought by its diplomacy, defense, and development components."[52] With over 100 military and civilian personnel on a team (including a security detachment), PRTs rapidly grew in number. PRT interaction with local leaders both increased local commitments to engaging the national government and tied the populace to reconstruction projects. Reports of Afghans defending PRT-enabled projects against attacks by Taliban or Al Qaida insurgents validated both the money and effort spent in creating these teams and making their efforts an integral part of the overall campaign plan. By mid-June 2005, the number of PRTs reached 22, with nine provided by other coalition countries.[53]

Civil Affairs was not the only element of "soft" power applied by the Army. Psychological Operations (PSYOP) elements rotated into the country with the missions of explaining American operations, isolating Al Qaida and the Taliban, discouraging interference with coalition operations, and soliciting support for the Afghan national government. The need to produce culturally appropriate messages in multiple languages increased the difficulty of these tasks. The negative opinion that many Afghans maintained about the United States and its involvement in their country complicated matters. This opinion originated in Taliban and Al Qaida claims that the United States was engaged in a religious war against Islam. While the creation of many PSYOP products took place in the United States, Kuwait, and the British Indian Ocean Territory of Diego Garcia there nevertheless remained a requirement for a sizable presence in Afghanistan. Like their CA counterparts, PSYOP soldiers—often Army Reservists—found themselves scattered across the country and operating in small teams with very little protection.[54]

Tactical operations acquired a rhythm as the American-led commitment to Afghanistan continued. Driving the rhythm was the rotation cycle of Army forces and Afghanistan's weather. The country's general weather patterns meant decreased operations through the winter months. The commitment of many American troops to the fighting in Iraq and Afghanistan's hostile geography initially meant the Army's light infantry formations would take on the Afghan mission. The 10th Mountain Division returned to Afghanistan as the headquarters for CJTF-180 in May 2003. 10th Mountain Brigades replaced those of the 82nd Airborne. The 1st Brigade assumed the combat mission while the 2nd Brigade took on training the Afghan National Army. Designated Task Force Phoenix (TF Phoenix), the training mission would remain an enduring one. The first Army National Guard conventional combat brigade, Oklahoma's 45th Infantry Brigade Combat Team, assumed the TF Phoenix mission in November with the 2nd Brigade, 10th Mountain Division, redeploying to its home station. The 1st Brigade, 25th Infantry Division, arrived in February 2004, followed by the arrival of the division's headquarters in April as the replacement for the 10th

Mountain. Accompanying the headquarters was the 25th Infantry Division Artillery and its Combat Aviation Brigade. When the 25th assumed CJTF responsibilities CJTF-180 became CJTF-76. August 2004 saw another TF Phoenix rotation with the Indiana Army National Guard's 76th Infantry Brigade Combat Team assuming this mission. The cycle repeated itself in April 2005 with the Southeast European Task Force taking over the CJTF-76 mission. Arriving from Vincenza, Italy, the 173rd Airborne Brigade replaced the 3rd Brigade, 25th Infantry Division. The 1st Brigade, 82nd Airborne Division reinforced CJTF-76. The rotation plan identified units up to 18 months before their deployment. This facilitated a rigorous training program building individual and collective skills before conducting a full-scale mission rehearsal exercise with a stable chain of command. This methodology represented a marked departure from the Vietnam War and the individual troop rotation. As a result, units arrived in the theater at full strength and proficient in the tasks trained before deploying. [55]

All types of Army units—conventional, Special Operations, Civil Affairs, or Psychological Operations—took casualties through their rotations. The pre-deployment training program, while beneficial, did not fully insulate units from battlefield risks. The requirements of combat and civilian engagement provided insurgents with frequent opportunities to observe American units and identify their vulnerabilities. Such observations facilitated an evolution in enemy tactics that resulted in increasing casualty rates. As a result, the cycle of action-adaptation-reaction-adaptation became a constant of operations in Afghanistan.[56] The desire to try and adapt faster than its opponents helped provide the ultimate impetus for the Asymmetric Warfare Group.

In Iraq, Baghdad's fall did not mean an end of combat operations. For Wallace, the "true transition was from major combat operations to something nebulous to counter-insurgency operations."[57] The environment itself was highly dynamic with four major groups of actors. The first group was the coalition and included military units and civilian administrators. The Iraqi civilian population represented the second. The third was the insurgents whose motivations were highly varied. The Iraqi security services, both pre- and postwar represented the fourth. Violence characterized the interaction between the coalition and the insurgents. Friction, misunderstanding, and at times outright deceit, defined coalition relations between the Iraqis and the security services.

The fundamental problem for the coalition lay in the lack of a clear vision and operational concept for post-conflict reconstruction. Retired Army Lt. Gen. Jay Garner's Office of Reconstruction and Humanitarian Assistance (ORHA) initially had this responsibility. Appointed to the position in January 2003, Garner shouldered responsibilities that included civilian administration, humanitarian affairs, and reconstruction. In theory, ORHA's objective was to "rapidly work itself out of a job within 90 days."[58] This objective, however, rested on two assumptions: that the military campaign would take far longer than it did, and that the overall damage to Iraq's government infrastructure would be far less than that actually encountered. Exacerbating the situation was ORHA's lack of situational awareness as the invasion unfolded. Largely excluded from the CLFCC headquarters and thus unaware of changes in the plan, ORHA entered Iraq ill-postured to assume its responsibilities. Wallace and his staff knew the ORHA existed but had no contact with it and assumptions made about its capabilities proved incorrect. Even before ORHA had a chance to fully establish itself, the Coalition Provisional Authority (CPA) replaced it. Headed by former Ambassador L. Paul Bremmer, the CPA represented the long-term governing structure of an occupied state. For Wallace, this transition seemed rapid and uncoordinated.[59]

Not appreciated before the war was the high level of control the Iraqi government exercised over ordinary citizens. With that control fractured, and with the ORHA and then CPA making things up "on the fly," conditions for an insurgency ripened. The failure of either the ORHA or CPA to provide any truly stabilizing influence meant a flourishing insurgency and an increasingly disaffected population that validated General Shinseki's original remarks. The Iraq security services might have successfully performed that role, but the Coalition Provisional Authority order disbanding the army and police removed that option and contributed to both the insurgency and popular discontent. The combination of the ORHA and CPA decisions created a vacuum—one soon filled by a multi-faceted insurgency. Former Regime Elements (FRE), Al Qaida foreign fighters, Sunni groups fearful of Shi'a domination, Shi'a groups seeking dominance, and Salafist groups battled the coalition—and each other. Interspersed among the mélange were criminal groups exploiting opportunities. This kaleidoscopic threat would challenge the Army in unimagined ways and force major changes in organization, training, equipment, personnel policy, and unit stationing around the world.[60]

Baghdad's fall meant an expansion of the Army's footprint as units deployed across the country. By late April 2003, units occupied not only Baghdad but the major population centers in all of Iraq's 18 governorates. The 3rd Infantry Division centered its operations on that portion of Baghdad under Army control and would remain in the country until relieved by the Germany-based 1st Armored Division (1AD). The 4th Infantry Division moved into Iraq in April and established its headquarters in Tikrit, some 150 kilometers north of Baghdad. The 173rd Airborne Brigade operated under 4th Infantry Division control. Farther north, the 101st Airborne Division re-established its headquarters at Mosul. To Baghdad's west, Al Anbar Governorate saw several units perform occupation duties. The 3rd Armored Cavalry Regiment initially performed the mission but then transferred responsibility to the Marine Corps Task Force

Scorpion. In September, the 82nd Airborne Division replaced the Marine Corps Task Force Scorpion and controlled operations from Ramadi. To the south of Baghdad, a British Army division centered its operations on Basra, while a Polish division established its headquarters in Diwaniyah in September. Dividing the country into six divisional areas, units assumed the operational designation of one of six multi-national divisions (MND). The multi-national division designation reflected the presence of coalition forces of varying size within each division's area. Coalition forces performed a variety of roles with one of the most important reducing requirements for additional Army units and thus helping sustain the Army's unit rotation model.[61]

The Army followed President George W. Bush's May 1, 2003, speech announcing "mission accomplished" in Iraq with a May 27 statement modifying its controversial "stop-loss" policy that prevented soldiers from leaving the service. June 14 saw a structural change with V Corps assuming the title Combined-Joint Task Force Seven (CJTF-7). This change made Wallace the senior commander for all ground forces—American and coalition—in Iraq. Wallace's tenure as Commander, CJTF-7, was brief as Lt. Gen Ricardo Sanchez replaced him effective July 10. Recognizing that Iraq would join Afghanistan as a long-term American commitment, the Army announced a rotation plan on July 23 predicated on units spending a full year in the combat zone before returning home. Sustaining such a schedule required significant mobilization of Army National Guard and Army Reserve units and added to the controversy already attached to the war in Iraq. It also required detaching brigades from Regular Army divisions and deploying them independently of their parent units. As part of the rotation plan, the Army created Task Force OLYMPIA (TFO). Largely built on units from I Corps at Fort Lewis, Washington, TFO began deploying to Iraq in November.[62]

For the units already in Iraq, 2003 proved a difficult year as the Army tried to simultaneously contain an expanding insurgency and restore essential services across most of Iraq. One major problem area, critical to any counterinsurgency campaign, was producing usable intelligence to drive small-unit operations. The Army's Center for Army Lessons Learned (CALL) observed in September that the tactical intelligence teams (THT) responsible for developing human intelligence failed to perform adequately. The THT was theoretically the best intelligence asset available to small units as it collected information from the local population. In September, the 69 THTs in Iraq should have produced 120 actionable reports a day. Instead, they averaged around thirty and these were often of "poor quality." "Poor quality" meant there was little useful intelligence in the report with this creating two problems. First, it could not drive operations. Second, it could fill databases with incomplete or inaccurate information which would complicate subsequent analysis of later reports.[63]

The albatross of reconstruction also created unforeseen demands on units—and helped fuel a steadily rising level of anger towards the American presence in Iraq. This anger, in turn, further strengthened the insurgency as Iraqis turned frustration into action. The Coalition Provisional Authority's failures meant troop units had to step up and try and fix problems or they would get worse. Looters largely destroyed what infrastructure remained untouched by military action. Units struggled to simultaneously provide security, restore electrical power, reopen schools and markets, ensure food distribution, and establish civilian government. Making this task more difficult was the removal of the Iraqi managers who could assist in such efforts as part of the CPA's de-Ba'athification efforts. A vicious circle emerged with the Army at its center. As insurgent attacks increased through 2003, the Army's challenges multiplied.[64]

On December 13, 2003, the 4th Infantry Division accomplished a strategic objective when soldiers from its 1st Brigade captured Saddam Hussein. Formerly accustomed to living in gold-plated luxury, Saddam found his freedom stripped from him as soldiers pulled him from an underground burrow near Ad Dawr, not far from his Tikrit birthplace. Still claiming the title of Iraq's president, Saddam sought to negotiate with his captors. The capture achieved nothing decisive, however, as attacks against coalition forces increased in number and violence.[65]

As 2004 unfolded, the 1st Infantry (1ID) and 1st Cavalry Divisions (1CD) deployed as part of a robust unit rotation schedule including Army National Guard brigades. The Guard's 30th (North Carolina) and 155th (Mississippi) Armored Brigades, and the 39th (Arkansas), 81st (Washington), and 256th (Louisiana) Infantry Brigades all served under Regular Army or Marine Corps division-level headquarters. To address personnel strength issues, the 30th Infantry received a battalion from New York's 27th Infantry Brigade, while the 39th received a battalion from Oregon's 41st Infantry Brigade. The attachment of "new" battalions to brigades illustrated a new reality—the Army National Guard severing traditional organization alignments to deploy full-strength units. The demands of homeland force protection (Operation NOBLE EAGLE) and peacekeeping missions in the Balkans and the Sinai coupled with soldier readiness issues meant many Guard soldiers found themselves in new units.[66]

Both the 1st Infantry and 1st Cavalry Divisions began deploying units into Iraq in January 2004. On February 5, TFO relieved the 101st Airborne as the Multi-National Division, North in Mosul. In March, the III Corps staff replaced that of the V Corps in Baghdad, while 1ID relieved the 4th Infantry Division in Tikrit and 1CD relieved 1AD in Baghdad. Both 1ID and 1CD saw major changes in their standing task organization. The 1st Infantry Division received both the 30th Armored Brigade and the 2nd Brigade, 25th Infantry Division. The 1st Cavalry Division created a provisional brigade based on its divisional artillery and received the 39th and 256th Infantry Brigades as it became the Multinational

Division—Baghdad. In Al Anbar, the I Marine Expeditionary Force assumed the duties of Multi-national Force—West (MNF-W) from the 82nd Airborne. In August, MNF-W received the 2nd Brigade, 2nd Infantry Division, previously stationed across Gyeonggi-do Province, South Korea, as the relief unit for 1st Brigade, 1st Infantry Division. The juggling of units in both Afghanistan and Iraq vividly illustrated a fundamental problem: the Army lacked the proper structure for the long war in which it found itself. The unfortunate reality is events proved Shinseki correct: properly controlling the vast areas now theoretically under Army control required far more troops than were readily available.[67]

While units rotated in and out of the theater, the mission's basic requirements remained unchanged: defeat the insurgency and reconstruct the country, which would supposedly place Iraq on the road to national sovereignty. To accomplish the first task, the Army sought to improve its intelligence collection and analysis efforts. Despite Army efforts to expand the Arabic linguist pool through a new enlistment program and the employment of contractors, linguists remained critically short. To compensate for this, units employed a variety of technical solutions including aircraft, surveillance cameras, and even the Joint Land Attack Cruise Missile Elevated Netted Sensor System (JLENS). Originally developed to detect incoming missiles, the JLENS proved useful in identifying attack locations. Tactical intelligence production is comparable to constructing a mosaic. Each small piece, when fitted with other small pieces, helps develop an intricate picture that ultimately allows identifying actors, locations, resources, and, most importantly, vulnerabilities. These high technology aids allowed analysts to identify patterns, tip and cue sensors, and ultimately direct maneuver forces against locations associated with insurgent activity. In the words of Col. Mike Formica, commanding 1CD's 2nd Brigade, "It's all about intel (…) all about intel."[68] Even with more reconnaissance and surveillance support, the task of intelligence production remained difficult. This captured two interrelated facts: the rapid attrition of insurgents incapable of adapting to a changing environment, and the increasing sophistication of attacks made by those who did adapt. Unsurprisingly, there is an element of Charles Darwin's theory of evolution highly applicable to combat.[69]

The continuing failures of the Coalition Provisional Authority, Department of State, and other government agencies to provide personnel for Iraqi reconstruction meant that troop units continued filling important positions in Iraq's various ministries and agencies. At the local level, this same paradigm degraded overall unit effectiveness by increasing the tasks a unit performed without a corresponding increase in troop strength. Then-Lt. Col. Frank Hall, commanding 1ID's 101st Military Intelligence Battalion, found himself concurrently directing an Iraqi school district. Other field grade officers who did not hold commands served as de facto cabinet officials in the faltering CPA administration. Still other soldiers found themselves serving not with the U.S. Army, but with the Iraqi Army as part of the effort to create a force responsible not to a dictator, but to the nation's citizenry. Initially undertaken by the Spanish Army and American contractors, Army training teams fell in on new Iraqi Army units in March 2004.[70]

July 1, 2004, ushered in a major change in leadership of Army efforts in Iraq. Gen. George W. Casey, Jr. assumed the duties of Commanding General, of Multi-National Force Iraq (MNF-I). Replacing CJTF-7, Casey's command operated in a different legal environment than Sanchez as Iraq had regained its sovereignty with the Coalition Provisional Authority's dissolution on June 28. An Iraqi Interim Government (IIG) exercised national authority until a permanent one could form following elections in 2005. This government's authority lay in the transitional administrative law (TAL) that would remain in force until a constitution could be composed and ratified. Multi-national Force—Iraq would not occupy Iraq but would partner with the IIG under the terms of a new legal code. For Casey, the greatest risk to strategic success was the Iraqi government's failure. Iraqi Interim Government's success would hinge not only on how it partnered with the coalition but also on how it engaged tribal, especially Sunni tribal, leadership.[71]

Drawing on his previous experience in the Balkans, Casey sought Iraqi participation in everything MNF-I did in the belief this would help realize five important objectives. These were enhancing the new Iraqi government's legitimacy, driving a wedge between the insurgency and the population, changing perceptions of the coalition, engaging the populace in the counterinsurgency effort, and improving the image of Iraq's new security services. Partnership did not mean any slackening of American operations as Casey appreciated the need for aggressive operations to prevent the insurgency from overwhelming political objectives both in Washington and Baghdad. To facilitate this partnership, Casey's operational plan emphasized "full-spectrum counterinsurgency operations."[72] This meant not only combat, but also information, stability, and support operations. Killing or capturing the enemy through patrols, raids, or attacks (the major tactical tasks of combat units) was critically important, but also meaningless if the Army could not concurrently accomplish the campaign plan's political objectives.[73]

The year 2004 proved much more violent than 2003 as the insurgency grew in strength and responded to coalition and IIG action. Army and Marine Corps units found themselves conducting major battles in Baghdad, Samarra (north of Baghdad), Al Najaf (south of Baghdad), and Fallujah (west of Baghdad). Fighting in Sadr City and Al Najaf involved Shi'a militias loyal to cleric Muqtada Al Sadr. In Samarra, the insurgents represented a combination of criminal elements, Sunni tribes vying for control, and foreign fighters. In Fallujah, insurgents included FREs, foreign fighters, Sunni extremists, and even deserters from the newly created Fallujah Brigade, an Iraqi unit responsible for the city's security. As the year progressed, the multi-faceted nature of Iraq's insurgency made itself obvious with escalating violence. March witnessed the first increase in total insurgent attacks since November 2003. The combination of the fighting in Baghdad and Al Najaf and the release of photos showing prisoner abuse at Abu Ghraib in April resulted in attacks that month that almost doubled the number sustained in March. May logged a small increase in attacks before a small drop in June and July. August proved the year's most violent month with attacks nearing 3,000 before a drop in September and significant increases in October and November. December saw attacks drop to a three-month low, but included a suicide bombing of an Army dining facility at Forward Operating Base Marez that killed 22 and wounded over 70. Perpetrated by the Sunni insurgent group Ansar Al Sunna, this was the first major attack against Americans in Iraq's Kurdish region.[74]

As the insurgency grew in intensity, the Army prepared for another round of unit rotations to take place between February and September 2005. With MNF-I's creation, III Corps had assumed the duties of Multi-National Corps—Iraq (MNC—I). XVIII Airborne Corps replaced III Corps as the new MNC-I headquarters. The 11th Armored Cavalry Regiment and 172nd Infantry Brigade, the Regular Army's only combat unit in Alaska, replaced TFO in Mosul. The New York Army National Guard's 42nd Infantry Division (42ID) replaced 1ID and controlled a mix of Regular Army and Army National Guard brigades. The latter included the 278th Armored Cavalry (Tennessee) and the 116th Cavalry (Idaho and Oregon). The 3rd Infantry Division found its 1st and 3rd Brigades serving under the 42ID as it replaced 1CD in Baghdad. The II Marine Expeditionary Force assumed the duties of MNF-W and had the Army National Guard's 2nd Brigade, 28th Infantry Division (Pennsylvania) under its control. The 29th Infantry Brigade (Hawaii) and the 56th Infantry Brigade (Texas) deployed as theater security brigades, operating across Iraq and focusing on securing transportation convoys that increasingly became insurgent targets. The 56th was unique in that it consisted only of Texas Guardsmen. The issues of unit organization and alignment remained a problem as 42ID and every other Guard brigade deployed in 2005 required reinforcement from other states. The 116th represented the extreme case, drawing on soldiers from 22 different states to bring its seven battalions and supporting elements to full strength. This approach ensured full-strength units but seriously complicated the planning and conduct of necessary pre-mobilization training.[75]

The struggle for Iraq reached a milestone in 2005 with three significant elections shaping events and American objectives. On January 4, Multi-National Force-Iraq described seven objectives it sought to attain with two of them inextricably linked to the year's ambitious election cycle: establishing a "federal, democratic, pluralistic and unified Iraq" and defeating the "campaign of intimidation and bringing the insurgency and foreign terrorism to a level that can be contained by the ISF (Iraqi Security Forces)."[76] The first contest, conducted on January 30, elected members to Iraq's interim National Assembly. The second, held on October 15, was a referendum on Iraq's proposed constitution. Finally, an election on December 30 determined membership of the country's permanent Council of Representatives. Iraqis often viewed election-related questions from tribal or sectarian perspectives, and the country's religious and ethnic fault lines once again became clear. Insurgents and political rivals increasingly resorted to violence against Iraqi civilians to influence voter participation and choice as voting dates drew near. Foreign fighters posed an increasingly determined threat across Iraq's Sunni regions as Al Qaida in Iraq grew in strength, adding to the threat posed by FRE. By mid-June, the estimated size of the Sunni insurgency alone was between 17,000 and 20,000 combatants and supporters. Former regime elements continued benefitting from existing tribal relationships. Iran also exerted its influence in training and supporting several Shi'a groups. The result was a year of greater violence, even as American troop levels remained high and the ISF grew in both numbers and capabilities.[77]

While the overall increase in violence was discouraging, one bright spot emerged: the distribution of attacks. Those parts of Iraq experiencing violence remained exceptionally violent with four governorates (Al Anbar, Baghdad, Nineveh, and Salah al-Din) accounting for over 80 percent of all attacks. By comparison, the rest of the country was almost placid. By year's end, the Arbil, Dahuk, and As Sulaymaniah governorates had gone months without a single reported attack while several others experienced only a few.[78]

Through 2005, another pattern emerged—the unwillingness of insurgents to fight the sorts of pitched battles that characterized 2004. Learning that the cost was high (the major 2004 battles resulted in over 2,200 insurgents killed and another 1,300 captured) and the return small, insurgents largely moved to stand-off attacks involving less risk. This shift capitalized on insurgent strengths while complicating both intelligence collection and operational planning. Such challenges, however, did not go unmet. A focus on identifying Al Qaida leadership led to Army or SOF forces killing or capturing nine senior Al Qaida in Iraq leaders. As 2005 ended, three things were clear. First, the insurgency was not ending any time soon as foreign fighters continued to arrive in Iraq. Second, the ISF was proving itself increasingly capable—the partnership was working. It might have been difficult at times, particularly when issues of trust developed, but it was working. Finally, enemy successes—no matter how small—overshadowed American successes, no matter how important. At day's end, the insurgency's asymmetric approach continued exacting a cost in American lives and equipment.[79]

Tactical Problems: Strategic Implications—Improvised Explosive Devices and Information Operations

Insurgents employed a variety of weapons and methods including ambushes, snipers, and mortar or rocket fire. The two most serious threats, however, soon became apparent: IEDs and Information Operations (IO). In Afghanistan, the number of casualty-producing IED incidents was relatively small. Out of 94 battlefield deaths of American soldiers between 2002 and 2005, 27 resulted from IEDs. In Iraq, IEDs accounted for 546 of the 1,141 US. Army fatalities caused by hostile action. In 2005, the IED-related fatality rate climbed to almost 66 percent of all those either killed in action or died of wounds. Army fatality rates, however, do not tell the full story. IEDs also claimed the lives of other American and coalition service members as well as local civilians. Amputations and traumatic brain injury (TBI) amplified the human costs exacted by IEDs. In March 2006, the Army's Surgeon General released statistics on both injury types. IED incidents resulted in 282 amputations. Soldiers sometimes suffered the loss of two or more limbs due to blast effects, penetrating wounds, or severe burns. More than 1,560 soldiers suffered traumatic brain injuries with 17 percent of these assessed as "serious" or "penetrating." Most such injuries resulted from IED detonations.[80]

The seriousness of an IED detonation depended on several factors. These included the amount of explosives used, a soldier or vehicle's distance from the IED at the time of detonation, and the level of protection provided by the soldier's vehicle or personal equipment. A major problem for the Army was the mismatch between its vehicle fleet and operational requirements. While armored combat vehicles offered protection, the Army's vast fleet of wheeled transport lacked any real protection. The Army designed its wheeled vehicle fleet to meet the needs of a traditional battlefield with clearly defined "front" and "rear" areas. In Afghanistan and Iraq, those vehicles (and their crews) operated on a battlefield lacking the theoretical protection of the "rear" area devoid of any enemy. Successful IED strikes thus frequently proved catastrophic to vehicles, cargo, and most significantly, soldiers. In Iraq alone, over 30,000 vehicles required hardening. Some units took to fabricating armor with locally available materials. Dubbed "hillbilly armor," this improvisation offered psychological comfort more than actual protection. Starting in 2003, the Army began a large-scale program to provide vehicles better suited to conditions in Afghanistan and Iraq. This included providing armored protection to its cargo vehicles, replacing its HMMWV fleet with protected passenger compartments, and introducing mine-resistant, ambush-protected (MRAP) vehicles.[81]

Improvised explosive devices became an increasingly difficult challenge as operations continued in both Iraq and Afghanistan. The proliferation of explosive material in both countries coupled with the widespread transmission of information about constructing IEDs meant insurgents from every group possessed the ability to build them. More importantly, they developed the ability to continuously improve such devices in response to American efforts to counter them. The combination of readily available materials and technically skilled individuals meant the cycle of action-reaction-counteraction continued with IED incidents increasing in frequency. In Afghanistan, the number of fatality-producing explosions went from two in 2003 to 20 in 2005. Iraqi insurgents executed seventy such attacks in 2003, more than doubled that number to reach 163 in 2004, and then almost doubled the number again in 2005 with 313 such fatal attacks against American soldiers. This roughly correlates with the total number of reported IED-related incidents in Iraq in 2004 and 2005. The former year ended with 9,372 such events, while 2005 recorded 17,485. In contrast, the total number of non-IED attacks in 2004 was 15,387 and 15,569 in 2005. This meant IEDs represented almost 38 percent of all attacks in 2004 and over 52 percent of 2005's total reported attacks. Successfully employed or not—and used by all insurgent groups—IEDs were clearly the most significant tactical problem facing the Army.[82]

Improvised explosive device technology continuously evolved with most changes found in the means insurgents employed to detonate their devices. Triggering mechanisms included cell phones, garage door openers, radio-controlled toys, infrared sensors, and a variety of pressure plates. Such devices also included the most dangerous detonator of all—the suicide bomber. The Army and its partners invested heavily in creating measures to counter command and pressure-plate detonated IEDs, but the insurgent willing to end his (and increasingly her) own life to ensure a successful detonation remained a significant problem. The suicide bomber also presented another problem—the ability to inflict massive civilian casualties. While a command-detonated IED astride a road used by American units could reduce the risk of unwanted civilian casualties, the suicide vest killed or maimed great numbers depending on when and where detonation occurred. Such IEDs created a tremendous concern for Army forces as their appearance demonstrated an inability to protect the population, an essential task in extinguishing an insurgency.[83]

The specifics of the IED threat in both Afghanistan and Iraq varied based on what groups operated in which country, and who group members could include as benefactors. One of the most dangerous IEDs was the explosively formed penetrator (EFP). These entered Iraq through Iranian Quds Force operatives aiding Shi'a groups. An EFP consisted of a machined concave copper disc backed by a high-explosive charge. When the device detonated, the copper's shape changed from a disk to a molten metal slug that essentially burned through a vehicle's armor. This behaved in the same way as the conventional high explosive anti-tank (HEAT) round. First introduced in 2004, the EFP relied on penetration rather than blast effects. As a result, such devices proved devastatingly effective against both unarmored and protected vehicles. The wide variety of triggering devices coupled with the EFP's ability to defeat armored protection and concurrently inflict traumatic amputation greatly complicated the Army's IED problem by making armored protection less effective.[84]

The consequences of fighting in Afghanistan and Iraq extended across and well beyond the boundaries of these countries. Insurgent groups commonly video-recorded their activities, particularly those involving IEDs, and then posted them to the world wide web. In the Defense Department's language, IEDs became "weapons of strategic influence" when exploited in this manner. Such videos broadcast insurgent capabilities for several purposes. These included demonstrating to the local population insurgent abilities to inflict casualties on American forces and counter coalition displays of strength. The latter was particularly important in areas under insurgent control by discouraging individuals from supporting coalition or national government efforts. This complicated efforts to re-establish local governance by exposing American vulnerabilities and undercutting efforts to establish effective relationships with the local population. Striking symbolic targets, such as Abu Ghraib prison and the Al Rashid Hotel, used by the foreign media, both communicated insurgent strength and reinforced claims of power.[85]

Insurgents also deliberately targeted civilians with vehicle-borne IEDs or suicide vests to either intimidate a local population or respond to attacks by a rival religious or political faction. Such violence reduced the victims to mere props in an information campaign. Video footage of insurgents successfully fighting coalition forces also aided in recruiting foreign fighters. Combining Islamic themes with images of blows against infidel invaders helped draw insurgents from across the Muslim world. Insurgents also took advantage of such American failings as Abu Ghraib or targeting mistakes resulting in civilian casualties to aid recruiting efforts.[86]

Finally, such imagery sought to target a western—and especially American—audience and demonstrate the futility of continuing operations. This audience included the families of American soldiers killed or wounded in insurgent action who received e-mails with attached video footage of the event. Propaganda as a weapon is as old as war itself and the conflicts in Afghanistan and Iraq found insurgent groups undertaking sophisticated efforts on a scale not seen previously. Leveraging technology, Al Qaida's information operations literally reached around the world and added another layer of complexity to the asymmetric battlefield. The coalition required the support of each participating nation's populace. Such support was difficult to sustain when official statements from Kabul or Baghdad had to compete with well-produced messages from insurgent groups presenting a different version of events and circumstances.[87]

Ultimately, information operations complicated Army operations by creating a military problem that spanned the tactical, operational, and strategic levels of war. In weaponizing information, insurgents sought to seize the strategic initiative, and in so doing compel a change in Army operations. Information operations might not force Army forces from Afghanistan or Iraq, but they could—and did—make the political and diplomatic costs of staying higher than would otherwise have been.

Conclusions

A popular saying within the Army in the early 2000s was, "Sometimes you fight the Jetsons, sometimes you fight the Flintstones." "You fight the Jetsons (…) and the Flintstones" was more accurate. The Army deployed to both Afghanistan and Iraq was the correct instrument to perform the tasks initially assigned. Special Operations Forces precipitated the Taliban's collapse by facilitating the marriage of local troops and American precision weapons. In Iraq, V Corps' performance again validated the utility of the "Big 5" and concurrently demonstrated the force enhancements placed following the Persian Gulf War. The Revolution in Military Affairs did not fully live up to its expectations, but it gave Wallace and his soldiers capabilities that helped compress an envisioned 120-day-long campaign into one lasting only a few weeks. Investments made based on the Army's self-conceptualization as a mechanized force staffed by skilled soldiers dominating the battlefield proved sound.[88]

Counterinsurgency operations represented the other side of the effectiveness coin. The Army's armored and mechanized units were initially neither trained nor equipped for that form of warfare. The publication dates of the Army's pre-war manuals dealing with counterinsurgency demonstrate the Army's dwindling interest in that form of warfare. *Field Manual 100-20, Military Operations in Low Intensity Conflict*, released in 1990, reflected the Army's experiences in Vietnam and Latin America in the 1980s. *Field Manual 71-100, Division Operations*, which also appeared in 1990, did not describe counterinsurgencies. The 2001 edition of *Field Manual 3-90, Tactics,* also made no reference to counterinsurgency. Similarly, the 2001 edition of *Field Manual 3-0, Operations,* contained only four paragraphs describing counterinsurgency operations with those focusing on foreign internal defense missions. What the Army had encountered in Somalia now seemed unimportant, as did the increasing rumblings of terrorist action in the Middle East. The Army also took great comfort in the so-called Weinberger doctrine. This doctrine precluded committing forces except in those circumstances where it could employ fully its technological and firepower advantages. While light infantry units had great utility in counterinsurgency, the ugly reality is that 9/11 caught the Army off guard and largely unprepared for the fights it did not want. [89]

While the Army did not draw extensively from its historic experiences with asymmetric warfare, it did capitalize on one lesson learned from the Vietnam War: the need to cycle units through a combat theater. Sustaining the rotation schedule was an ongoing challenge, but the effort paid dividends. Units executed a comprehensive pre-deployment training program, deployed to the fight with stable leadership, and then returned to home stations for rest, refitting, and retraining. As a result, organizations enjoyed better leadership, cohesion, and training than their Vietnam War counterparts. Sustaining this approach imposed a colossal burden on the Regular Army and Army National Guard, but the Army remained committed to this model through both wars.

The "coalition of the willing" stood in marked contrast to the largely unsuccessful "Many Flags" effort of the Vietnam War. Employing coalition forces, both Special Operations and conventional, created opportunities for employing Army forces. Even the smallest deployments tying small, allied units to fixed positions benefited tactical operations by eliminating the need for Army units to perform such duties. The size, capabilities, and restrictions cooperating governments imposed on coalition forces varied, but the presence of over 23,000 additional servicemembers from more than 58 countries meant American soldiers could address high-priority areas and problems. This proved critical as the tempo and size of the insurgencies increased and the Army scrambled to field the units required to effectively respond. The quality of foreign units deploying to both Afghanistan and Iraq was particularly noteworthy as countries frequently deployed their premier formations. Equally important was the growth of Afghan and Iraqi forces willing and able to take on insurgents. [90]

By conducting effective information operations, the Taliban, Al Qaida, FRE, and Shi'a extremists created additional challenges for the Army by expanding the battlefield. Such operations fueled both insurgent numbers and questions about the Army's ability to defeat it. Already burdened with the requirement to restore both effective governance and public services, insurgent information operations created additional distractions and requirements for an Army increasingly stretched thin. Defeating propaganda efforts became essential not only to win in Afghanistan and Iraq but also in those countries that had joined the coalition. Success in this battle would take more than psychological operations against insurgent groups, it would require major information efforts both in theater and in the United States. The propaganda successes insurgents enjoyed made American counter-efforts significantly more difficult as the campaigns dragged on with no clear end in sight and magnified the effectiveness of the insurgents' best weapon: the IED.

Finally, IED attacks—increasing in both numbers and sophistication—meant the Army needed to change its approach to operations if it was to stem rising casualty rates. These rates imperiled not only soldier morale but the public's willingness to support ongoing operations in both Afghanistan and Iraq, a problem magnified by insurgent propaganda. Newly assigned Army Chief of Staff Gen. Peter Schoomaker, known for colorful phrases, summed up his goal in this area quite simply: "Dorothy and Toto can tip-toe down that road [Route Tampa, the Army's main supply line in Iraq] naked and not be touched."[91] Schoomaker and a small number of officers and Department of the Army civilians would both manipulate long-standing processes and recast Army priorities to create organizations able to meet his objective.

Chapter 4

Creating the Asymmetric Warfare Group

...getting to the left of the boom.[1]

Colonel Bruce Jette describing an approach to the IED problem in 2003.

Setting the Stage

Whatever questions existed over the decision to topple first the Taliban regime in Afghanistan and then Saddam Hussein in Iraq, one fact is clear: initial operations succeeded dramatically. The Taliban lasted only 60 days. Saddam Hussein's regime fell in only 21 days. The former operation represented a combination of American and allied skills in unconventional warfare, integrating the elements of combat power, and diplomacy. The latter demonstrated a mastery of joint and combined operations. Taken together, optimistic projections regarding how the American effort would end seemed entirely justified. [2]

Things began going wrong for the Army once the Afghan and Iraqi regimes fell, and occupation and rehabilitation became the Army's larger tasks. It seems absurd in retrospect, but no military planner or civilian policymaker seriously considered a post-conflict environment that significantly challenged American military power, particularly in Iraq. In Afghanistan, a combined American-Afghan force removed a repressive regime from power in a matter of weeks. In Iraq, the population recognized overwhelming American ground and air power had twice shattered their army. U.S. and allied forces ranged the entire country inflicting crippling damage irrespective of time, terrain, or weather conditions. An actor viewing things from a western perspective would realize the utter futility of attempting to contest American power and accept the new political order, no matter how alien it might be. In fact, such acceptance was essential if the United States was to realize its objective of creating states able to serve as regional models of democratic government.[3]

Western perspectives, however, were the exception in Afghanistan and Iraq. Both countries had long-standing religious and cultural traditions and structures that shaped popular attitudes. These countries also possessed experienced soldiers as well as disaffected populations lacking a robust attachment to any central authority. They also tended to be mistrustful of foreigners—especially when they perceived those foreigners to be Christian, imperialist, or both. These two populations also absorbed an influx of dedicated foreigners imbued with a religiously inspired belief that their duty lay in contesting American power anywhere it presented itself. Finally, years of conflict meant both nations contained all the material resources needed to sustain a bloody fight. Under "normal" conditions, the defeated populace would remain quiescent: this was the American experience following both world wars, Korea, Panama, and the small portion of Iraq occupied following the First Gulf War. Normal, however, is a cycle in a washing machine. [4]

First Efforts

The realization that operations would continue in Afghanistan for the foreseeable future, coupled with the discovery of various equipment shortfalls, led the Army to a rather obvious realization: while combat experience validated many of the Army's approaches to training, organizing, and equipping the force, "many" was not "all." The Army would need to make some rapid changes if it was to provide units with the institutional support needed for battlefield success. This realization provided the impetus for creating the first organization that would play an increasingly important role as the American presence both grew in Afghanistan and expanded into Iraq—the Rapid Equipping Force (REF). Established in 2002 under Gen. Eric Shinseki's tenure as Army Chief of Staff, the REF's mission was a straightforward one: quickly procure and issue needed equipment not currently in the Army supply system. Initially, the REF oriented on robotic technologies offering the potential to reduce casualties by extending the physical reach of units while concurrently improving their protection. By focusing on squads and companies—the small units engaging in combat—the REF sought to provide the best equipment possible through the "buy, try, decide" model. The REF would purchase equipment, give it to the operational force to put it through its paces, and then, based on feedback, make decisions regarding additional acquisitions. It would work with the Army's Test and Evaluation Command and its research and development center at Aberdeen Proving Ground. The focus, however, would lay with the deployable soldier and his or her feedback. The REF was one of three new, small, and focused organizations that would play a disproportionately important role in counterinsurgency operations in the coming years, but its focus was on material procurement. [5]

The REF represented a first step in adapting the institutional Army to a changing environment. The second laid in finding solutions to the expanding problem of improvised explosive devices. Before 9/11, the Department of Defense established the Combatting Terrorism Technical Support Office (CTTSO) and the Joint Staff Anti-Terrorism Directorate (J34). The Assistant Secretary of Defense for Special Operations/Low-Intensity Conflict established the former in 1999 to consolidate research and development programs previously administered by the Office of the Assistant Secretary of Defense, Command, Control, Communications, and Intelligence. The latter produced information for the services with the publication of a periodical titled *The Guardian.* It also provided training for individual service members, unit commanders, and installation commanders and served as the proponent for several Department of Defense directives and instructions. Equally important, it supported targeting efforts against known terrorists.[6]

The J34, however, suffered from two significant weaknesses. First, most in the military in general—and the Army staff in particular—were not directly concerned about the terrorist threat before September 11, 2001. While President William J. Clinton made countering terrorism a national priority in 1995 with his signing Presidential Decision Directive 39 (PDD-39), primary responsibility fell to the Departments of Justice and State. Second, the events of 9/11 largely gutted the J34 and moved many of its components to the newly established Department of Homeland Security. What little existed was inadequate for the Army's needs. In October 2003, the Army established the IED Defeat Task Force. Under the direction of Brig. Gen. Joseph Votel, and positioned under the Army's Deputy Chief of Staff for Operations, Plans, and Training (G-3/5/7), the task force served as a coordination and synchronization agency. In centralizing its diverse counter-IED efforts, the Army established priorities, generated cost efficiencies in training and procurement, and, most importantly, provided accurate information. As a purely Army activity, the IED Defeat Task Force was short-lived. Realizing the benefits provided, then-Deputy Secretary of Defense Paul Wolfowitz directed the Defense Department to create a similar team under Army leadership. This action effectively transferred Votel and his efforts from the Army G-3/5/7 to the Joint Staff, ultimately returning lost capabilities to that higher jurisdiction. Votel continued his close coordination with senior Army leadership, even as the scope of efforts expanded. The IED Defeat Task Force proved an effective actor, but, as was the case with the REF, it operated with a comparatively narrow focus.[7]

In concert, the REF and IED Defeat Task Force enjoyed successes in both better equipping the soldier and reducing the risks posed by IEDs. Despite the Army fielding a variety of countermeasures and improving training, however, casualties continued to mount. While the REF and IED Defeat Task Force represented steps towards a solution to the problems posed by asymmetric threats, they were insufficient. The Army required more, but what "more" looked like and the form it should take were undefined and facilitated a great deal of discussion at the Army's most senior levels. Army Chief of Staff General Peter Schoomaker's earthy statement about Dorothy, Toto, and the yellow brick road lay at the heart of those deliberations. As discussions and debates continued, ideas began to coalesce and point toward a specific course of action.

The only clearly defined elements were the need to consider a broader approach than the REF or the IED Defeat Task Force and the recognition that any effort would benefit from the ability to operate outside the Army's standard institutional and organizational structures. The Special Operations Forces experience of some involved provided insights, particularly concerning materiel and structure.[8]

The Initial Concept

Identifying the need for a new type of organization represented one challenge. Determining its form and functions constituted two more. The path to answering those questions lay in the operational and organizational concept, often referred to as the "O&O." Creating an O&O provided answers to the many DOTMLPF-related questions inherent in establishing a new type of unit. Getting the O&O "right" meant the organization would have the proper structure and capabilities for performing its mission. Getting it "wrong" meant the group would fail to meet expectations, needlessly squander time and money, and undermine the credibility of the very idea of a new unit. Crafting the O&O required both an understanding of the Army's mainstream organizational and business practices and the unique practices of its Special Operations community. Their specialized roles and requirements meant Special Operations forces had developed different procurement and acquisition processes from the larger Army. The Defense Department's 5000-series regulations governed both conventional and Special Operations Forces procurement. Special Operations procurement differed from conventional acquisition in three ways. It delegated decision authority to the appropriate level, tended towards simultaneous vice sequential development, and, partnered the different individuals within the process to accelerate programs. The Army did not need another Special Operations unit, but the discussions established the need for a unit capable of operating in special ways with the general-purpose forces operating in Afghanistan, Iraq, and elsewhere. Perhaps the best metaphor for describing the process of creating the O&O is that of making sausage: it would be neither clean nor pretty, but it would produce the desired product.[9]

Fortuitously for the Army, both Chief of Staff Schoomaker, and the Vice Chief of Staff, Gen. Richard Cody, possessed a wealth of experience in both Special Operations and general-purpose forces. These materially aided efforts to build the capabilities required. Recalled to active duty as General Eric Shinseki's replacement in 2003, Schoomaker started his Army career as an Armor officer. While he occasionally returned to the general-purpose force after qualifying as a Special Forces (SF) officer in the 1970s, he spent much of his career in Special Operations assignments. In the words of Col. Robert Shaw, who served as the

AWG's first commanding officer, Schoomaker considered it critical to change the Army from "the inside out by providing solutions to the latest threats."[10] Failing to do this would lead to others imposing answers, an approach capable of providing the Army with multiple complications for every "solution" provided. Schoomaker, however, could not focus solely on emerging threats. As Chief of Staff, his duties included serving as a member of the Joint Chiefs of Staff, formulating policy, presiding over the Army staff, and coordinating with the Secretary of the Army. While Schoomaker might have desired to focus on specific concerns, federal statute meant he spent much of his time resolving other issues.[11]

Recognizing the need for a Vice Chief of Staff capable of both understanding his vision and having the temperament to drive the Army to realize it, Schoomaker selected Cody (then serving as the G-3/5/7) to replace outgoing Vice Chief of Staff, Gen. George W. Casey, Jr. Filling the role described by Stephen Rosen, Cody embodied the quintessential activist leader positioned to drive a host of necessary changes within the Army. While the statute governing the Vice Chief of Staff said very little about his specific duties, actual Army practice meant the Vice Chief of Staff focused on the day-to-day business of running the Army. With only two erstwhile brakes on his freedom of action—the Secretary of the Army and the Chief of Staff who selected him for the position—the Vice Chief maintained an unusual ability to operate across the Army. Cody spent much of his career in Special Operations aviation. He also commanded multiple conventional aviation units, the 160th Special Operations Aviation Regiment, and was the first Aviation branch officer to command the 101st Airborne Division (Air Assault). His background gave him an understanding of both the conventional Army and Special Operations community's bureaucratic processes and how to work both through and around them to achieve objectives. Among the processes Cody had mastered was Special Operations materiel acquisition, especially the acquisition of aviation-related items. Cody's experience as the 101st Airborne Division's commanding general expanded his knowledge of asymmetric threats and the assignment of then-Lieut. Col. Christopher Hughes as one of his battalion commanders proved fortunate. Having served with J34, Hughes possessed intimate familiarity with the Department of Defense's knowledge of the terrorist threat and the efforts underway to mitigate it. Equally important, Cody's existing relationship with Schoomaker meant the two effectively communicated with each other. Finally, his background in two different communities allowed him to create and maintain a list of officers capable of solving the most vexing problems.[12]

Schoomaker was not particularly interested in the details involved in creating the O&O. His technique was straightforward: provide the initial concept and then "hand it off to the real hands-on folks."[13] The "hands-on folks" included Cody and several officers whom Cody believed possessed the ability to both think creatively and work collaboratively with each other and with the units and agencies their efforts would impact. Cody issued these officers "mission-type" orders–orders that described the fundamental problems—and then trusted they would not lock themselves into existing methods or solutions to resolve them. Cody gave them the first 50 percent and then empowered and resourced them to solve the still very amorphous problem. Reflecting on the AWG's creation, Cody observed it was the majors, lieutenant colonels, sergeants major, and staff sergeants "who were the brains and gave the organization life."[14]

The first task to accomplish was defining an actual objective. Translating General Schoomaker's quote into a practical and achievable mission represented the initial step in creating the O&O. In a "spit-balling session" with Cody in 2003, Rapid Equipping Force commander, Col. Bruce Jette, suggested a better concept lay in getting "to the left of the boom."[15] Moving to the left proposed a fundamentally different approach to the problem by expanding the conceptual universe beyond narrowly focused material solutions to specific threats. Material would remain an essential part of the solution effort, but changes in how units would act and react would join them. Recognizing the need to expand beyond IED detonations planted the seed for the creation of the AWG.[16]

Efforts to develop an organization that realized Schoomaker's intent meant Cody and the others involved would continue spending long hours considering concepts. That placed them in extremely close contact for several months. James J. Lovelace, then a major general serving as the Army's Deputy Chief of Staff for operations, recalled that many of those involved were Roman Catholics who would meet on Sunday, discuss the problem, attend Mass, and then go back to their discussions. Occasionally, ongoing discussions would interrupt Mass itself. For Cody, an essential element of the solution included creating a unit able to perform three crucial tasks simultaneously: inform the current fight, identify the emerging one, and identify DOTMLPF-based solutions to both. This capability would provide a unit able to function in the training base and on the battlefield with the capacity to identify capability gaps and develop doctrinal and organizational solutions. Ideally, the unit would serve as "scouts." These scouts would allow the Army to become more agile by decreasing the time between identifying emerging threats and deploying appropriate responses. Concurrently, the unit would also function as a sort of "directed telescope" able to provide Cody and the rest of the Army's senior leadership information beyond that typically provided through existing channels. Officers who worked for Cody were not surprised by this as he routinely employed this method to attain situational awareness rapidly. Importantly, the unit would *not* report on how other individual outfits performed. Instead, it would provide information on the environment that would encourage both greater and more rapid understanding. While the issues were complicated, the objective was ultimately simple: move from mitigating problems to providing longer-term solutions. Yet as Carl von Clausewitz famously said, "Everything in war is very simple, but the simplest thing is difficult." Those working on the task would spend a great deal of time defining and refining concepts as the AWG's form began taking shape between the initial 2003 discussions and its January 2006 activation.[17]

By April 2004, the long hours began paying off. Both the Vice Chief of Staff and the Deputy Chief of Staff G-3/5/7 published memoranda defining some guiding principles. After providing direction to the planning team on February 18, 2004, and receiving the latter's feedback, Cody directed the U.S. Army Training and Doctrine Command (TRADOC) Force Design Directorate to design an "Asymmetric Warfare Regiment" on April 12. Cody's initial guidance placed the unit under the U.S. Army Forces Command Chemical, Biological, Radiological, Nuclear, Explosives Command. Later designated the 20th CBRNE Command, this organization was the Army's largest dealing with unconventional munitions and their effects. Cody's guidance included establishing the unit as a table of organization and equipment (TO&E) capable of augmentation. Directing establishment as a TO&E unit was significant as the Army maintained different policies for TO&E units, table of distribution and allowances (TDA) units, and special mission units (SMU). He also stated that the unit should achieve initial operating capability by January 2005 and that it should achieve full operating capability by January 2007. Finally, Cody provided the first, albeit tentative, approved mission statement.[18]

On April 28, the Army G-3 published a memorandum in response to questions from TRADOC's Force Design Directorate following meetings at Fort Leavenworth, Kansas, to work on specific force design issues. As a result of this session, the Army G-3 further clarified the AWR's mission as encompassing the basic design of the AWR headquarters, the support required for a Combined Joint Task Force, liaison teams, and advisory teams. This memo also confirmed the AWR would remain subordinate to the FORSCOM.[19]

While Cody, the G-3/5/7 staff, and TRADOC tackled issues associated with mission statements and force design, other elements of the Department of the Army addressed issues associated with creating the AWG. Foremost among these were obtaining the required legal authority and money. Funding a unit that neither previously existed nor appeared in any Program Objective Memorandum (POM) submissions meant the Army needed to shape the legislative process. While the Army receives billions of dollars annually, it has little discretion in how it uses those funds: Congress allocates funds for specific purposes and it is illegal for the Army to use them for other purposes. This constraint means ideas, no matter how good, are irrelevant if the funding required to turn them into action is not available. Army efforts ultimately bore fruit with Congress passing *Public Law 109-148: Department of Defense,*

Emergency Supplemental Appropriations to Address Hurricanes in the Gulf of Mexico, and Pandemic Influenza Act, 2006. Finally approved on January 4, 2005, Section 8403 (c) (2) provided the Secretary of the Army the authority to form and fund the AWG.[20]

While some staff officers worked on the legislative issue, others undertook the detailed coordination required to move the AWG concept forward with the expectation that congressional authority would soon arrive. The means for moving forward lay in a process created by Cody early in his term as the G-3/5/7: the Army Strategic Planning Board (ASPB). The ASPB provided a weekly forum for adjusting priorities at the Department of the Army level. This board reassessed the Army Campaign Plan (ACP) in light of emerging requirements, particularly those flowing from the combat theaters. The 2004 version of the ACP consisted of 49 pages of text. Embedded in this were eight ACP objectives with 69 supporting objectives at the Department of the Army level arrayed on 22 different lines of operations (LOO). Each LOO represented "functions that enable supported MACOM (major command) commanders or HQDA staff principals to link multiple tasks with the logic of purpose (cause and effect) to achieve ACP objectives."[21] Simply put, the LOO grouped the activities of both the Army staff and the commands to realize a specific objective.[22]

Every Friday saw a working level by the colonels representing various commands and staff sections. These officers came to the ASPB thinking for their bosses—but not necessarily thinking like their bosses. Since a simple priority change in the ASPB could affect months (or years) of work, not to mention millions (or billions) of dollars, the sessions tended toward acrimony and acquired the nickname "Friday Fights." The Friday sessions, however, became the precursor to the actual ASPB conducted each Saturday with general officers (including three- and four-star generals) participating. The Saturday sessions included both an open and a closed session. Co-chaired by the G-3/5/7 and the Army Deputy Chief of Staff for Resource Management, these sessions changed priorities—and the associated flow of money within the Army to the extent permitted by the National Defense Authorization Act.[23]

Predictably, ASPB decisions generated hostility among some participants. While the objective was making smart decisions to best support deployed forces, the process created "winners" and "losers" with the losers having to sometimes make significant changes to plans and programs—a requirement rarely welcomed. The ASPB demonstrated Cody's ability to bend the bureaucracy, but it concurrently embittered some who made the bureaucracy work.[24]

As the original ideas and TRADOC's work matured into something more defined the maintenance of momentum fell to three individuals and a team of retired generals working for Cody. Then-Col. Joseph Votel, Lieut. Col. Chris Hughes and a retired Field Artillery lieutenant colonel named Michael Bowman comprised the original team. Votel concurrently led the IED Defeat Task Force, so much of the day-to-day work fell to Hughes and Bowman. Hughes had just returned from Iraq. He has also served on the Strategic Initiatives Group, a Department of the Army planning organization and supported the IED Defeat Task Force. Bowman, who was also a former Department of the Army G-3/5/7 staff officer, played a significant role both in the AWG's development and operations, moving from the IED Defeat Task Force to the AWG itself upon its activation. These individuals handled the staff coordination tasks associated with developing the AWG. This included identifying and then contracting with a civilian firm, the Burdeshaw Group, to translate proposed objectives into the O&O concept. Then headquartered in Bethesda, Maryland, the Burdeshaw Group was a firm possessing the ability to marshal some of the Army's retired talent and focus it on resolving pressing problems. Even with the demands of leading the IED Defeat Task Force, Votel made time to provide guidance and remain engaged with those working to create the O&O.[25]

Retired Gen. William Hartzog, who formerly commanded TRADOC, led the Burdeshaw Group, while retired Lieut. Gen. Harold "Tom" Fields served as the AWG team lead. Fields' selection represented something of an inspired choice. Commissioned an Infantry officer after graduating from the Citadel in 1960, Fields qualified for Special Forces in 1962 and spent two years as a Green Beret before an automobile accident hospitalized him with a broken femur. After convalescence at Walter Reed Army Medical Center, he reported to Fort Benning, Georgia, where he participated in the reorganization of the 11th Airborne Division (Air Assault). This effort culminated with the Army re-designating the unit the 1st Cavalry Division (Airmobile) and then deploying it to the Republic of Vietnam in 1966. Fields followed this tour with an assignment as a Training and Counseling (TAC) officer at the U.S. Military Academy. As a TAC, he helped shape then-Cadet Richard Cody. Subsequent assignments as an Infantry officer saw Fields return to Vietnam, complete a graduate degree, and then serve in a variety of assignments in the United States. After attaining star rank, he served as the assistant division commander and then division commander of the 6th Infantry Division (Light) in Alaska, as the Army G-3/5/7 during the Persian Gulf War, and as the deputy commander of the U.S. Indo-Pacific Command. Fields possessed a deep understanding of Special Forces, insurgency and counterinsurgency as well as the cultural norms of the conventional Army, the Army staff and one geographic combatant command.[26]

Under Fields' direction were retired Maj. Gen. John Greenway, Maj. Gen. James A. Guest, Lieut. Gen. Charles Otstott, Lieut. Gen. James C. Riley, and Maj. Gen. Steven Silvasy. In terms of a basic branch—the arm or service into which the Army commissions an officer—this group represented a mix of backgrounds with Fields and Guest providing SF perspectives, Greenway possessing Armor expertise, and Otstott and Riley possessing Infantry backgrounds. All had extensive combat experience. Perhaps more importantly, each also had experience in both senior Army and joint staff positions emphasizing training or doctrine development. Guest commanded the Army's John F. Kennedy Special Warfare Center and School at Fort Bragg while Greenway served in several doctrine or force design assignments and directed the Army's 1990 PROJECT VANGUARD study exploring post-Cold War changes to the institutional Army. Otstott commanded TRADOC's Combined Arms Combat Development Activity at Fort Leavenworth. Riley's assignment history included command of the Combined Arms Center and service as Commandant, the U.S. Army Command and General Staff College. Silvasy brought the experience of service as TRADOC's Deputy Chief of Staff for Doctrine and Development.[27]

As the O&O would drive both the AWG's employment and DOTMLPF elements, it was essential to state the AWG's enduring mission cogently. The enduring mission statement broadly described a unit's fundamental purpose and provided the foundation for resourcing it. For the AWG, converting the language from "getting to the left of the boom" to something less abstract constituted the first task. Doing this required defining what asymmetric warfare was—and was not. IEDs represented an asymmetric tactic, but asymmetric warfare required a broader definition if the AWG was to develop into the organization Schoomaker envisioned. The contemporary focus on terrorism and the ongoing IED problem shaped the working definition:

> Asymmetric Warfare: Operations conducted by terrorists, guerrillas, militias and paramilitaries to limit US military effectiveness while achieving their political objectives. The operations are conducted at the strategic, operational, and tactical levels to attack military and civilian targets. Objectives of these operations are to undermine US/Allied political, military, economic, and psychological strengths and the will to win.[28]

Aligned with this definition was the latest revision of the AWG's tentative mission statement, which underwent several modifications in a few weeks before its release on April 3:

> JTF Spartan (the proposed AWG) integrates, coordinates, deploys and provides trained and ready forces, and is prepared to exercise command and control of full spectrum asymmetric warfare operations to Joint and Army Force Commanders. Maintains technical links with appropriate Joint, Federal, and State counter-emerging threats assets, as well as research, development and technical communities to assure Army counter-emerging threat response readiness. Provides or assists in the

> training and readiness oversight of counter-emerging threat assets (Active, Guard and Reserve).[29]

This mission statement emphasized command and control and liaison with other organizations and highlighted its proposed role both overseas and in the United States. By April 12, action officers had crafted a revised mission statement recognizing two different missions based on both the unit's TO&E foundation and reinforcement with a TDA:

> Asymmetric Warfare Regiment (AWR) Mission (TOE): AWR integrates, coordinates, deploys and provides trained and ready forces, and exercises command and control of assigned forces in support of Joint and Army Force Commanders full spectrum asymmetric warfare operations. Observes, collects, develops, validates, and disseminates emerging Tactics, Techniques, and Procedures (TTPs) to TRADOC and the Joint Warfare Fighting Center, USJFCOM. Provides Advisory Training Teams to conduct pre-deployment Train-the-Trainer training on counter asymmetric threats. Conducts and provides prediction modeling, forensics, and trend analysis on asymmetrical threats to the supported commander.
>
> Classified TDA Augmentation Mission: Recruits, assesses, and selects unit members. Conducts liaison, on combating terrorism, with Joint Services, multi-national, and interagency work groups. Identifies and fields Rapid Prototyping and/or commercial off-the-shelf (COTS) solutions to counter asymmetrical threats. Conducts military utility assessments and initial integration of prototypes and/or COTS equipment to the force. Deconflicts, synchronizes, and integrates materiel solutions for countering asymmetric threats with HQDA G-3, PEOs (Program Executive Officers/Program Managers) and AMC (Army Materiel Command).[30]

This mission statement provided greater detail on both the tasks performed to support the operational Army and those Army elements involved in equipment design and procurement. Clearly absent from both mission statements was any direct reference to IEDs. While the IED issue provided the initial impetus, thinking had moved beyond a single type of asymmetric threat.

General Cody's April 12 memorandum reflected yet another change in the tentative mission statement by inserting the phrase "and exercise command and control of assigned forces" into the TO&E mission statement, thereby emphasizing the fact that the unit would control its elements even after deployment.[31] Cody's April 12 change, however, saw extensive modification on May 26. This revision defined the mission as:

> The AWG conducts operations in support of Joint and Army force commanders to mitigate and defeat specified asymmetric threats.
>
> Key tasks:
>
> - Provides, deploys, integrates, coordinates and executes C2 of trained and ready forces.
> - Observes, collects, develops, validates and disseminates emerging tactics, techniques and procedures.
> - Assists in the exploitation and analysis of asymmetric threats.
> - Provides advisory training for in-theater or pre-deployment forces.
> - Provides identification, development and integration of counter-measure technologies[32]

Far more descriptive than the previous versions, this edition essentially "binned" tasks the unit would perform and concurrently provided a great deal of conceptual space in determining how to perform those tasks by not setting specific limits on how to approach them. The key tasks identified in the mission statement coupled with the intellectual freedom of action provided made clear the AWG required individuals with unique skills.

The Mission Statement

With the Department of the Army making the IED Defeat Task Force a standing capability in February 2004, the proposed AWG enjoyed a largely unconstrained conceptual mandate: a dedicated counter-IED organization meant the AWG could—and should—focus elsewhere. The absence of an existing conventional Department of Defense organization with responsibility for asymmetric warfare also meant that those crafting the O&O could be as creative as they desired. It possessed the Army Chief of Staff's intent, the Vice Chief's support, TRADOC's design ideas, a descriptive mission statement, and few bureaucratic restrictions. As a result, General Fields' team possessed a great deal of intellectual freedom in considering a final restated mission.[33]

The O&O's restated mission, approved on March 5, 2005, demonstrated an understanding of General Schoomaker's intent and General Cody's guidance. In its final draft form, the O&O stated the AWG's mission as:

> The Asymmetric Warfare Group conducts operations in support of Joint and Army force commanders to mitigate and defeat specified asymmetric threats. Serves as the global U.S. Army expert in irregular warfare and conducts predictive modeling and trend analyses concerning global asymmetric threats. Maintains liaison with Regional Combatant Command HQ, JFCOM (Joint Forces Command), and USSOCOM (U.S. Special Operations Command). Establishes linkages with all internal, Combatant Command and National intelligence agencies. Deploys, integrates, coordinates and executes battle command of assigned and attached forces. Conducts or assists in advisory training for in-theater or pre-deployment forces. Assists in identification, development and accelerated integration of lessons-learned, TTPs (tactics, techniques and procedures), and countermeasure technologies.[34]

The proposed mission gave the AWG five basic tasks. It would provide the Army subject matter experts on asymmetric threats, coordinate activities with geographic and functional combatant commanders, coordinate activities with both the Army's and other intelligence agencies, deploy and control forces forward, and train the rest of the Army. This mission statement presented a massive statement of work for an organization with no precedent that would compete with the larger Army for increasingly scarce personnel and resources. Implicit in "internal agencies" was continuous coordination with both the IED Defeat Task Force and the REF. These three organizations—IED Defeat Task Force, REF, and the AWG—would work hand-in-glove to ensure the Chief of Staff's intent was satisfied. If the AWG could successfully execute its mission, it would represent a "one-stop source of expertise" for understanding and countering emerging asymmetric threats. Countering these threats would profoundly change the tactical and strategic landscapes by degrading or eliminating the enemy's most effective tactics and returning the Army to the dominant position on the battlefield.[35]

While a mission statement represented a significant milestone, that by itself could not adequately drive operations or organization. Translating the mission statement into reality required linking that document with two critical concepts. Gaps—gaps in understanding, gaps in capabilities—were the first. The counter-asymmetric warfare cycle was second. The counter-asymmetric warfare cycle consisted of five elements: identify and predict, prevent, react, eliminate, and, assess. Framing activities against the counter-asymmetric warfare cycle's five elements both identified and positioned gaps for resolution. Resolving the gaps, in turn, meant an adversary's ability to fight asymmetrically would suffer. Resolving the gaps included an essential task: ensuring units received and understood *effective* solutions to the challenges they faced. It might seem superfluous to qualify the word "solutions" with "effective," but the Army's long history included multiple examples of solutions being anything but practical, particularly when it sought to resolve asymmetric requirements. If deployed units could not receive the AWG's theoretical benefits, then the effort would become another one of the proverbial "self-licking ice cream cones" the Army often created when problems became bureaucratized. To truly meet the Army's needs, the operational concept had to ensure that the larger Army benefitted from what the AWG learned in the combat zone. Implicit in this was ensuring the AWG possessed a high degree of operational and organizational agility. If the AWG could not rapidly produce results it would have no real value.[36]

Each of the original mission statement's five basic tasks carried with it an extensive set of operational requirements possessing specific DOTMLPF requirements. Providing the Army subject matter experts on asymmetric threats required developing a cadre capable of operating anywhere in the world and under virtually any set of weather, illumination, or tactical conditions. Tactical conditions are the amalgam of terrain, weather, friendly, enemy, and battlefield effects. As an example, the use of industrial chemicals, such as chlorine, to create a weapon of mass effect would require changes in how explosive ordnance disposal (EOD) teams dealt with an IED containing such chemicals while concurrently changing tactical conditions. The enemy's ability to manipulate or "ramp up" the difficulty and risk of tactical conditions provided a source of leverage that the AWG would need to negate. The inherent uncertainty meant the AWG required individuals possessing a wide range of technical, tactical, and analytical skills. Complicating the issue was the simple fact that some required skills did not really exist in the contemporary Army. Developing these skills meant the AWG would look beyond the Army to acquire the baseline skills required to understand and effect emerging threats.

Coordinating activities with geographic and functional combatant commanders meant the AWG required effective communications paths with Central Command, European Command, Indo-Pacific Command, Northern Command, Southern Command, Joint Special Operations Command, and, Strategic Command. The first five commands are responsible for the vast majority of American military activities within their boundaries. With few exceptions, if there was activity taking place, the geographic combatant commander was responsible. Not surprisingly, each of these commands possessed a unique organizational culture reflecting the attitudes of the combatant commander, the dominant service within the command, and the cultures of the military services of other countries within its boundaries. The last two commands are functional combatant commands. Like the geographic combatant commands, they also have their own cultures. Their operations span geographic regions and are the source of most of the exceptions to a geographic combatant commander's responsibility. As the Global War on

Terrorism continued, it became common for USJSOC to establish multiple Joint Special Operations Areas (JSOAs) within a geographic combatant command.[37]

Also required was the ability to effectively communicate with both the various Army Service Component Command headquarters supporting these organizations and their Marine Corps counterparts to ensure all elements of a Land Component Command had access to the same information and the ability to function collaboratively. Effective communication paths did not merely mean the technical means used to pass information. Far more important was developing the interpersonal relationships that gave an organization a meaningful and authoritative voice. Absent such a voice, the AWG would become an irrelevant actor unable to influence operations in any meaningful way. For the AWG, this requirement meant developing a robust capacity for liaisons able to understand and effectively interact with these very different organizations. Recognized early on was the need to ensure liaison officers had enough rank to both cultivate required relationships and have legitimacy when they spoke.[38]

Coordinating activities with both the Army and other military and civilian intelligence agencies would put the AWG in the unique position of not only consuming intelligence but also directly providing it to the higher levels of the intelligence community. Most tactical units did not engage such agencies directly, but the AWG's mission meant it would coordinate directly with a host of agencies, a task that would have been impossible only a few years earlier. The attacks of 9/11 profoundly shaped intelligence culture. After generations of emphasizing "the need to know" as the means of restricting the flow of sensitive information to ensure its security, the post-attack analysis led the intelligence community to realize "the need to share." Restrictions still existed, but the realization that one organization might hold information useful to other agencies fueled a new approach stressing collaboration across agency and departmental boundaries. This new attitude helped establish the AWG's operational environment.

The task of deploying and controlling forces overseas partially constituted a "mainstream" Army task. The Army's largest major command, U.S. Army Forces Command (FORSCOM) existed for one fundamental reason: to provide combatant commanders with trained and ready tactical units of corps and smaller size. FORSCOM units exist only in the United States and only for one reason: to deploy in support of a geographic combatant commander. The difference between the AWG and FORSCOM units laid in the task of controlling deployed units. When FORSCOM deployed units to a combatant command, the units temporarily severed their command and control relationship with FORSCOM: the unit worked for either the geographic combatant command or the Army Service Component Command supporting him. The AWG would deviate from this model by maintaining command and control of its teams while they operated in support of a geographic or functional combatant commander. This exception would give the AWG commander tremendous flexibility in how and where he deployed elements.[39]

The last tasks (conduct/assist in advisory training and assist in the development, and integration of lessons learned) represented the most critical ones. That was because they were involved in training the rest of the Army. For General Schoomaker, the AWG's success would reside in its ability to impart to the wider Army the skills necessary to understand and defeat the emerging threats it would face in combat. Even with a unit built around a cadre of experienced trainers, this task was simultaneously the easiest and the most difficult. Knowing how to train units would not present a problem, knowing *what* to train would—and the answer to that question could only come from effectively performing the other tasks in the AWG's proposed mission statement. That would allow the AWG to provide the relevant and timely training deployed units needed. The AWG *had* to successfully perform its training tasks to truly benefit the Army. If it failed to do so, it could find itself imperiled during the next POM cycle and the transfer of funds to other requirements. Put simply, the Global War on Terrorism was proving expensive, requirements generally outpaced resources—and the Friday Fights attracted many participants.

The Operational Concept

The draft operational concept presented a multi-faceted description of the environment in which the AWG would operate, the objectives it sought, and the methods it would utilize to achieve them. Very much a product of its time—the influence of Iraq and Afghanistan was unmistakable—it presented an ambitious approach to understanding the contemporary circumstances and acting to change those circumstances for the better.

The operational concept neatly linked the "how" of the mission statement to the "what" described as well as the conditions under which it would accomplish the myriad tasks the Army would task the AWG to complete. If the operational concept successfully enabled the outcomes desired by generals Schoomaker and Cody, then it was viable. Exercises or simulations could not prove viability. This concept would require actual testing on the battlefield, coupled with an objective analysis of the outcomes.

As initially envisioned, the operational concept demonstrated a sophisticated understanding of the contemporary battlefield. Beginning in the 1980s, the Army's tactical intelligence doctrine emphasized intelligence preparation of the battlefield or IPB. IPB provided an analytical framework for determining and then identifying enemy courses of action (COAs). COAs reflected the combination of terrain, weather, and the capabilities associated with a given enemy unit's strength and equipment. The first step in IPB was the Battlefield Area Evaluation. This evaluation determined both the area of operations (AO), defined as the area in which a commander has the authority and responsibility for military operations and the area of interest (AI), defined as the location of all enemy activities which might affect the friendly force. Doctrinally, the AI was larger than the AO but nevertheless comprised both a relatively small volume of space and a small number of actors—the friendly command and the enemy force he sought to defeat. The draft O&O acknowledged war's new reality: the global nature of the AI and the fact that actors extended to include non-combatants thousands of miles removed from the battlefield.[40]

This change emphasized a vital element of the asymmetric threat: leverage. Threat actors did not have to commit to a decisive engagement with Army units on the ground; they only had to create the impression of dominance in the minds of citizens and their allies to erode national will and realize their objectives. For the enemy, the AO and AI merged into a single entity. This conflation was not an entirely new phenomenon: the Army experienced this in 1968 with the Vietnamese Tet Offensive that saw the temporary seizure of portions of the American embassy in Saigon by the Viet Cong. What made the latest expression of the expanded battlefield different from earlier examples was the plethora of ways to communicate the message and influence non-battlefield actors. The continuous nature of the news cycle in the age of cable coupled with the Internet gave contemporary actors capabilities far beyond those of the Viet Cong or other earlier groups that sought to use information as a weapon. The new battlefield's depth in both time and space meant the AWG's operational concept had to ensure that no dead space (areas the AWG could not address) or seams (where elements or capabilities came together) were left behind for the enemy to exploit. Failing to incorporate these ideas into the operational concept would result in an organization incapable of addressing asymmetric threats through the depth of the battlefield and failing to meet the Chief of Staff's objective.[41]

The paragraphs discussing the operational environment were not purely theoretical. Also included were assessments on asymmetric weapons and tactics, non-state actor command structures, and the vulnerabilities identified in various non-state actor groups. In stating "adversaries are increasingly able to conduct operations with minimal use of traditional communications means and have demonstrated the capability to be creative and innovative in causing both mass destruction and far-reaching mass effects," the O&O emphasized a critical point.[42] IPB's third and fourth steps, evaluating the threat and developing threat courses of action, required producing doctrinal (DOCTEMP) and situational templates (SITEMP). These posited understanding the details of how an enemy organized and deployed his units. The SITEMP sought to graphically present how the enemy would appear at different points in time: on his approach march, when deploying his artillery, when attempting a penetration, et cetera. While doctrine recognized the need to conduct IPB against terrorists, its discussion of how to do this was minimal. IPB methodology meant units conditioned themselves to operate against organized forces whose battlefield behavior foreshadowed subsequent actions. The ugly reality of the asymmetric battlefield was a loosely organized force whose behavior was far more difficult to accurately template. [43]

The paragraphs referencing non-state actor command structures called attention to both their relatively flat structure and the near-universal presence of three tiers of membership: the ideological leadership; the mid-to-low level planners, trainers and technicians; and the "soldiers" who carried out operations. Noted were the varying motivations between the tiers and the role social structures played in their composition. Highlighted were both the cellular nature of the tiers and the potential for widely varying motivations between strategic directors on the first tier and soldiers on the third. Implicit in this description was how variations in motivations and resources would affect efforts to engage elements on each tier. [44]

In considering the inherent vulnerabilities in non-state actor groups (especially the first tier), the O&O paid particular attention to the need for freedom of movement—both individuals and funds—and for support from recognized states or other sources of security. It also observed the relative benefits between attacks against elements on each tier. First-tier elements were fundamentally more difficult to target but offered a much higher return on the efforts expended against them. Second-tier elements, particularly trainers and technicians, also represented higher pay-off targets owing to the loss of vital skills or capabilities on which both the first and third tier depended. [45]

The operational concept also recognized the need for a consistent and standardized model to analyze asymmetric threats and actions. Absent such an approach, the group's efforts would fail to address the many facets of the asymmetric threat consistently and could contribute to the fundamental problem the group sought to eliminate: gaps in friendly understanding and capabilities. The O&O emphasized the incident investigation model as the group's fundamental methodology. Championed by retired Major General Silvasy, this called for developing a viable hypothesis and then empirically testing it to provide both actionable intelligence, and revised tactics, techniques, and procedures units could use to counter those threats.

Creating actionable intelligence, intelligence that could effectively drive tactical operations such as raids or deliberate attacks, was the most challenging task the Army faced. It required rapidly sifting through enormous volumes of information, identifying specific elements of information useful in making time-sensitive decisions, *and* concurrently providing useful and focused information to the individuals undertaking the operation. This task took place under significant time constraints, with these reflecting the competition for scarce resources. These resources included the intelligence collection assets necessary to maintain an accurate intelligence picture as well as the air and ground combat units conducting operations against a specific target. Complicating the problem was the need to support the shaping operations that isolated the target or otherwise set the most favorable conditions possible for the decisive operations that physically placed soldiers or weapons on the assigned objective. These operations frequently required the same level of support as the primary mission to be effective. The incredible size of both Afghanistan and Iraq, coupled with low American troop densities, and increasingly sophisticated opponents in both countries served only to magnify the challenges of providing actionable intelligence and contributed to the duration of operations in both locations.[46]

Closely linked with the incident investigation model were technical countermeasures. The original O&O called for the AWG to "to develop and implement a full spectrum analysis capability that considers and applies multiple materiel, doctrinal, and training strategies to effectively counter the various means of attack that will be used by the asymmetric threat."[47] The incident investigation would drive a comprehensive analytical effort incorporating the efforts of technical experts from across the defense community. Viewed from the Army's various major commands, the broader intelligence community, and the Army's sister services, the O&O sought the broadest engagement possible. The incident investigation model provided the linkage between events on the battlefield and the efforts to provide useful counters. This requirement acknowledged an unpleasant battlefield reality: asymmetric methods extended beyond the IED threats that provided the AWG's initial impetus. The AWG would not compete with the IED Defeat Task Force but would work closely with it to provide solutions to that problem. It would concurrently work the technical aspects of the suicide bomber, counter-Man Portable Air Defense (MANPADS), counter-sniper, counter-unmanned aerial vehicle, and subterranean infrastructure missions.[48]

Time also affected identifying and promulgating tactics, techniques, and procedures. A relationship existed between an asymmetric threat's persistence or duration and the tactical and strategic effects created. As lieutenants and sergeants, the Army's senior leaders learned the factors of "METT-TC" (mission, equipment, time, terrain, troops available, and civilian considerations) as an aid to mission analysis, but for the AWG time assumed even greater importance.[49] The operational concept thus sought to give the AWG both a high degree of operational agility and the ability to impart that to the larger Army. Concurrently, it endeavored to remove the elements of agility and persistence from the enemy's array of tactics, techniques, and procedures.

The O&O observed the "AWG is organized and equipped to assess and understand insurgent organizations/capabilities, and to assist commanders in defeating the insurgents at all levels." [50] The overarching focus resided in a layered approach that aligned specific components of the AWG against discrete requirements to attack not only the threats in various tiers but also the linkages between them. The headquarters provided necessary staff estimates and similar products for deploying teams even as the teams themselves would provide the headquarters, training support, and concept integration components the information necessary to close the gaps the enemy exploited. This strategy allowed the various pieces of the AWG to focus on specific items and concurrently reduced the time required to break out requirements and identify those responsible for satisfying them.

The operational concept also articulated fundamental concepts for operational employment—the basic "how" of executing its mission. The overarching goal described was allowing conventional units to rapidly adapt through training assistance, building their capacity to identify enemy vulnerabilities, and improving both their speed and ability to attack enemy vulnerabilities throughout "the depth of enemy capabilities."[51] Using a series of vignettes describing the existing challenges of convoy security, counter-mortar, counter-man portable air defense, and offensive operations, the document linked the actions of various parts of the AWG with the actions of the institutional Army and the deployed units dealing with the problem.[52]

Finally, the operational concept addressed DOTMLPF implications for not only the larger Army but the joint force as well. The concept emphasized the need to address both the asymmetric threat's foundational tier structure in the Army and joint doctrine as well as the issues of supported and supporting relationships between the AWG and other forces. Highlighted was the need to further develop capabilities in information and psychological operations, military police, explosive ordnance disposal, civil affairs, and forensic science. Expanding these capabilities established the conditions for earlier identification of specific threats and both reduced the time required and increased the effectiveness of methods to counter them.

Ultimately, the operational concept presented a sophisticated description of the environment in which the AWG would operate and the tasks it would perform. Capturing the extensive experience of the authors, it linked the elements of time, space, agility, and method. The operational concept did not do this in a theoretical vacuum: it recognized the demonstrated commonalities of the asymmetric threat, the ever-present realities of DOTMLPF, and organizational and institutional culture. Most importantly, it provided the essential background for building and validating the proposed organizational concept.

The Organizational Concept

However well considered the mission statement or operational concept, neither would amount to anything without proper organization. The complexity of the AWG's mission, the uncertainties of its operational environment, and the requirement to work at all levels of the Army and joint force required a robust, flexible, and tailorable organizational concept. Robustness would provide the AWG itself with a core set of capabilities to reduce the need to rely on other organizations and allow it to maintain a high operational tempo. The highly dynamic operational environment meant the unit had to possess the ability to rapidly organize and deploy teams from across the group to ensure it worked effectively with units both deployed and preparing to deploy. As both the larger Army's mission changed and the threat evolved, the AWG's requirements changed with these deviations outpacing the institutional Army's ability to provide the required personnel.

An additional requirement resided in providing the AWG not just with an individual—but the right *type* of individual. The ideal was someone possessing extensive tactical and technical skills, the ability to interact with all levels of the command effectively, exceptional communication skills, and most significantly comfortable working in an environment best described as "uncertain."[53]

General Fields' team considered a range of organizational structures before settling on the final proposal. The proposed structure called for a group headquarters, a headquarters detachment, an operations squadron, a concepts integration squadron, and a training advisory and assessment squadron. Each of these elements maintained discrete responsibilities and capabilities. An embedded mechanism for allowing this was constructing these units from a mix of soldiers and civilians. Civilians would include both U.S. government employees and contractors. As a practical matter, most contractors would be retirees from the Army and Navy's Special Operations Forces. By employing high-skill retirees, the AWG retained the ability to both rapidly employ and terminate human resources as circumstances changed. This concurrently reduced the need to draw personnel from the small number of active-duty soldiers qualified for special operations.[54]

The group headquarters exercised command and control with a colonel as the commander. Reflecting the AWG's mixed military-civilian structure, a Department of the Army civilian (DAC) would serve as the deputy to the commander. The deputy was not in the chain of command but performed a range of administrative tasks. While the headquarters would contain the standard coordinating staff structure of personnel and administration (S1), intelligence (S2), operations (S3), logistics (S4), and communications (S6), it would also have an assigned lawyer and command liaison teams to facilitate operations with the various combatant commands. While the coordinating staff performed the functions common to all units, the S2 executed some unique tasks. These included providing asymmetric warfare intelligence to units outside the AWG chain of command, providing much of the direct support provided by a military intelligence company, and assisting the S3 with information operations. Rounding out the headquarters were a forward support team to assist in the control and support of deployed elements, a plans cell to assist combatant commands in integrating AWG capabilities into operations, and a Department of Army staff element. As the AWG was not subordinate to any of the Army's existing commands, the Army Staff element provided the interface between the AWG and the Army Staff and ensured the unit received both the requirements and support necessary to operate. [55]

A small headquarters detachment provided both administrative and logistical support to the headquarters and ensured weapons, vehicles, and other maintenance for the group. The detachment also maintained responsibility for nuclear, biological, and chemical (NBC) training and equipment support. The importance of this laid in the various ways the various facets of NBC could integrate into asymmetric attacks. [56]

The operations squadron represented the AWG's functional heart. If its proposed mission, "conducts operations in support of Joint and Army force commanders to mitigate and defeat specified asymmetric threats" sounded a bit vague, its organization was more explicit.[57] The squadron contained four field troops, each comprised of three field teams (four people per team) that would deploy and embed with conventional units and provide expertise and assistance in identifying and defeating the asymmetric threats faced. Anticipating growth, the O&O proposed a second squadron to provide support beyond the fights in Afghanistan and Iraq. The civilian element of the operations squadron was significant, comprising almost 40 percent of the squadron's strength. Within the field teams, this increased to 50 percent. There was a stroke of genius inherent in this model. It provided a mechanism to remove military rank from the equation when necessary.[58]

While its field troops and teams worked closely with operating units, the operations squadron performed several additional roles critical to success. The first was collecting, developing, validating, and disseminating emerging asymmetric warfare tactics, techniques, and procedures. As a subset of that task, the squadron identified tactical asymmetric warfare innovations for input to the TRADOC Futures Center and the Center for Army Lessons Learned and supported asymmetric warfare predictive modeling and trend analyses. The second was assisting in threat identification, providing tactical training, and assisting in the integration of emerging countermeasure technologies.[59]

Though small (the TDA called for 62 individuals), the operations squadron ultimately became the best-known element of the AWG by its ubiquitous presence in both the training base and on the battlefield.[60]

If the operations squadron's focus was "downward" on deployed units and ongoing operations, the concepts integration squadron would focus "upward" on the institutional Army, most importantly the Army Acquisition Executive (AAE) and the Army Materiel Command (AMC). Directed by the Assistant Secretary of the Army for Acquisition, Logistics, and Technology (ASA ALT), the AAE's primary function is developing, acquiring, fielding, and sustaining equipment and services to meet Army needs. AMC performs a variety of missions, including operating the Army's installations and both managing its global supply chain and synchronizing its logistics and sustainment activities. With its "upward" focus, the concepts integration squadron coordinated asymmetric warfare materiel requirements and innovations with the AAE and AMC. It would liaise with the Army's Program Executive Offices (PEO), acquisition Program Manager (PM), and acquisition communities to assist in developing solutions to identified material gaps. The overarching objective was simultaneously simple and complex: identify the right equipment and then get it to the right people as quickly as possible. The squadron consisted of three distinct cells to accomplish these tasks: program and concept development, requirements analysis, and operations analysis.[61]

Program and concept development would orient on the development of identified requirements to close operational gaps through technology vetting, test support and coordination, and training material development. The requirements analysis cell would work with the combat training centers, installations from which units deployed, test sites and areas, and inter-agency work groups. It would also work with the TRADOC Systems Managers responsible for managing DOTMLPF elements associated with specific equipment items. This squadron represented the vehicle for meeting General Cody's objective of providing the larger Army DOTMLPF-based solutions to existing and emerging asymmetric warfare problems. Finally, the operations analysis cell would test and perform operational analysis and assessments of equipment, tactics, techniques, or procedures (TTPs) to validate performance. Completing these tasks allowed the AWG to meet the objective of serving as the "one-stop shop of asymmetric warfare expertise."[62]

The last subordinate unit, the training advisory and assessment squadron, would perform three distinct missions. The first mission lay in conducting the AWG's selection course, while the second was ensuring members received proper and relevant training before deployment. The third mission was the analysis of intelligence, operational TTPs, training, and materiel solutions, and then articulating an overarching strategy to combat asymmetric threats. Part of this analytical effort meshed with the concept's integration squadron and its assessment role. The training advisory and assessment squadron would develop or propose solutions with the concept's integration squadron lending its expertise to validate them.[63]

The AWG represented a unique approach to solving complex problems under exceptionally demanding conditions. Operators—both uniformed and civilian—required special skills and an unusual temperament: comfort in dealing with uncertainty. If the AWG received its personnel in the same manner as the conventional Army its utility—and existence—would likely suffer. To preclude this, the AWG would operate an assessment course designed to weed out those individuals who might have possessed the desire to serve with the group but who lacked the capacity. The selection course requirement imparted a certain mystique to the unit, but more importantly, it mitigated the risks inherent in creating a specialized unit. For the AWG, the greatest risk would reside in establishing credibility with both the operational and institutional Army. Failing to do this would condemn the unit as a failure and force the Army to consider a different pathway to developing the asymmetric warfare skills so desperately needed.

As proposed, the O&O met General Cody's requirements and the document received initial approval on January 6, 2005. The basic concepts found expression in more concrete terms, terms that allowed assigning specific tasks to specific individuals. Public Law 109-148's passage meant the Army now had the authority to move the AWG from idea to functioning unit. Building the unit was the next step.[64]

Subsequent Actions

The first substantive administrative action was the August 11, 2005, memorandum that provided a proposed Table of Distribution and Allowances (TDA) for an Asymmetric Warfare Group. This reflected both the O&O and ongoing discussions within the Army staff. Published by the G-3/5/7 Force Management Office, the memorandum provided a point of departure for subsequent staff work regarding the actual staffing of the unit. The Army required 16 months to get from an initial idea to a proposed unit. This was a short time span considering the many issues requiring resolution. It would require considerably less time to complete the most critical follow-on actions.[65]

January 31, 2006, marked a major milestone in the AWG's development when the Army published General Order Number 2 (GO 2). GO 2 formally established the AWG and directed it to "execute missions, responsibilities, and functions required to eliminate, mitigate, or counter the effects of improvised explosive devices and, as determined by the Secretary of the Army, other similar threats."[66] This general order represented the third of three key actions—formalizing the basic concept, obtaining legal authority to obligate funds, and establishing the organization itself—necessary for transitioning the AWG from amorphous ideas to an actual unit possessing institutional legitimacy.

While the O&O called for assigning a high number of civilians to the unit, a serving officer would exercise command. With the AWG breaking new ground, its commander needed to possess not only outstanding leadership and tactical skills, but also the ability to foster a unit culture emphasizing initiative, adherence to standards and procedures, and commitment to unit objectives. Individual successes would be important but would prove counterproductive unless the Army viewed them as unit accomplishments. If this did not happen, if the unit did not prove its value to the wider Army, it would likely become a short-lived experiment and historical footnote. Finally, the first commander required the capacity to understand and negotiate the range of bureaucratic hurdles that would largely define his or her command tour. These included not only navigating the Army's personnel and logistics systems, but also dealing with a variety of external stakeholders, some of them members of Congress, who had opinions of how, what, and, where the AWG would function. In the judgment of Col. Joseph Votel, Col. Robert "Bob" Shaw embodied the qualities necessary for the job.[67]

Commissioned an Infantry officer through the Army Reserve Officers Training Corps program at the Citadel, Shaw's career included extensive Infantry and Special Forces assignments. These included assignments with the 172nd Infantry Brigade in Alaska, a tour with the 3rd Ranger Battalion, service as a Ranger instructor at the Infantry School, a tour with a Special Mission Unit, and service with JSOC. In addition, he served as a plans officer with I Corps at Fort Lewis, Washington, and commanded the V Corps Special Troops Battalion. A graduate of the School of Advanced Military Studies, he returned to Fort Leavenworth, Kansas, as a fellow in the Advanced Military Studies Program. Shaw was an officer equally at home in conventional and Special Operations Forces units and had practical experience in accomplishing complex missions and solving difficult problems. Concurrent with General Order 2's publication, Shaw assumed command on January 31, 2006.[68]

With a clearly defined concept, authorization, a funding stream, and the assignment of a commander, the only things the AWG lacked were a home—and the personnel required to perform the mission. Providing both would present Shaw with significant challenges over the coming months.

Conclusions

Analysis of the AWG's creation allows one to draw several conclusions. The first of these is the pivotal role of senior leaders in driving institutional change. In seeking to create a new solution to the asymmetric warfare problems bedeviling the Army, Generals Schoomaker and Cody opted to create highly specialized units—the IED Defeat Task Force and the Asymmetric Warfare Group. Cody embodied the "activist leader" described by Stephen Rosen in *Winning the Next War: Innovation and the Modern Military.* In *Forging the Sword,* Benjamin Jensen considered the importance of both personality and bureaucratic power in shaping change. While Jensen focused on doctrine, these themes resonate in the AWG's creation. To better exploit technology, automation, and systems-based business practices, the Department of the Army bureaucracy changed in 2002. Nevertheless, it fundamentally reflected the Army's peacetime requirements, requirements vastly different from those existing since 2001. Cody understood how the bureaucracy worked, and how and where he could influence, bend, or break it to achieve the outcomes he desired. Cody's establishing the ASPB reflected his understanding of institutional practices as well as his ability to impose change upon them. It is unlikely that an officer lacking Cody's deep knowledge of conventional and Special Operations, not to mention his acumen in working through (and sometimes breaking) the Army's bureaucracy, would have succeeded in establishing the AWG. [69]

The second is that the AWG, while not a Special Forces organization, embodied many aspects of Special Operations culture as it developed. The imprint of Special Operations permeated its O&O. This involved more than designating AWG troop elements "squadrons" or "troops." It extended more fundamentally into the "what" and "how" of the mission—including the group's role as a "directed telescope" focusing on asymmetric methods. The essence of Special Operations manifested itself in the utilization of a selection course to determine membership and in employing small teams composed of highly skilled individuals possessing a high degree of initiative with little direct supervision. The O&O built this into the new unit. Interestingly, this also represented something of a potential liability. The AWG's first commander, Colonel Shaw realized this and would ultimately seek to draw applicants from across the Army to acquire individuals possessing the required aptitudes but with perspectives, methods, and skills not found in Special Operations.[70]

The third is that the AWG was, from its very beginning, envisioned as a permanent unit—one intended to survive beyond the end of the conflicts in Afghanistan and Iraq. In considering a sophisticated vision of the operational environment, building in a robust but flexible structure, in focusing on DOTMLPF approaches, and in developing the capacity to function in any combatant command, the AWG did not become a short-term program focused on the challenges of Afghanistan or Iraq. Instead, its design provided an *enduring* institutional solution to the *global* challenges of asymmetric warfare that would remain long after any withdrawal from Iraq or Afghanistan.

Chapter 5

Growing the Asymmetric Warfare Group

Normal is a cycle on a washing machine.
Motto of the Asymmetric Warfare Group

Assembling and Shaping the Group

The January 2006 publication of General Order 2 created the AWG on paper. Turning it into an operational unit involved a long process, initially characterized by competing demands, sub-standard facilities, and the pressure to translate the O&O into tangible results. For Colonel Shaw, the days leading to the AWG's activation presented daily challenges as he juggled two important duty positions—senior fellow at the Army's School of Advanced Military Studies (SAMS) at Fort Leavenworth and command of the AWG. As a SAMS fellow, he led a seminar of field grade officers immersed in a year-long study of military history, theory, and doctrine. This was a full-time position since seminars met daily and fellows had the role of moderator, facilitator, and mentor. Additionally, they served on committees supervising student monographs and conducting the comprehensive oral examinations required for graduation. Shaw found himself devoting weekdays and nights to his SAMS duties and weekends to the AWG as he moved into command.[1]

This unusual set of circumstances reflected both the nature of the nascent organization and Shaw's capabilities. This highly unusual arrangement persisted until Shaw's students graduated in May, allowing him to join his unit on a full-time basis. Satisfying all the AWG's DOTMLPF elements should have presented Shaw with more than enough work, but the command's unusual mission meant additional requirements appeared. In common Army parlance, Shaw, like other unit commanders, had to "keep the plates spinning." For a conventional armor or infantry brigade, the fundamental missions were already defined and most DOTMLPF issues were resolved. In contrast, Shaw had to start spinning the plates while reporting directly to Headquarters, Department of the Army as the commander of a field operating agency. He also had to spend a great of time interacting with elected political officials and their staffs, not a normal task for an armor or infantry brigade commander. All these tasks transpired against the backdrop of combat and the inherent risks it posed to Shaw's subordinates.[2]

An initial task for the command was finding an actual home. Working out of the Pentagon was infeasible, especially as the unit started receiving personnel and began actual operations. The AWG established an interim headquarters at Fort Meade, Maryland, and concurrently developed a plan for permanent stationing. To this end, the unit invested considerable time determining the criteria for objectively recommending a suitable installation to serve as its permanent home station. Among these was access to the senior leadership of both the Army and other government agencies. Equally important was the availability of firing ranges and training areas, availability of building space for offices and secure storage of equipment, access to major airports to facilitate travel and communications support. Course of action (COA) development considered options on two separate tracks: within the National Capital Region (NCR), which extended from northern Maryland to northern Virginia; and outside the NCR. The fundamental challenge facing the unit was a simple one: competing requirements.[3]

Within the NCR lay Aberdeen Proving Ground and Fort Meade in Maryland; Fort Lesley J. McNair in Washington, D.C.; and Forts Myer, Belvoir, and A. P. Hill in Virginia. Outside the NCR, the analysis identified nine different installations ranging from Fort Irwin, located in the high Mojave Desert of California; to Fort Bragg, in eastern North Carolina. In between lay Fort Carson, Colorado; Fort Hood, Texas; Fort Polk, Louisiana; Fort McCoy, Wisconsin; Fort Knox, Kentucky; Fort Benning, Georgia; and Fort Jackson, South Carolina. These installations reflected a variety of missions with Forts Irwin and Polk hosting the Army's two most important training centers in the United States and Forts Jackson and Benning hosting large initial entry training populations. Forts Carson, Hood, and Bragg hosted one or more divisions. Bragg and Hood also hosted both a corps headquarters and many of the associated troops. Fort McCoy was an Army Reserve installation focused on deploying reserve component units from across the country. Fort Knox was undergoing a complete realignment as the Army both moved the Armor Center from there to Fort Benning and relocated several personnel-related commands to the post. Sorting through the options meant the AWG worked with the Installation Management Command (IMCOM) and the Army G-3/5/7 training staff to identify the strengths and weaknesses of each one. This collaborative effort sometimes led to conflicting opinions as these parties could not reconcile their perspectives.[4]

Despite this friction, the AWG developed five permanent stationing options for the Secretary of the Army for approval. COA 1 was an outside-the-NCR option with all AWG elements located together at either Fort Hood or Fort Knox. COA 2 was an inside-the-NCR option with all AWG assets on the same post within the NCR. COA 3 was a plan with the AWG stationed at two locations: a headquarters inside the NCR with headquarters at Fort Meade and other elements located at Fort A. P. Hill. COA 4 offered a hybrid arrangement with the headquarters in the NCR and other elements stationed at either Fort Hood or

Fort Knox. COA 5 was the non-plan of deferring the decision to a later date. None of these schemes was optimal, but COA 2 and COA 3 best met the critical criteria of access to senior leadership, suitable installation support (facilities, firing ranges, and communications), and airport access for travel.[5]

Courses of Action 2 and 3 reflected the reality of the NCR—too many occupants vying for too little space. While multiple installations existed within the area, their unique patterns of development meant no one installation met the requirements. This situation stemmed from two long-past decisions. The first was the transfer of much of Fort Meade to the U.S. Fish and Wildlife Service in 1991 to expand the Patuxent Research Refuge. The second was the Base Realignment and Closure (BRAC) decisions of the 1990s and early 2000s. BRAC meant that much of Fort Belvoir's former range and training areas saw conversion to office space as several Army and Department of Defense commands moved from outlying locations to the post. Much smaller than Belvoir, both Fort Lesley J. McNair and Fort Myer lacked suitable office space—and the room to build any. If the AWG were to locate within the NCR, it would require a split-basing plan to meet the command's divergent requirements with Forts Meade and A. P. Hill representing the best compromise. The AWG was already validating this approach as it conducted fieldwork at A. P. Hill even as the command element worked from its interim headquarters on Meade.[6]

Fort Meade was relatively close to the Pentagon (less than 30 miles) and possessed existing (if woefully inadequate) building space meeting the organization's requirements. It also had the physical space required for constructing purpose-built administrative facilities should the Army decide to make the post the unit's home. Additionally, it occupied a convenient location near both Baltimore-Washington International Airport and Aberdeen Proving Ground, with the latter serving as the home of the 20th Support Command (20th SUPCOM) the organization envisioned as ultimately becoming the AWG's higher headquarters. The most glaring deficiency was the post lacking adequate space for the critical tasks of threat replication and training. Threat replication consisted of taking the information available on emerging threats and then completely reconstructing them in a non-hostile environment. This reconstruction facilitated the developing and validation of countermeasures. While the institutional Army periodically engaged in such experiments, the AWG's mission required it to routinely conduct such activities on a much larger scale.[7]

Fort A.P. Hill is an installation best known as a field training center for the Army units stationed within the NCR and National Guard and Reserve units in the mid-Atlantic states. Located 70 miles from the Pentagon and 90 miles from Fort Meade, A.P. Hill offered several advantages as a stationing option, most notably firing ranges and training areas enabling the full range of threat replication. It also had enough space for the new construction necessary to meet the AWG's requirements fully. Absent, however, was good airport access and suitable office space for initial operations. Additionally, the surrounding area was mostly rural with limited options for soldier and contractor housing, an issue that could prove problematic as the unit grew.[8]

Over the short term, Fort Meade was the most suitable location for the AWG headquarters. The unit established an interim headquarters in the former installation stockade and the basement of adjacent Snowden Hall as the Army deferred a final decision on its ultimate station. Concurrent with this, the unit would train at Fort A.P. Hill. The issue of locating elements on Meade and A.P. Hill would increasingly consume Colonel Shaw's time as the unit grew. As the AWG became an increasingly visible element in the Army's construction budget, United States senators from Maryland and Virginia took increasing notice as they sought to influence final stationing plans through the May 22, 2008, decision-making Forts Meade and A. P. Hill the unit's permanent homes.[9]

If Fort Meade represented a good "general" location for the AWG, the first facilities occupied were not. "Terrible" would be a charitable term to describe the AWG's original headquarters, the former Fort Meade Confinement Facility. Primarily constructed in 1954 as the "Fort Meade Stockade," with auxiliary buildings dating to World War II, the facility reflected the demands of a massive Cold War Army. By the late 1980s, a much smaller all-volunteer Army found it no longer needed the prison space previously required. As a result, the Army abandoned the facility in 1991. It lay derelict and decaying as the structures were unsuitable to meet existing requirements. With the AWG in search of a home, the facility gained a new lease on life. Despite its lacking much of the required infrastructure, it did possess two characteristics in abundance: it was in the right location and it was available. Colonization of the new home began in February 2005 with AWG members Randy Brumit and Jim Mallon moving from their Pentagon office to the former jail. Mallon was a retired squadron sergeant major from Special Forces Operational Detachment Delta (SFOD-D), the famed "Delta Force." His association with the AWG began in July 2004 while working for the IED Defeat Task Force. Having served in five different combat zones and the Balkans, Mallon first acted as an advisor to Colonel Votel on the IED Defeat Task Force. His background made him the natural choice to start working on AWG-related issues as the initial concepts unfolded. He soon found himself as the *de facto* original (and only) member of the AWG. This distinction entailed performing a wide range of tasks for several months before the arrival of additional personnel allowed him to distribute the workload. These actions included advising on both the Operational and Organizational Concept (O&O) as it underwent development and the initial personnel decisions as well as working on the issues associated with providing the unit a physical duty station. The latter ultimately occupied much of his time.[10]

Eliminating the mess that accumulated over 14 years on an abandoned post took some time, but even a good cleaning did not make the facility suitable. Lights and doors did not work correctly, roofs and plumbing leaked, and summer's arrival caused an explosion in mold that compelled the AWG Command Sergeant Major to move his office outdoors. All these issues took time to correct, but by August the unit could fully occupy its own, albeit decrepit, headquarters. Concurrent with this, Mallon acquired additional building space on post. Working the facilities element of DOTMLPF was an important task, but even more critical were the elements of personnel, training, and material.[11]

While the O&O called for a rigorous selection process under the AWG's direction, it took several months to put together the multiple pieces required to make that process effective. Bringing in soldiers who had not undergone a selection process contained an element of risk. Accepting volunteers might rapidly fill spaces on the organizational chart, but it could both fill those spaces with unqualified individuals and concurrently saddle the AWG with these soldiers for a period ranging from several months to several years. This tactic could severely hobble the AWG's most critical operations—the first ones undertaken to validate both the O&O and the unit's ability to perform them.

To mitigate this risk, Shaw successfully lobbied for the opportunity to hand-pick the original military personnel cohort. Once again, Shaw's Special Forces background stood out with his initial preference for personnel from the Special Operations community. All these soldiers had completed airborne school, the psychologically demanding Special Forces Assessment and Selection Course, military occupational specialty training, and then successful service in one of the Army's Special Operations units. Always in demand, these units had maintained an exceptionally high operational tempo following September 11, 2001. Comfortable with the battlefield's uncertainties, possessing an unusually deep repertoire of tactical skills, and accustomed to training and advising others, the initial reliance on Special Operations soldiers ensured the AWG built on a solid professional foundation.[12]

A critical position was that of the Command Sergeant Major (CSM). Army Regulation 600-20, *Army Command Policy*, defined a CSM's duties as "carr(ying) out policies and standards, and advises the commander on the performance, training, appearance, and conduct of enlisted soldiers."[13] The first individual to hold this position would fundamentally shape the unit's growth and development. Holders of the Army's highest enlisted rank, CSMs are neither capriciously selected nor assigned. There is a cultural hierarchy both in units and within the CSM ranks. Selecting the AWG CSM meant looking through the ranks of some of the Army's most elite non-commissioned officers and identifying that individual with the right skills mix. More than his peers, the incoming CSM would need to understand the operational environment in its totality, and that included both the battlefield and the Army's bureaucracy. He also needed to have an extensive Special Operations combat record to ensure credibility with both the unit's enlisted and civilian members. Finally, he needed exceptional skills as a trainer, a training developer, and a human resource manager. The last skills would ensure both the assignment of the right soldiers to the unit and their acquiring the skills essential to performing their duties.

The Army's senior leadership selected Command Sgt. Maj. Alfred G. "Greg" Birch to fill the CSM position. Birch's career included service as a military police soldier before joining the 2nd Battalion, 75th Infantry (Ranger). He served with SFOD-D for 15 years before returning to the Rangers as the CSM of the 3rd Ranger Battalion and then as the Ranger Regiment CSM. A recipient of the Silver Star Medal, the Bronze Star Medal with "V" (valor) device, and the Purple Heart for wounds received in Afghanistan; Birch brought solid judgment rooted in years of realistic training, extensive operational deployment experience, and actual combat to the mission at hand.[14]

Birch was an unconventional thinker who thrived in the closed world of Special Operations, where individual soldiers traditionally enjoyed far greater freedom of action than their counterparts in the conventional force. He also understood that changing circumstances required modifications in how units operated and that meant not having a high degree of tolerance for old methods unbacked by sound justification. "Because we always did it this way" was never the right answer for Birch. His changing the Ranger Regiment's haircut standards provide one example of his approach. Before joining the regiment, Rangers always had the so-called "high and tight" haircut—virtually no hair on the sides of the head and very short on top. The haircut immediately marked a member of the regiment when off duty but did not serve a useful purpose. Despite encountering a great of resistance, Birch changed the standard to that of the rest of the Army and used the change as a messaging vehicle. Within the regiment, the message was, "Hang on to shit; things are about to change."[15] To the rest of the Army, this change said, "The regiment is changing; this is not the same old Ranger Regiment." Comfortable with change, uncertainty, and challenge, Birch provided much-needed stability as the tumultuous process of growing the unit unfolded. Much more than the typical command sergeant major, Birch spent considerable time arbitrating and refereeing the discussions that culminated with specific actions as the unit executed the long list of tasks required to move it from "idea" to operational unit.[16]

Birch's arrival predated the AWG's official activation by several months. This allowed him to gain a broad understanding of the existing—and developing—operational environments and associated challenges. He quickly immersed himself in the tasks at hand, which included getting the AWG's activation orders finally published. This task involved unofficial consultation with a former member of the Army's Office of Legislative Liaison, who ensured the incorporation of the AWG's legal authorization into Public Law 109-148. Once armed with the activation orders, tangible proof of victory over the bureaucracy, Birch was able to start bringing order and structure to the group.[17]

A firm believer in the importance of marksmanship, physical fitness, battle drills (standardized actions made in response to common battlefield occurrences), and first aid, Birch also ruthlessly enforced performance standards. More importantly, Birch understood success for the AWG required credibility, not only within the unit but with the deployed elements with which members would serve. Ensuring credibility meant that even with the high bar established for unit membership, Birch returned several members, sometimes longtime friends, to other duties when he determined they ultimately lacked the capacity for fruitful service. Ensuring credibility also meant Birch spent considerable time moderating the AWG's organizational culture. These actions ensured members understood that while the Army viewed the AWG as a Special Mission Unit for purposes of manning (assignments would not follow conventional Army guidelines), its non-standard mission did not make it a Special Mission Unit. Several "wannabe Ninjas" had attitudes adjusted as the unit's realities began coming into sharp focus—with Birch doing much of the focusing. What Birch sought was an organizational culture rooted in the unique aspects of Special Operations. These included a high degree of innovation, the ability to be self-starting and "think on one's feet," and pushing to one's physical and mental limits every day. The AWG's center of gravity ultimately resided in those individuals serving as operational advisors to deployed forces. Only the most capable individuals are needed to fill those positions.[18]

If relying on Special Operations soldiers solved several vexing problems, it also created a thorny bureaucratic one: providing adequate compensation. Cody's June 23, 2004, memorandum designated selected positions on the TDA as qualifying for Assignment Incentive Pay (AIP) without providing additional details. Special Operations soldiers could and would serve with the AWG, but initial service outside a designated position meant a reduction in pay. All soldiers receive basic pay, a salary determined by their rank (pay grade) and years of service. In addition to basic pay, soldiers performing special duties receive special duty assignment pay (SDAP) and, if on jump (parachute duty) status, also receive Hazardous Duty Incentive Pay (HDIP). Depending on the individual's rank and length of Special Operations service, a soldier received anywhere from $315 to $575 per month in SDAP and HDIP. The AWG represented a new field of endeavor for Special Forces soldiers, but participation could result in hefty pay cuts as service with the AWG would mean forfeiting entitlements to this pay.[19]

Both the Defense Department and the Army had policies for awarding SDAP, HDIP and AIP. Department of Defense Instruction (DODI) 1304.27, "SUBJECT: Award and Administration of Special Duty Assignment Pay," described three requirements for SDAP, while DODI 7000.14 outlined considerations for AIP. If the duty assignment met just one of the three, the servicemember qualified for the extra pay. Shaw argued that AWG service met all three requirements listed. These included extremely demanding personal effort while serving with the AWG, a higher-than-usual degree of responsibility based on the soldier's rank, and completing the rigorous screening and special schooling to serve with the unit. Army Regulation 614-200, *Enlisted Assignments and Utilization Management*, governed the Army's procedures for determining which soldiers qualified for SDAP. The regulation called for the commanding general of the soldier's major command to determine if SDAP was appropriate and to then forward the recommendation to the Deputy Chief of Staff for Personnel (G-1). The G-1 would make a recommendation and forward it to the Assistant Secretary of the Army for Manpower and Reserve Affairs (ASA M&RA). The ASA M&RA then forwarded the recommendation to the Secretary of the Army who approved the request. This created a bureaucratic hurdle for AWG members. As the unit was a field operating agency, there was no major command to initiate the process. To solve the problem, the Army G-3/5/7 acted in the major command role and initiated the process with one major change: requesting AIP rather than SDAP. Substituting AIP for SDAP would set a standard amount for all recipients, and prevent creating groups of "haves" and "have-nots" with respect to additional pay tied directly to AWG service.[20]

Interestingly, neither General Schoomaker nor General Cody was directly involved in the process and the decision would stand on case the AWG made. These efforts, undertaken by the AWG Pentagon Cell, bore fruit with AWG personnel receiving a monthly AIP payment of $400 per month starting in 2006. While this meant some of the group would see a decrease in pay, the amounts were not as high as they would otherwise have been. This action both improved morale and provided an incentive for Special Forces or Special Mission Unit soldiers to seek an assignment with the group. This also dissuaded non-SOF individuals from seeking assignments for financial reasons. More significantly, the AWG had successfully engaged the Army bureaucracy to achieve a specific objective.[21]

Filling the AWG's civilian slots proved as challenging as filling the military ones. The designation of many positions as contract positions, and the high number of recently separated Special Operations personnel interested in continuing service yielded many candidates to serve as operational advisors (OA). Willingness, however, did not necessarily equate to suitability. The company that received the AWG contract, Wexford, employed a rigorous screening process to ensure the quality of each OA. This process included competency and behavioral screening, a criminal records check, a security background check, and, most critically, psychological testing. Despite most applicants having extensive Special Operations combat experience, Wexford's hiring rate admitted only 8 percent of those applying. Those hired came from four basic backgrounds—Army Special Mission Units, Special Operations Forces, the Ranger Regiment, and civilian law enforcement. Former Rangers entered the AWG with an average of 5.9 years of experience under their belts and Special Operations veterans boasted 13.7 years. Interestingly, some of the former did not come from the Army, but from the Special Operations communities of other services. OAs with civilian law enforcement experience generally had military experience as well. Their real value, however, stemmed from the ability to apply law enforcement investigative techniques to sensitive site exploitation, maintaining evidentiary chains of custody, training local national law enforcement, and assessing criminal and terrorist networks. One other facet of contracted OAs warrants discussion: the OAs' motivation to serve. Most OAs were retired sergeant majors or chief warrant officers who felt the Army had "been good to them." They had a sense of *noblesse oblige* and wanted to assist in the development of the current generation of soldiers.[22]

Similarly, there was a wealth of talent available within the Pentagon for critical leadership positions. Like Jim Mallon, Mike Bowman left the IED Defeat Task Force and joined the AWG's operations staff. In Bowman, the unit gained an individual intimately familiar with each step taken thus far in creating the AWG. Equally important, he also understood the myriad details and backstories associated with them. As a former Department of the Army G-3/5/7 staff officer, he also possessed an intimate understanding of how bureaucratic processes played out within the Pentagon. Bowman provided the unit the effective continuity it required as its growth continued to impinge on the institutional Army's bureaucracy. [23]

A key position within the AWG was that of the deputy to the commander, envisioned as a Department of the Army civilian. Frank Bohle, a retired Special Forces colonel possessing extensive experience with both operational units and higher staff, assumed that position. Fortuitously, much of his career coincided with major transformations within Special Forces and he possessed a wealth of first-hand experience in managing change. This experience proved a valuable attribute for a unit breaking new doctrinal, organizational, and training ground. Bohle's ability to visualize, describe, and direct activities complemented Shaw's, while his status as a civilian meant he would remain with the group after Shaw completed his command tour.[24]

Supervising the AWG

While the AWG continued the transition from idea to functioning unit, the G-3/5/7 underwent changes showing a new focus on asymmetric warfare. Seeking to improve its ability to direct the AWG and REF and restore capacity lost with the IED Defeat Task Force moving from the Army to the Department of Defense, the Army created the Army Asymmetric Warfare Office (AAWO) under the G-3/5/7 in March 2006. April and May saw the publication of Department of the Army Execution Order 158-06 and the "AAWO Charter" with both documents amplifying the AAWO's responsibilities. These included providing oversight and operational direction to both the AWG and REF until the AWG moved to its intended "organizational home," the 20th Support Command. The AAWO also sought to "change the culture of the Army to a mentally agile and adaptive force."[25] Ultimately, the AAWO ensured that even as the AWG and REF performed their respective missions, the benefits gained accrued to the larger Army.[26]

For the AWG, the AAWO's most visible mission lay in providing oversight—and that included the funds that would sustain the unit until it fully entered the Army's budget. Cody's June 2004 memorandum provided the information required to perform that function. For fiscal years 2004 and 2005, the AWG would receive funding through a re-prioritization of sums already appropriated by Congress through the Army Strategic Plan (ASP) budgeting process. Created by the G-3/5/7, the ASP nests the Army's institutional planning with national and Department of Defense guidance. This activity includes budget planning and programming over future year defense programs—determining where, and at what rate, to apply funds against requirements. In re-evaluating budget priorities, funding for the AWG would come from budget lines assessed as having a lower priority. This process imparted dynamism into the Army's cumbersome budgeting process. Without the ASP budgeting process, the service would lack the ability to move money to meet developing requirements.[27]

The fiscal year 2006 funding required the Army Budget Office to submit an appropriate Program Budget Decision (PBD). The PBD represented another way to reallocate funds after the Secretary of Defense approved the POM as part of the Department of Defense's budget review process. Like the ASPB, it provided flexibility in the Army's budget. Unlike the ASPB, which only involved the Department of the Army, the PBD demonstrated the Secretary of Defense's concurrence in reprioritizing funds. As a practical matter, this meant that a PBD could allow the Army to receive funds for a specific program at the expense of a sister service. The PDB also meant an additional set of bureaucratic actors could influence decisions as the process involved both the other military departments as well as the Office of the Secretary of Defense's staff.[28]

Finally, fiscal years 2007 through 2011 saw a "mini-POM" submission of $60,000,000 per year for a total of $300,000,000 to completely meet the unit's funding requirements. The mini-POM represented an abbreviated POM process complying with regulatory guidance. A core element of the POM is the Program Evaluation Group (PEG), which is a group of senior officials considering discrete elements of the POM. This division of labor ensures that the correct individuals are evaluating requirements and providing recommendations. Cody's memorandum aligned the AWG under the Training PEG for POM purposes. This arrangement was not necessarily ideal given the AWG's mission set straddled multiple PEGs (a PEG represents each of the Army's specified functions under Title 10 United States Code), but the Training PEG represented the closest bureaucratic fit.[29]

These budget-related actions reflected a simple fact: Public Law 109-148's passage provided the Secretary of the Army the legal authority to establish the AWG, but it did not allocate any funds for that purpose. Cody's plans thus provided the bureaucratic solution to the problem at hand: authorization without resources. Once the POM included the AWG's full integration, the need for such actions would terminate as the Army would fund the unit as part of its permanent force structure. [30]

The AAWO's role in providing oversight of the AWG makes it noteworthy. It is also remarkable as another example of Cody bending the bureaucracy to foster the AWG's success. With Lieutenant General James Lovelace replacing Cody as G-3/5/7, the latter had an officer who shared his vision in a critical position. In creating the AAWO and placing it under Lovelace, Cody institutionalized the AWG's unusual standing as a field operating agency reporting to a Department of the Army staff section. In laying out the funding mechanisms for the near future, Cody also ensured the AWG would receive the required resources. Cody, in effect, established a bureaucratic bubble around the AWG, giving Shaw and his subordinates the necessary institutional positioning. Their ability to remain in position, however, would rest on their ability to accomplish their daunting mission. [31]

Finally, Cody instituted one final bureaucratic adjustment, the "JAAR" review. "JAAR" stood for Joint IED Defeat Organization, AAWO, AWG, and REF, and the review was a recurring conference focusing on these organizations. Cody employed the JAAR to receive feedback from the field, gain situational awareness regarding issues on which he could take action, and influence the Army Requirements and Resourcing Board (AR2B). Created in December 2004 as a replacement for the Army Strategic Planning Board, the AR2B served as the vehicle for presenting critical operational needs to the Army's senior leadership for rapid decision-making. Consisting of the Army's primary staff; Assistant Secretaries of the Army for Acquisition, Logistics, and Technology and Financial Management and Comptroller; the heads of the Army National Guard and the Army Reserve, and the Assistant Chief of Staff for installation management the board provided a means for adjusting priorities. While Cody could veto AR2B decisions, shaping them saved both time and effort.[32]

Building Capacity

The training represented a particularly thorny problem for the AWG. With no fully qualified personnel on hand, the unit had to train its initial cohort. Additionally, it needed to concurrently develop the selection course through which follow-on applicants would pass. Complicating the issue was the unit's Special Operations culture. Shaw recognized that while this element brought tremendous tactical capabilities to the AWG, there were many soldiers in the conventional Army with the temperament, intellect, and desire to serve with the unit. Importantly, he understood that the perspectives and experiences of these soldiers could profoundly enhance the unit's capabilities as it grew. Additionally, he saw a potential danger in keeping the AWG closed to the larger Army as a "Special Operations only" club: the unit becoming a dumping ground for Special Forces soldiers no longer possessing the capacity to remain in operational units. Shaw's objective was a straightforward one: cast the widest net possible to "capture" the highest number of the *right* soldiers from across the Army. This objective was impossible to realize, however, until the unit both fully developed and validated the selection course and the subsequent syllabus for awarding the requisite additional skill identifier.[33]

The overriding task lay in creating the required AWG selection course. Here again, the Army turned to civilian contractors. The Wexford Group's AWG contract included developing a course capable of challenging an applicant physically, mentally, and psychologically. In giving this task to Wexford, the Army leveraged years of carefully nurtured experience without placing additional demands on the existing selection courses experiencing higher throughput as the Army sought to expand Special Operations forces in response to ever-increasing demands for these units. Wexford benefitted from its own hiring practices that brought on board several retired senior noncommissioned officers who had previously supervised the Special Forces Operational Detachment-Delta selection course.[34]

To preserve the course's utility, the AWG released few details of its structure or the methods used to assess a candidate's suitability. Selection for the Ranger Regiment required the ability to pass the Ranger qualification physical fitness test, completing a 12-mile march with a 35-pound rucksack, and the water survival assessment. Selection for Special Forces, by contrast, required passing a selection course placing a much higher demand on an applicant's mental and psychological abilities. The difference in the two requirements sets vividly demonstrates the differences in how Ranger and Special Forces units operate.

Rangers operate as companies or battalions. In contrast, Special Forces, while organized into groups, battalions, and companies, really functions at the A-Team level with an A-Team consisting of 12 soldiers. The AWG's selection course required a high degree of physical fitness but emphasized mental and psychological abilities with the unit citing a rough 75% mental-25% physical ratio. Using a series of scenarios, it sought to identify the applicant comfortable with uncertainty and confident in their ability to determine and implement solutions to complex scenarios. This approach made sense as the deployed AWG member would both serve as part of a two-person team and interact with a deployed unit's most senior leadership as well as its most junior members under a broad range of battlefield conditions.[35]

The AWG selection course also included additional elements: a psychological assessment, and an interview with the commander and Command Sergeant Major. The former sought to identify individuals who made it through the course despite possessing personal characteristics that might prove detrimental once deployed. The latter sought to ensure the individual was both a good fit and possessed a clear-eyed understanding of what service with the unit entailed. In the words of Command Sgt. Maj. Birch, "the unit's reputation was too important to risk."[36]

In recognizing the AWG as a Special Mission Unit for purposes of manning, the Army's Human Resources Command (HRC) eliminated many potential staffing problems. HRC was cooperative in dealing with the AWGs and did not place any obstacles to manning. HRC, however, did not control any soldiers attending the AWG selection course. Those soldiers belonged to units in FORSCOM, TRADOC, or one of the overseas commands. While HRC could direct movement, the soldier's current organization could stop or delay an assignment to the AWG. This issue frustrated the AWG as it sought to fill all its personnel authorizations and establish a sustainable overseas deployment schedule for its military members. The irony lay in the fact that the AWG would employ these soldiers' unique talents while the current unit saw him or her as part of a crew, squad, or section. This view meant that instead of influencing the battalion, brigade, or division's operations, the soldier by remaining in his or her tactical unit could only directly influence the smallest elements on the battlefield.[37]

As members of the command assembled, the organization underwent the normal process of "form, storm, norm and perform." [38] Establishing norms, both formal and unstated, was essential for the unit's long-term success. Shaw and Birch were tasked with establishing the formal norms for those areas the O&O did not address.

Shaw established a baseline with his "bumper stickers," statements that sought to set some guiding principles for the unit. In response to a request that he develop policies on a variety of subjects, he reduced them to a single sentence:

"If it is stupid, it isn't my policy."[39] This approach initially violated the policy letter requirements of Army Regulation 600-20, *Army Command Policy,* but it neatly communicated the unit's focus on operational readiness. Administrative correctness followed, but only after the unit resolved operational issues. Shaw's priority was "solving problems and finding solutions for the soldiers who needed them, getting our guys/gals down range, getting our AWG folks into the right agencies and units, departments to operate (...) build the AWG in flight, so admin policy posting wasn't a high priority. I felt I could do that because we hired mature soldiers who knew already what the Army policies were and didn't need to have something signed by me telling them what they already knew."[40]

In helping ensure everyone understood the environment, Shaw also frequently reminded all concerned, "there is no box." Shaw recognized that unnecessary constraints would impair the unit's ability to accomplish its primary mission and engage the Army's institutional bureaucracy. From the very beginning, Shaw also communicated a point neatly summarizing the operational environment, the unit's mission, and the member's relationship with them: "get comfortable with being uncomfortable."[41]

Birch emphasized the mastery of fundamentals and sought to incorporate practical aspects of training into all the unit's activities. Understanding both the importance of effective shooting and shortfalls in how the Army trained and equipped soldiers, Birch leveraged the Rapid Equipping Force and obtained $2,000,000 worth of weapons, ammunition, and equipment to facilitate advanced small arms training with AWG members. The objective was not only ensuring AWG members possessed essential combat skills but also confirming their ability to train others as part of their role as OAs. Birch viewed this as critical as the then-current edition of the Army's governing regulation, AR 350-38, *Standards in Weapons Training,* reflected pre-2001 conceptions of how the Army would fight. This regulation meant the individual soldier received enough ammunition to qualify with a weapon, but not enough to become truly expert in its battlefield use. Most combat support or service support soldiers had not experienced realistic live-fire exercises. Most combat arms soldiers had experienced such exercises, but the new dynamics of the battlefield—urban clusters, significant differences in shooter-target elevation, and a high concentration of non-combatants—represented challenges not replicated on most ranges. In obtaining the additional resources, Birch ensured AWG members could assist other units in developing the improved marksmanship skills demanded on the contemporary battlefield.[42]

Physical training also presented opportunities for creatively integrating training in fundamental skills. The AWG's activation day saw a physical training session after formal ceremonies concluded. Unit members were divided into two teams and then donned protective (gas) masks and body armor before engaging in a game of soccer. The only distinction between the two teams lay in the hands: one team wore gloves, the other did not. Watching the hands became essential, just as it was on the battlefield where understanding where an adversary focused his eyes and positioned his hands could make the difference between life and death. "Combat soccer" was not a routine event but the incorporation of battlefield elements into training whenever possible was.[43]

Capacity building extended beyond the AWG itself and included recurring engagement with the Army bureaucracy. Periodically, this meant engaging elements outside the Department of the Army staff and included both TRADOC and various program managers. While the AWG enjoyed success when it properly executed bureaucratic processes, the demands of time sometimes precluded the expected staff work accompanying actions. The need for rapid action sometimes meant employing the expedient of saying, "Schoomaker said" or "Cody said." While this approach would generally produce the desired outcome, it did not endear the AWG to those elements that had to respond to its demands. This tactic resulted in occasional "self-inflicted wounds" requiring follow-on action by AWG staffers to assuage feelings and re-generate or maintain support. The assignment of civilians to the unit went a long way toward repairing relationships by providing a long-term cadre capable of understanding the damage done and possessing the ability to repair it.[44]

Initial Operations

A new unit's normal progression is to move from non-operational to initial operational capability (IOC), and then to full operational capability (FOC). How long it takes a unit to reach these milestones depends on its type and the priority it enjoys within the Army for personnel, equipment, and training support. Enjoying a high priority within the Army and leveraging its ability to influence both the REF and its unique procurement authority, the AWG rapidly achieved IOC with the first operational advisors embedding with deployed units in June 2006. Within 12 months, OAs, most often operating as teams with a mix of soldiers and contractors, had touched 29 of the Army's maneuver brigade combat teams either in the combat zone or as units completed pre-deployment preparations.[45]

In the combat zone, advisors engaged both downward/outward with the units they supported and upward with the Concepts Integration Squadron and AWG headquarters. The downward/outward focus was on the asymmetric tactics enemy forces employed. This effort, however, went beyond describing enemy tactics, techniques, procedures, or countermeasures that could offset them. Instead, the OAs took a more expansive view based on a broader understanding of both the counter-asymmetric warfare cycle (identify and predict, prevent, react, eliminate, and assess) and the structure of insurgent groups. Operational advisors assisted units in understanding not only those pieces of the puzzle they saw but also the associated pieces not readily visible. In doing this, units uncovered tactical gaps degrading their ability to operate. These gaps included intelligence, reconnaissance, surveillance, incident investigation, and small-unit tactics. Understanding the asymmetric threat model, and then applying that understanding to the unit's area of operations, mission, organization, equipment, and available support led to the closing of gaps and better battlefield outcomes.[46]

Even as OAs assisted units, they provided relevant information about new or evolving threat tactics back to Fort Meade and Fort A. P. Hill. After receiving this information, the Concepts Integration Squadron could start developing, modeling, and refining viable responses. This event chain epitomized Colonel Jette's idea of "getting to the left of the boom." The OA effort required investing significant amounts of time in unit command posts. It required even more time outside the protective wire of forward operating bases or combat outposts where the real business of counterinsurgency operations transpired. OAs performed another important role—serving as a commander's "directed telescope." Contract OAs excelled in this role as their many years of experience allowed them to gain insights on unit capabilities in general—and training shortfalls in particular—as they accompanied units on operations. Wexford contract OAs participated in thousands of patrols and hundreds of missions in Iraq, Afghanistan, and elsewhere. Such engagement, coupled with their prior experience, allowed them to provide objective, unfiltered feedback to commanders from company to division that facilitated improvements in the most critical area of all: training.[47]

If deploying to Afghanistan or Iraq as an OA represented the "glamor" mission, perhaps more important was the pre-deployment support OAs provided to units getting ready for combat. This support focused explicitly on the threats present or emerging in a unit's pending area of operations with those related to IEDs remaining an area of interest. Focusing reconnaissance and data collection activities in this manner facilitated both tailored training and the development of standing operating procedures and refined battle drills by the small units slated to operate in those areas. Concurrent with this, it presented openings for realistic multi-echeloned training rooted in the latest (and most effective) battlefield methods. Affording every echelon and unit involved in a simulated attack in training before deployment presented opportunities to rehearse and refine responses. As an example, confronting an IED or sniper situation offered a training squad the opportunity to execute its battle drill, which included notifying its higher headquarters of the incident and requesting medical evacuation as required. Reporting to higher headquarters allowed both the intelligence and operations sections the opportunity to update situation maps, ascertain additional meanings of such an attack, and forward that information to adjacent and supporting units. This also facilitated practicing techniques of the incident investigation model: properly executing such actions was critically important in identifying other threats.[48]

Such integrated training meant the unit providing medical evacuation learned to plan and execute that function while the fire support element could call in artillery fire as appropriate. The functional benefit of such a training strategy laid in its compressing time. Units gained the benefit of in-theater experience as they underwent mission rehearsal exercises at the Army's various combat training centers. By the time a unit arrived in the combat zone, its soldiers knew the threat, and how to respond to it, and had rehearsed contingency plans, often with the very units they would work with on the battlefield, under a variety of conditions. While this would not eliminate the fear associated with a sniper or IED attack, it did mean better responses by leaders and soldiers alike, with this helping to both suppress attacks and reduce their effectiveness.

Even as the AWG sent its first members overseas, the unit received missions allowing it to demonstrate its ability to solve problems for the larger Army. Throughout the fiscal year 2007 (October 1, 2006, through September 30, 2007), the AWG unveiled several programs, products, or recommendations either in response to requests for assistance, to taskings from the Army G-3/5/7, or most tellingly from observations made by deployed OAs. These included providing the framework for an organic intelligence cell at the maneuver company level, training methods of personality targeting, instruction on the methods of tactical site exploitation, developing a Combat Applications Training Course (CATC), and providing solutions to the increasing sniper problem. Two of these—CATC and providing solutions to the sniper problem warrant exploration as they demonstrate how the AWG executed its mission.[49]

The CATC originated with two of AWG's contract employees—Morgan Darwin and Scott Flanagan. Hired by the Wexford Group, both possessed extensive shooting experience. They also had strong training backgrounds and recognized marksmanship training, correctly done, could not only improve technical skills but also training ones. Excellent training skills translated into a host of both tangible and intangible benefits for the Army's tactical formations. CATC reflected two hallmarks of Special Operations: awareness of one's self and awareness of the environment. From this foundation, CATC taught situational awareness and how to improve it, and the process of purposeful action. Understanding and integrating these principles allowed leaders to learn how to train in new ways. A believer in this approach was Maj. Gen. Jim Schwitters, commanding general of the Army's largest Basic Combat Training (BCT) installation, Fort Jackson, South Carolina. The AWG had put the CATC "on the road" around the Army, and Schwitters realized it could help him solve several challenges. [50]

Schwitters was an anomaly within the Regular Army: a former enlisted soldier who neither went back to college as an ROTC cadet nor attended Officer Candidate School. Instead, he received an appointment to second lieutenant directly from the ranks. Like so many involved with the AWG, Scwhitters had an extensive Special Operations background, but that was not what drew him to the CATC. Instead, he drew his motivation from his prior experience at Fort Jackson as a BCT brigade commander and his observations after serving a year as the commanding general. Schwitters was "very fascinated with the business of transforming people and acculturating them to the profession," and also very frustrated with the basic model for BCT, which he described as "formulaic, dumbed down to the lowest common denominator, risk-averse, and hidebound."[51] For Schwitters, CATC allowed recasting the BCT mold. Schwitters believed BCT should not just produce soldiers who could meet the minimum standards. Instead, the experience should create soldiers possessing both a soldier's heart and mind as well as the ability to think critically and creatively, function as part of a team, and most importantly—think for themselves.[52]

For Schwitters, CATC created opportunities for his drill sergeants—the non-commissioned officers directly responsible for transforming the raw recruit into a soldier—to reimagine their basic understanding of training. While Schwitters recalled that many drill sergeants told him, "They'd never had more fun training, the truly smart ones talked about using CATC as a vehicle and adapting it to train others skills by applying its basic principles." [53] CATC used marksmanship as a vehicle to make a significant point: one could train routine tasks differently to realize different objectives or characteristics within the training audience. With the conflicts in Afghanistan and Iraq returning the lowest-level non-commissioned officers to positions of battlefield prominence, a training methodology giving them the battlefield leadership skills of more experienced or mature soldiers was highly desirable. Unfortunately, the benefits of CATC reached only a few as the Army did not incorporate it into the basic combat training (BCT) syllabus. The challenges of resourcing such training and the need to train a broad range of skills in the short amount of time allotted for BCT meant wide-ranging change proved infeasible.

The counter-sniper mission represented a more encompassing challenge that involved deployed OAs and the Concepts Integration Squadron. The AWG's efforts in providing solutions to this problem present an example of how the unit addressed both specific DTLMOPF elements and deployed and non-deployed units. The Army's underlying problem was its lack of a standard against which it could either train personnel or develop equipment. General Shinseki's legacy force remained organized around the requirements of Air-

Land Battle, an approach presumed to limit the use of, or exposure to, snipers. Despite the use of snipers during the Balkans conflict, the Army did not place great emphasis on such a threat. While the Army established a sniper course at Fort Benning, Georgia, in 1986 and followed with a National Guard course at Camp Robinson, Arkansas, in 1993, these schools focused on training shooters, not on training the operational Army to eliminate them. As late as 1979, the Army had a formal doctrine for countering this threat, but it had long passed into oblivion. By 2003, the Army was starting to address the threat, but it downplayed its seriousness with publications noting "little evidence of well-trained and qualified snipers" and the use of "shoot and spray" tactics.[54] The spike ineffective enemy sniper use forced the Army to reconsider its position and provided the AWG its first opportunity to demonstrate its full potential. The AWG operated on two distinct lines in executing this mission. One focused on doctrine and training, while the second focused on materiel. Integrating both lines would drive changes in leader development.[55]

Doctrine and training focused on what units were doing—or, more importantly, not doing—concerning snipers. An early effort in improving unit responses to snipers was the June 2006 symposium hosted by the AWG. Attended by approximately 40 former or serving snipers, the symposium sought to harvest years of thinking and action by individuals repeatedly performing this mission. The insights gained informed both how to train units and provided additional material considerations. Adding to this was the information OAs fed back to the Concepts Integration Squadron. [56]

The fundamental problem was that most soldiers generally did not understand how skilled snipers functioned, nor did they understand that simple actions could reduce vulnerability. Using a variety of videos, including those produced as propaganda by insurgent groups, AWG members trained soldiers to think about the sniper threat as they prepared positions or engaged in patrols. Training meant understanding the geometry of the fight—the time and space relationships between elements on the battlefield—and the immediate actions required upon detecting a sniper. Units that gained a better understanding of sniper operations changed their behaviors to both minimize exposure times and concurrently present a more difficult target array for a sniper to attack. In addition to developing a training program using videos, the Concepts Integration Squadron also produced a field guide, akin to a graphic training aid, for soldier use. Ultimately producing 140,000 copies, the aid emphasized a key truth: the most important thing was to kill the enemy sniper. It also made a second point: the need to immediately find concealment to deny a sniper the time needed to acquire and engage a target. This counter-sniper doctrine was the AWG's first mass-produced training support product. While its blunt language generated concerns in some circles within the Army (it expressly stated "Kill

Snipers"), the information provided proved useful to units in Afghanistan and Iraq. The field guide also provided an additional benefit—it introduced the AWG to those brigade combat teams that had yet to receive any OAs. Not intended as a marketing or branding aid, the guide nevertheless presented the AWG and its mission to the larger Army.[57]

Deployed AWG members also performed other tasks. Command Sgt. Maj. Birch made it a point to train by example by personally participating in combat patrols with the units supported by OAs. With skills honed by years of real-world experience, Birch both knew how to read the environment and concurrently train others to do likewise. He also knew another battlefield truth—that leadership required full engagement and could not occur in a forward operating base or combat outpost. It required a leader presence outside the wire. While Birch never occupied an OA position, he nevertheless periodically provided a graphic example to those who needed reminding of that fact. To the extent Birch retained any regrets over his AWG service, it was in his having to impart that lesson to more than one commander or senior non-commissioned officer who had not grasped that simple truth.[58]

The material line led to a new device, the Fast Obscurant Grenade, better known as the "FOG grenade," as a counter to snipers. The Army's inventory of munitions and related devices included a variety of smoke pots or smoke grenades intended to screen or conceal a friendly force from enemy observation. There was, however, a fundamental problem with all these devices. They all required some amount of time to generate enough smoke to cloak their users. Part of the problem lay in the fuses used to initiate the device. The interval between the moment a soldier pulled the safety pin and the device began dispensing smoke could run from two to five seconds. Once activated, the grenade would dispense smoke for roughly one to two and one-half minutes, depending on the specific model. The "quality" of the smoke—how dense, how stable, and how large the area covered—varied by both grenade type and atmospheric conditions. The latter included wind speed and direction, air temperature, and relative humidity. For a unit employing smoke to mask its movement and possessing the ability to wait for the cloud to form, the delay was not an issue: they could delay action until the cloud fully developed. For a unit trying to employ smoke in response to a sniper, the grenade's vagaries could result in additional casualties.[59]

Assigned to the Concepts Integration Squadron, retired Sergeant Major George "Eric" Zipperer recalled developments undertaken at Fort Bragg to produce a fast-acting screening grenade while assigned to the John F. Kennedy Special Warfare Center as a combat developer. The FOG grenade capitalized on a fast-acting fuze to reduce the time between the soldier pulling the pin and the grenade activating and a different compound, titanium dioxide, to produce the smoke.

The result was a grenade that activated in 0.5 seconds and created a dense cloud, described by many as an exploding fire extinguisher. This grenade offered near immediate concealment and better protection, especially for the soldier moving to aid a sniper's victim. [60]

When it came to material, the AWG enjoyed another unique capability— the authority to "buy, try, and decide" on new equipment items. Cody granted the AWG the ability to buy low numbers of items that analysis indicated would have battlefield benefit. Once bought, the AWG could try them under a broad range of conditions either at Fort A.P. Hill or another testing/training site. After carefully considering the item, including how it compared to similar equipment, it could coordinate with the Rapid Equipping Force for procurement. This ability meant that instead of having to work with combat developers at one of TRADOC's service schools, the AWG could (within limits) identify items, determine their utility, and then initiate the process of fielding such items to the larger Army. The introduction of new camouflage systems and effective ballistic shields for vehicles (an outgrowth of the counter-sniper mission) reflected early AWG experiments. When coupled with the FOG grenade, they provided units with a variety of passive and active means of negating the sniper threat. The AWG also considered items enhancing soldier lethality. Examples include the Trijicon M150 rifle scope and adapters for the Army's M68 close combat optic. These provided soldiers with two greatly improved capabilities. The first was the ability to engage targets at much longer ranges precisely. The second was the ability to accurately engage targets in densely populated areas under a variety of lighting conditions. The improved M68 denied the enemy the safety of hiding in the shadows.[61]

Nevertheless, "buy, try, decide" represented the proverbial slippery slope. Theoretically, the unit could squander enormous amounts of money on esoteric, "sexy," or questionable equipment. The combination of mature leadership, close coordination with the Rapid Equipping Force, and AAWO oversight instead demonstrated the soundness of this approach. As executed, "buy, try, decide" significantly reduced equipment procurement time and delivered materiel that nested with elements of doctrine, training, leader development, organization, personnel, and facilities.

Heraldry

Unlike its sister services, the Army embraces unit distinctions in both service and dress uniforms. Alone among the services, its dress uniforms bear distinctive unit insignia. Unit-specific insignia also adorn the service uniform. The Army maintains strict policies for developing and approving such insignia with the view towards ensuring such insignia properly represent the unit. Most often, such insignia show elements of a unit's branch assignment and elements of its history. Many of the shoulder sleeve insignia of the Army's divisions reflect their

World War I origins while the distinctive unit insignia of battalions or regiments display specific chapters in the unit's history.

For the AWG, however, there was no history to memorialize. Like the unit, the AWG's insignia would start with a blank slate, and ultimately the heraldic devices developed were unusual. Like so much of the AWG's early development, the path to obtaining authorized insignia included breaking some long-standing bureaucratic norms. While some might think the efforts trivial, Shaw recognized the importance attached to a shoulder sleeve insignia.[62]

The Army's final authority on insignia is the Institute of Heraldry (TIOH) at Fort Belvoir, Virginia. Subordinate to the Office of the Administrative Assistant to the Secretary of the Army, TIOH ensures all insignia conform to heraldic rules and Army policy. With an authorized strength of only 231, the AWG required an exception to the existing policy prohibiting distinctive unit insignia for units with an authorized strength of less than 250. While it would take several months to work the AWG developed ideas for insignia for the TIOH to validate. Foremost in Shaw's mind was insignia that both accurately conveyed the unit's purpose and provided a rallying symbol. The latter was particularly important as it would reinforce perceptions of the AWG as an effective Army asset even as it put more pressure on the unit to demonstrate its importance, utility, and value.[63]

The AWG's original proposal shoulder sleeve insignia called for a Blackfoot warrior on a black circle. The envisioned distinguishing unit insignia was a black circle defaced with a red horizontal arrow and the motto, "*Otatsiihtaissiistakio Pilski mako ta paawia,*" Blackfoot for "Normal is a Cycle on a Washing Machine," displayed around the insignia's edge. Command Sgt. Maj. Birch suggested the motto. Drawn from John P. McAfee's *Slow Walk in a Sad Rain,* Shaw and Birch, believed the motto accurately captured the unit's mission. The references to the Blackfoot paid proper homage to an individual who embodied the sort of soldier the group desired, retired Master Sergeant William "Chief" Carlson. Carlson's 20-year-long Army career included service with the 2nd Ranger Battalion, Special Forces, and SFOD-D. Retiring in 2003, he served with the Central Intelligence Agency's Special Operations Group before his death to enemy action in Afghanistan later that year. The memorialization of Carlson reflected the strong cultural influence of Special Forces within the AWG. Further discussion among the AWG's senior leadership led to a change in the shoulder sleeve insignia design from a Blackfoot warrior to the red arrow originally envisioned for the distinguishing unit insignia. The revised design remained rich in symbolism. The black circle symbolized both the AWG's global mission (the shape) and the unknown nature of emerging asymmetric threats (the color). The red arrow connoted the transition of the battlefield from two to three dimensions and concurrently alluded to the Indian Wars in which the Army gained early experience with asymmetric threats. Additionally, the direction—

always facing forward—portrayed forward thinking, vision, and action while the color connoted aggressive action. Not stated in the official description was another reference to Special Forces whose branch insignia consisted of two crossed arrows. Demonstrating the dark sense of humor common among experienced combat arms soldiers, the patch acquired the nickname "the shot in the dark."[64]

The Institute of Heraldry recommended a change to the original unit insignia since the AWG's request to wear a distinctive patch was winding its way through the Pentagon's halls. While TIOH recommended disapproving a new insignia in keeping with set policy, it continued working with the AWG to ensure that if the Army granted the exception, the items authorized met existing heraldic standards. Rather than the red arrow, TIOH proposed the Blackfoot pictographic rendering of the coyote's paw print. The pawprint displayed three unique elements. First, it alluded to some traits supposedly common to the coyote and AWG members, including cunning and resourcefulness. Within Blackfoot tradition, the coyote often served as a trainer or instructor, sometimes tricking the student into the lesson and teaching him that things are not always as they seem. This element of Indian folklore neatly paralleled elements of the O&O and emphasized the AWG's training role. Second, it again tied the legacy of Master Sergeant Carlson and his example to the AWG's aspirations. Finally, it specifically paid tribute to General Schoomaker, whose nickname was "Coyote."[65]

The Institute of Heraldry's case against granting the AWG distinctive insignia was rooted in Army Regulation 670-1, *Wear and Appearance of Army Uniforms and Insignia.* Governing all aspects of the Army's various uniforms, the regulation sought to prevent small units, particularly provisional ones or those not envisioned as remaining in the force structure for an extended period, from receiving such a privilege. The regulation represented something of a sound barrier against the unnecessary expenditure of government resources. The AWG's case *for* distinctive insignia emphasized the unit's permanency, the rigorous process that applicants had to complete for assignment to the unit, the morale boost and cohesion provided by such insignia, and the awarding of additional skill identifiers (ASI) to the unit personnel. Every soldier possessed a specialty—armored crewmember, imagery analyst, combat engineer, et cetera—but not every soldier possessed an ASI. The ASI further identified soldiers with additional specialized training or skills and facilitated better management of human resources at the higher levels of the Army. Despite TIOH's objections, the Deputy Chief of Staff for Personnel granted the AWG's request on August 29, 2006.[66]

The authority to wear the insignia was significant. Even more so was what approval represented: another victory over bureaucratic norms—and formal recognition of the underlying cultural norms Shaw and Birch sought to instill within the unit.

Conclusions

In 24 months, the AWG racked up an impressive string of successes—both in its efforts to negotiate the Army bureaucracy and in the conduct of tactical operations with various Army units. As a practical matter, the command appears to have executed two simultaneous efforts. The first was the direct engagement of those elements of the Army bureaucracy positioned to say "yes" or "no" to AWG efforts. The second was focusing resources to achieve and demonstrate the AWG had realized initial operational capability.

The AWG derived considerable benefits from having two senior leaders willing to serve as champions—General Schoomaker and General Cody. While their authority was neither limitless nor absolute, many within the bureaucracy, particularly the action officers (colonels, lieutenant colonels or majors on assignment to various Army staff offices) were more than willing to take the steps required to realize the aspirations of these two leaders. Nevertheless, there remained other individuals with different perspectives who neither saw the need for new structures such as the AWG or who believed the bureaucracy's requirements were sacrosanct and that satisfying its needs took precedence over realizing change. The latter were not necessarily hostile to change itself but believed that change required adherence to existing processes. This insistence on existing norms represented some of the stiffest resistance. Such recalcitrance illustrates a key point: it is not enough for a senior leader to want to change a military organization. He must possess the willingness to drive new modes through a bureaucracy organized to make rapid change difficult through processes optimized for the peacetime needs of recruiting, training, manning, and equipping the Army.

As with General Shinseki championing the interim armored vehicle, Schoomaker, and particularly Cody, undertook a conscious effort to ensure as much of the Army bureaucracy as possible supported the AWG and its counterpart organizations. On those occasions when the AWG could not negotiate the bureaucratic hurdles and employed the expedient of using Cody's name to compel action Cody supported those actions. Cody wanted the bureaucracy to work, but he did not want to prioritize bureaucratic efficiency over creating desperately needed organizations to prosecute the fight in which the Army found itself. Like taking money from other programs, dropping the Vice Chief of Staff's name did not win the AWG loyal supporters in parts of the Army. Nevertheless, it ensured the AWG obtained the resources and freedom of action required to accomplish the tasks Schoomaker and Cody assigned the unit.

Operationally, the command's efforts to achieve initial operational capability resulted in multiple successes. Perhaps the most celebrated early success came in countering the increasing problem of enemy snipers. The AWG used its resources to both train units and validate both the FOG grenade and how to employ it. Once the AWG accomplished this task, it coordinated with the REF to quickly secure a production contract to provide deployed units with the grenade. The FOG grenade demonstrated both the AWG's strength in solving tactical problems and the benefits of closely coordinating its efforts with the REF. As important as this was, however, it was just part of a more fundamental issue: determining if the AWG's O&O provided solutions to asymmetric problems.

Observers across the Army could look at the AWG by the end of fiscal year 2007 and realize the command achieved and then moved beyond initial operational capability. It was important the AWG obtained a home, but it was critical the AWG demonstrate the time, money, and bureaucratic effort committed to its development justified the expenditure. The effects of improved ballistic shields, the counter-sniper program, CATC, the maneuver company intelligence cell, and training personality targeting support offered conclusive proof of both the O&O's fundamental soundness and of the AWG's ability to deliver what the Army needed.

Equally important was establishing a functionally sound and reliable command. Shaw, Birch, and their subordinates successfully brought together the initial personnel cohort and established the policies and structures to sustain the recruiting and training of new members. This allowed the command to increase its reach across the Army. They also developed and validated the internal processes necessary for the unit's success. The benefits of high-skill, self-starting individuals to this process are important to recognize. It is crucial, though, to recognize the role of leadership in melding such individuals into a cohesive unit as opposed to independent clusters of individuals operating in an unsynchronized or unfocused manner. Anything but effective leadership would have produced markedly different outcomes.

Despite achieving initial operational capability, however, the AWG still faced challenges. The first was moving to full operational capability while the second was even more daunting—surviving as a separate command as the inevitable changes in Army leadership resulted in questions about both the command's long-term utility and its placement within the larger Army.

Chapter 6

Transitioning The Asymmetric Warfare Group

Think...Adapt...Anticipate
The second motto of the Asymmetric Warfare Group

Changes in Senior Army Leadership

The AWG owed its existence to the vision of the Army's most senior officers, but those officers rarely remained in place for more than four years. Gen. Peter J. Schoomaker relinquished the duties of Chief of Staff of the United States Army on April 10, 2007. Replacing him was Gen. George W. Casey, a career Infantry officer. Where Special Operations defined Schoomaker's career, the conventional Army defined Casey's. Commissioned in 1970 through the Army Reserve Officers' Training Corps program at Georgetown University, his initial assignment was to the 1st Battalion, 509th Airborne Infantry, then stationed in Europe. Returning to the United States, he served with the 4th Infantry Division (Mechanized) as a mechanized infantry battalion staff officer and as the commander of both a mechanized infantry and combat support company. After a tour of duty with the United States Military Observer Group Israel and Egypt, he returned to the 4th Infantry Division, where he served as a mechanized infantry battalion executive officer and battalion commander. He then served as the Chief of Staff 1st, Division, and then commanded the division's 3rd Brigade. Promoted to brigadier general, he served as the V Corps Chief of Staff, the Assistant Division Commander of the 1st Armored Division, and then assumed division command in July 1999. Following assignments with the Joint Chiefs of Staff and Joint Forces Command, he served as the Vice Chief of Staff of the United States Army before assuming command of the Multi-National Force in Iraq. Casey knew the Army, and most knowledgeable observers identified him with its heavy conventional force. More valuable than Casey's background was his understanding of the state of the Army, and the areas in which he focused his efforts.[1]

The Army had been at war for almost six years when Casey assumed the duties of Chief of Staff. In the Army's 2008 Posture Statement to Congress, Casey noted the Army was "out of balance" and identified four policy imperatives—broadly termed "Sustain, Repair, Reset and Transform"—to correct this condition.[2] Casey focused on turning these imperatives into concrete outcomes. The routine supervision of individual units—and that included the AWG—remained the responsibility of subordinate commanders and staff officers. Casey recognized the AWG's benefits, but his preoccupation with policy meant he would not take an active role in the group's issues or activities. With the AWG representing a micro-fraction of the Army's strength and budget, and already functioning effectively, there seemed little need for him to take a different approach—especially with more pressing issues competing for his attention.[3]

More significant for the AWG than General Schoomaker's retirement was General Cody's relinquishing his post. Gen. Peter J. Chiarelli assumed the latter's duties on August 4, 2008. Cody's departure produced greater consequences than that of many of his predecessors. In creating the AWG, the Rapid Equipping Force, and the Improvised Explosive Devices Task Force, Cody manipulated the bureaucracy and upended many long-held assumptions and practices. Akin to a tornado, Cody had swept in, dramatically changed the landscape, and then blew out. His personal observation was that "many in the building" (the Pentagon) were "happy to see me go," and there is strong reason to believe his self-assessment was accurate.[4] Cody maintained a deep interest in the organizations he created from their inception and through the length of his tour. Despite the demands that being Vice Chief of Staff imposed, he diligently maintained the practice of conducting the JIEDDO-AWG-REF (JAAR) reviews. Bob Shaw noted the meetings were never pro forma, that Cody was engaged, and the time spent in preparing and conducting them was time well spent.[5]

Like Casey, General Chiarelli received his commission through ROTC. A career Armor officer, Chiarelli's assignment history included service with the 9th Infantry Division, the 3rd Armored Division, the 1st Cavalry Division, and the 2nd Infantry Division. He also taught history at the U.S. Military Academy; acted as the executive officer to the Commanding General, Supreme Allied Commander, Europe; and functioned on the Army G-3/5/7 staff. He commanded both the 1st Cavalry Division and the Multi-National Corps—Iraq from 2003 to 2006. Chiarelli was intelligent, well-educated, and committed to the Army and its soldiers. Indeed, a hallmark of his tour as Vice Chief of Staff was the "war" he waged against the barriers to its people seeking assistance for post-traumatic stress.[6]

Three words that could describe Chiarelli's career were "not Special Forces." Some Special Forces officers, however, perceived him as possessing a marked antipathy for their branch, its units, and those entities closely associated with them—including the AWG. In contrast, some close subordinates would dispute that view, noting his support for several Special Operations Forces initiatives during his tour of duty. An old Army saying holds the view is different from every foxhole, and Chiarelli's case validates that adage. What one thought of him was where he was relative to one's foxhole—and the view from there.[7]

Another key figure involved in creating and shepherding the AWG, Lt. Gen. James J. Lovelace, turned over the duties of Army G-3/5/7 to then-Lt. Gen. James D. Thurman, in 2007. Lovelace enjoyed a strong relationship with Cody and understood the importance of ensuring the AWG's growth and survival. With Lovelace's move to command of the Third U.S. Army/Army Forces U.S. Central Command, one of the controls that Cody placed within the bureaucracy fell by the wayside. [8]

Thurman's career track differed markedly from both Cody's and Lovelace's. Like Chiarelli, he was originally commissioned as an Armor officer. After completing several assignments as an Armor officer, he attended flight school. He served as an aviator with both the 82nd Airborne Division at Fort Bragg and along the East German border with the 11th Armored Cavalry Regiment. His combat experience included both the Persian Gulf War and Operation Iraqi Freedom. During Iraqi Freedom, he first served as the operations officer of the Coalition Combined Land Forces Command. He then returned to the theater as the commander of the 4th Infantry Division/Multi-National Division, Baghdad. He also commanded one of the Army's premier training establishments, the National Training Center at Fort Irwin, California. Additionally, he had prior duty assignments within the Army G-3, including that of the director, the Army Aviation Task Force. He understood how the Army's bureaucracy functioned and his credentials as a soldier and leader were unquestioned. Thurman believed the AWG provided the Army capabilities needed as it fought a counterinsurgency, but he also saw the AWG as representative of a larger institutional problem—that of too many organizations reporting to the G-3/5/7. Thurman believed the AWG required assignment to an actual command to resolve its manning, training, and equipping issues. So long as the unit remained under the G-3/5/7, it would not develop the processes necessary to meet these requirements over the long term and would instead remain reliant on individuals championing its needs. The sooner the Army could place the AWG in an existing command the better it would be for all concerned.[9]

The Asymmetric Warfare Group's First Change of Command

By March 2009, Bob Shaw had served 36 months commanding the AWG, well beyond the 24-month standard tour length. As a matter of policy, the Army sought to execute changes of command for units of battalion size or larger during the summer months. This policy meant Shaw would serve an additional four months as the AWG commander before turning command over to his replacement, Col. James M. Mis. On the surprisingly cool and cloudy morning of July 14, 2009, Shaw relinquished command to his successor. As Mis took possession of the unit's colors, he had no way of knowing the clouds overhead represented the perfect metaphor for the position he was assuming. In looking back on his tour,

Shaw could enumerate a long series of successes both in the combat theaters and in the United States. His greatest accomplishment, however, was establishing a solid foundation for the AWG. More than anything, this would ensure both its battlefield utility and its longer-term survival as a unit.[10]

Like Shaw, Mis was a Special Forces officer. Also first commissioned as an Infantry officer, Mis served with both the 101st Airborne Division and the 1st Battalion, 75th Infantry Regiment (Ranger). His operational experience included the Sinai, Panama, Haiti, Bosnia, Colombia, Paraguay, and Afghanistan. He was also one of a small number of Army officers entitled to wear the small bronze star on his parachutist badge designating a combat parachute assault. His education included master's degrees from both the Naval War College and the Marine Corps War College. Like Shaw, Mis possessed the unique tactical and educational qualifications required to command the AWG successfully. Mis had an additional qualification for the AWG: he had previously commanded Fort A. P. Hill, the focal point of AWG training and experimentation. As garrison commander, Mis' duties included supporting the AWG. This provided an opportunity to develop relationships with unit members and concurrently observe both aspects of AWG training and work done to support operational units. From the time of the initial discussions surrounding creating an Asymmetric Warfare Regiment, Mis was something of an engaged observer.[11]

A break occurred between Mis' tenure at Fort A. P. Hill and his taking charge of the AWG when he first attended the Marine Corps War College and then served with Special Operations Command (South), better known as "SOCSOUTH." His master's thesis focused on the nexus of military operations and law enforcement in counter-terrorism/counterinsurgency operations. At SOCSOUTH, he served as both the Chief of Joint Operations and as the Director of Operations. These assignments provided both a year of focused academic study and research centered on strategic thought and additional operational experience within a combatant command's component staff. Together, these tours provided a clear understanding of both ongoing special and conventional operations as well as the opportunity to see where gaps existed between them—gaps that the AWG could help close.[12]

Mis possessed an additional advantage when he took charge of the AWG: an experienced Command Sergeant Major in Ray Devens. Enlisting on a whim in 1982, Devens spent much of his career with the Ranger Regiment before moving to the AWG. His first combat experience occurred in 1983 in Grenada, where he served as a heavy weapons infantryman. Following an assignment to the Ranger Regiment's Long Range Surveillance Detachment, he served in an Airborne surveillance detachment in Germany, and as an advisor to the Alaska Army National Guard. Devens subsequently returned to the Ranger Regiment, where he served as a battalion Command Sergeant Major before joining the

AWG. Significantly, he became the first sergeant major of the AWG's Able Squadron, one of the two providing operational advisors. Consequently, Devens was intimately familiar with the mission, the personalities, and the issues surrounding the unit before he assumed the duties of Command Sergeant Major in 2007. His long service with the unit, coupled with his having served for several years as a squadron Command Sergeant Major, provided a level of continuity infrequently found in the Army.[13]

In assessing his new command, Mis quickly came to several conclusions. First, Shaw had done an outstanding job of setting the unit up for success. In terms of task organization, structure, and basic operational concepts, there was no need for any immediate major change—even if some in the organization possessed a differing perspective.[14]

Lieutenant Colonel Blakeslee Connors spent seven years with the AWG, with his tenure culminating in the command of Charlie Squadron (Selection and Assessment). In his view, a structural weakness existed in the form of the number of staff positions authorized. The absence of positions at the squadron level reflected the decision to centralize staff functions as much as possible at the group level. The advantage of this resided in a reduced requirement for personnel, important during a time when the Army found itself critically short of commissioned officers. This shortage meant making hard decisions about faces and spaces. Implicit in this was recognizing neither all faces nor all spaces were equal, and thus the service needed to consider how it distributed its captains and majors carefully. The disadvantage laid in the challenge for the group staff to balance requirements, resources (equipment, money, and personnel), and time. In an organization as dynamic as the AWG, this represented an ongoing challenge. The group's staff routinely demonstrated the capacity to meet this challenge, but it represented an issue that remained through Mis' tenure as commander.[15]

In evaluating the AWG's performance, it was clear that despite maintaining a minimal physical footprint, the group had not only "touched" many units in Afghanistan, Iraq, and the training centers in the United States, but that these contacts encouraged improved battlefield performance. As the Army sought to assess corporate performance during this period, two theoretical concepts came to the fore: measures of effectiveness and measures of performance. The former sought to determine if units were doing the right things while the latter sought to determine if it was doing things right. Changes in enemy methods and friendly casualty rates following AWG engagement showed the group both properly prioritized its efforts and that they resonated within the Army.[16]

Second, Mis realized misperceptions concerning the AWG existed across the Army. These misperceptions served to obscure the unit's real capabilities. The AWG's unique placement under the Army Staff, its designation as a Special

Missions Unit for manning, granting certain exceptions to policy, the selection course, and the high percentage of Special Operations personnel within the unit contributed to these skewed views. Some outside the AWG believed the unit's designation as the "Asymmetric Warfare Group" was a cover name that allowed the unit to perform missions not covered in the operational and organizational concept. Even within the group itself, there remained a bit of confusion stemming from the Special Mission Unit designation as some struggled to make the distinction between such status for personnel management rather than operational purposes. The important, if seemingly unglamorous truth, was that the AWG was exactly what it said it was. There were no secret missions or agendas, only soldiers and civilians seeking to improve the Army's response to emerging or evolving threats. For Mis, an important task was rebranding the unit to allow the larger Army to understand its purpose and capabilities more readily.[17]

Finally, the combination of the unusual circumstances surrounding the AWG's creation and the high operational tempo—the continuous deployment/redeployment of OAs to both the combat theater and the training centers—meant the unit's focus had shifted a bit from the original charter. The shift was not dramatic but was enough to convince Mis that some adjustments were in order. A refocusing would enable the unit to better function in a role Mis considered vitally important: that of a customer-focused organization with Army or joint force commanders representing the customer.[18]

For Mis, the issues of external confusion about the unit and the need to internally refocus the command had an elegantly simple solution. All that was required was a rebranding of the AWG with the change directed outward. "Normal is a cycle on a washing machine" ably communicated the unit's aspirations and environment within the unit, but failed to resonate with the larger Army properly. Frankly, if it had, there would be no need for the AWG as the Army's combat formations would change organically in response to changing threats. A change in brand that unambiguously communicated the AWG's primary purpose would remove much of the confusion surrounding the unit. It would also reaffirm to all the unit's members their fundamental role within the Army's force structure.[19]

One way the AWG advertised its existence and mission was through its OA and the training materials it produced. "Normal is a cycle on a washing machine" remained as the unit's traditional motto, but the words "Think. Adapt. Anticipate." soon appeared on AWG products as Mis undertook the rebranding effort. Conceptually simpler than the old motto, the new one more clearly communicated to the *external* audience whose connections Mis saw as critical to fulfilling the unit's original charter. As events unfolded under Mis' command, external perceptions would prove crucial.[20]

Group Operations and Activities

When Mis replaced Shaw at the head of the AWG, that change of command caused a few ripples within the unit. Operations continued uninterrupted, and changes within the group remained mostly invisible to the units the AWG supported. This perception demonstrated two significant points. First, the group achieved sufficient maturity to transition between leaders without difficulty. Second, units receiving support from the AWG did not experience disruptions in that support. Despite some personnel shortages, the ability of operational advisors to continue operations without difficulty showed the operational Army that the AWG remained a fully functioning unit.[21]

The monthly JAAR reviews with the Vice Chief of Staff continued, but these had taken a very different tone with General Chiarelli replacing General Cody. In his first JAAR with Chiarelli, Bob Shaw noticed a fundamental change in the tone and outcomes of these sessions: Chiarelli did *not* view the AWG in the same way Cody did. Mis never participated in a JAAR with Cody, but like Shaw, he found the Chiarelli reviews mostly ineffective and, at times, even counter-productive.[22]

Arguably, these meetings reflected the dynamics of the Army's almost "tribal" structure. In the wake of September 11, 2001, soldiers assigned to the Military District of Washington transitioned from wearing a service dress uniform to the battle dress uniform (BDU), and then to the BDU's replacement, the Army Combat Uniform (ACU). Soldiers wore boots rather than dress shoes with such uniforms. While the Army issued one type of boot to its recruits, soldiers wore a variety of types that connoted membership in a given community. Those who heavily identified with the Army's airborne forces wore "jump boots." Modeled on the specialized footwear worn by World War II paratroopers, these boots instantly identified the wearer as an airborne soldier. Members of the Armor branch often wore "tanker boots." This footgear came in several styles, but all had straps rather than laces. Derided by many outside of Armor (such footwear was unsuitable for forced marching, parachute jumping, or other dismounted activities), many officers within the branch wore them as a symbol of branch membership. As the Army ended the practice of wearing branch insignia on the ACU blouse, boots served as an identification badge—one Chiarelli proudly wore. The boots, in essence, stated: "I am from the heart of the conventional Army—the tank force."[23]

In contrast, Mis appeared in a uniform bearing the distinctive trappings of a Special Forces officer. Many viewed such uniforms as communicating the message, "I am NOT from the conventional Army—and I decline to do the things the conventional Army way." The reality was that Special Forces came from the Army's mainstream, attended the same professional schools, understood the conventional Army, and most acted accordingly in their dealings with it.[24]

Reality and perception, however, did not always coexist. From the very beginning, the cultural differences between Chiarelli and Mis made themselves apparent, and these would burden the AWG-Army staff-Vice Chief of Staff relationship with the stresses most prevalent during the monthly JAAR review. The first meeting between Chiarelli and Mis generated tension early on when the former observed: "You SOF guys don't like it when you get taskings from Armor guys."[25] That sent a clear message to Mis: Chiarelli perceived the AWG as a Special Forces unit, rather than as a valuable conduit chartered to support the conventional Army. The irony here lay in Mis' belief that his unit existed to serve the conventional Army. It was not a Special Mission Unit or a Special Forces Unit: it was an *ARMY* unit.[26]

For the AWG's advisors, the tensions between the headquarters and the Army staff were of little immediate consequence. Support to the Army's deployed units continued uninterrupted. While the AWG rightly recognized its mandate extended far beyond IEDs, insurgent use of such devices meant they continued as a focus area. The unit also recognized that asymmetric threats were not only the province of the enemy—sometimes the Army itself did things benefitting the enemy. Three efforts conducted early in Mis' command demonstrate the broad range of AWG activities and the group's ability to consider DOTMLPF elements in executing the mission. Two of these initiatives are particularly noteworthy—the Soldier Load Program (SLP) and the evaluation of the Army's camouflage uniforms and their battlefield impact, which demonstrated the AWG's ability to identify self-generated threats within the Army. The third, countering the Al-Sadid air defense system, showcased the AWG's ability to develop comprehensive solutions to threats posed to both ground and air units.[27]

The SLP showed fundamental changes in how soldiers equipped themselves for battle. In *The Soldier's Load and the Mobility of a Nation,* Brigadier General S.L.A. Marshall observed the modern American soldier "is as heavily burdened as the soldier of 1000 years B.C."[28] Marshall's work referenced campaigns ranging from ancient Greece to World War II. He noted that the GIs who came ashore on Normandy's beaches in June 1944 bore an average of 80 pounds of equipment, ammunition, clothing, and rations. The infantryman of Korea frequently saw his load reduced by the Korean Service Corps and its 75,000 laborers—the vast majority of whom spent their days carrying required supplies up the peninsula's steep mountains in return for clothing, food, lodging, and a small salary. Vietnam saw a return to loads of approximately 80 pounds for combat soldiers—sometimes more depending on the soldier's weapon or duty position. Carrying such heavy loads reduced soldier effectiveness, but the load seemed an unsolvable problem. Reducing equipment weight only meant the soldier carried more water or ammunition.

The fundamental dynamics of the problem remained unchanged as the Army entered the Global War on Terrorism.[29]

For soldiers in Afghanistan, the problem of weight only increased. New equipment, such as night-vision goggles and ground positioning system receivers, also came with heavy batteries. The bigger problem, however, lay in the part of the Army's response to snipers and IEDs: body armor. From its introduction late in World War II, body armor continuously evolved. The Korean War saw World War II's 12-pound aluminum plates give way to laminates and nylon. In Vietnam, layers of ballistic nylon provided all protection in a vest weighing approximately eight pounds. By the Persian Gulf War, nine pounds of Kevlar replaced nylon. By the early 2000s, the latest thing in self-protection lay in ceramic plates sandwiched between Kevlar layers, a system known as improved body armor (IBA). This system offered much better protection and included Kevlar attachments to protect the groin and upper arms. All this protection came at a high cost in weight—the system of plates and carriers weighed between 16 and 33 pounds. The weight of the armor was in addition to the other items the soldier carried. In contrast to the heavily equipped infantryman, the Taliban insurgent carried little more than a weapon and ammunition. This disparity meant the insurgent enjoyed a tremendous advantage over his opponent in tactical mobility. The soldier might fly or ride to a point on the battlefield, but after exiting the aircraft or vehicle, he was a heavy, cumbersome actor enjoying little mobility. He could find the enemy, and he might fix the enemy, but the heavy loads he carried impeded every effort to finish off the enemy.[30]

Retired Sgt. Maj. Jose Gordon was one of the operational advisors tasked with considering the issue for the AWG. Unlike many other group members, Gordon did not have a Special Forces background. A non-commissioned officer who had spent most of his career with the Ranger regiment, he retired from the U.S. Military Academy's Department of Military Instruction in 2003. Gordon's active-duty experience included two combat parachute assaults—one in Grenada and one in Panama. He made his Grenada jump as a member of the Ranger's jump-clearing team. This specialized element-maintained responsibility for clearing an airfield of obstacles to facilitate the entry of follow on forces. Gordon's subsequent service with the 1st Ranger Battalion included duties as a master breacher and as an assault team leader. He was a leader who understood the challenges of managing the competing requirements of a soldier's load.[31]

To become better familiar with the dynamics of the load problem in Afghanistan, Gordon embedded with two different units: the 173rd Airborne Brigade and the 3rd Brigade, 1st Infantry Division. Gordon found the fundamental issue was one of leaders properly identifying and mitigating risk. In many portions of Afghanistan's Logar province where these two units operated, the terrain itself provided adequate protection, or cover, for soldiers moving about the area. This realization provided a crucial element in managing a soldier's load.[32]

Retraining commanders to reevaluate how they considered terrain allowed them to change a soldier's load. In eliminating the need for IBA, commanders significantly reduced load weights and restored a much higher level of tactical mobility and agility to their soldiers. The problem of the soldier's load illustrates several points. The first is that while the Army thought it addressed all the elements of DOTMLPF when it fielded IBA, it did not entirely do so. The second is that while commanders considered the terrain, they did not fully consider the relationship between the soldier, the equipment, and the environment. Third, it showed that commanders tended to view IBA as the optimum solution for soldier protection, even when it was not. In identifying the issue and solution, the AWG provided a straightforward solution to a vexing problem. Equally important, the solution (considering the terrain's impact) was one that leaders could apply in any tactical situation. Fourth, it demonstrated that the enemy did not present all asymmetries—the Army's solutions to its problems could sometimes generate them as well. Ironically, the technical solutions employed to improve soldier performance could impede it. Finally, it proved that not all solutions needed to be materiel-based: changes in thinking could create effective solutions.

The evaluation of camouflage stemmed from Department of the Army uniform decisions beginning in 1990 with the replacement of the then-seldom-seen six-color desert camouflage uniform issued to most soldiers deploying to the Persian Gulf during Operation DESERT SHIELD with a three-color version. This new pattern offered good concealment in the deserts of Southwest Asia but was unacceptable for wear in temperate or forested regions. The Army thus had two completely different camouflage uniforms—woodland and desert—and bore the costs of producing and maintaining them. To resolve this problem, the Army sought what it termed a "universal camouflage pattern" (UCP). Conceptually, this pattern would work equally well in forested, desert, and urban areas. Ideally, it would also improve concealment from image intensification devices. Such a uniform would allow the service to issue and maintain a single combat uniform series. After testing in 2003 and early 2004, the Army approved a UCP uniform later that year. It consisted of a "digital pattern" comprised of small squares of light sage and medium olive green over a khaki background. This represented a radical departure from prior Army camouflage. The UCP proved an effective pattern—in a very narrowly defined set of environments. For most of Afghanistan, Europe, and the tropics, it proved remarkably unsuitable. Rather than concealing the soldier, it tended to increase his/her visibility. In Afghanistan, this gave insurgents one of the battlefield's greatest gifts: time. Time to run, time to hide, or, time to fight. In contrast, American soldiers faced an opponent who lacked a uniform and who most often appeared no different from the many non-combatants found on the battlefield. Under such circumstances, time conveyed a significant tactical advantage. In addition to conveying a time advantage, the uniform also simplified target acquisition and engagement. The uniform just did not work over much of the country. In response to both complaints from soldiers and congressional pressure, the Army sought to determine the best camouflage pattern possible for wear in Afghanistan and similar environments. With members of the Program Executive Office-Soldier (the agency responsible for uniform design and procurement), the AWG identified camouflage elements best suited to Afghanistan's diverse conditions. This resulted in a new uniform that performed successfully not only in Afghanistan but also in many other regions.[33]

In August 2009, the AWG released an information paper for maneuver brigades detailing countermeasures for the Al-Sadid improvised anti-aircraft system. The highest casualty-producing events in Iraq stemmed from the loss of troop-carrying helicopters, most notably the large CH-47 Chinook and smaller UH-60 Blackhawk. Recognizing the need for such aircraft to operate in forward areas in the event of conflict against the Soviet Union, the Army invested millions of dollars in developing aircraft survivability equipment (ASE) and the tactics needed to protect against various hazards. Soviet air doctrine emphasized an integrated air defense combining radar-guided guns with infra-red guided missiles augmented by the heavy machine guns of tanks and armored personnel carriers. The doctrine also called for linking all air defense weapons to allow the massing of fires on targets either in transit or in battle positions (the locations from which attack helicopters would fire their guns and missiles).[34]

Army doctrine called for using ASE to "spoof" enemy missiles either by jamming controlling radars or creating false targets for the guidance systems. This doctrine also mandated the suppression of enemy air defense (SEAD). SEAD meant using field artillery, missiles, or electronic attack to destroy, dislocate, or degrade air defense weapons before Army aircraft entered their engagement ranges. A generation of Army aircrews and staff rehearsed these methods and most units were proficient in planning and executing SEAD.[35]

In contrast, Al-Sadid sought to engage Army aircraft not with missiles or guns, but with IEDs rigged to detonate when aircraft were most vulnerable—while they were about to land. Standard ASE did not affect IEDs. The placement of anti-aircraft weapons in densely populated areas meant SEAD was not possible without inflicting casualties that would undermine counterinsurgency objectives. This development meant the Army's well-rehearsed (and expensive) protective measures were ineffective. In identifying countermeasures that extended beyond the aircraft cockpit and included elements of the ground maneuver force, the AWG reduced the overall effectiveness of the Al-Sadid.[36]

All these actions took place early in Mis' command and had roots in AWG initiatives originated under Bob Shaw. Significantly, Mis both understood what was taking place and took virtually no action to change them. This decision did not demonstrate any disengagement on Mis' part. Instead, it demonstrated his satisfaction with the activities taking place. These products sent a signal to the AWG's membership: there would not be any "tactical pause" owing to changed leadership. Members of the command appreciated Mis' confidence in their ability to meet requirements. Communicating this message internally was as important as communicating it to the Army at large. Inevitably, a change of command brings about some changes, but the unit's demonstrated performance meant these need not be sweeping. Changes would take place, but they would illustrate new requirements rather than fundamentally new approaches.

The year 2010 brought about another leadership change in the AWG as Command Sgt. Maj. Devens relinquished his position to Command Sgt. Maj. Mike Akridge. Akridge was a career Special Forces soldier and among the first non-commissioned officers assigned to the AWG. He found the AWG appealing as it kept him in an operational assignment after completing a tour as a Special Forces "A Team" sergeant—the team's senior enlisted soldier. Having completed multiple deployments as an operational advisor as the AWG first established itself meant he possessed a first-hand understanding of both how units received AWG support and their concerns. His experiences as an operational advisor, coupled with his knowledge of the more unusual aspects of the AWG mission, gave him a firm understanding of the unit and how he could best support it as the senior enlisted soldier. [37]

As the year unfolded, an additional action reflected the AWG's ability to conduct operations at extended distances. Army leadership always envisioned the AWG having benefits extending beyond the Army and contributing to the efforts of the joint, interagency (all US military and other US government agencies) or combined forces. In 2009, the unit advised the Colombian military in its efforts against the *Fuerzas Armadas Revolucionarias de Colombia*, better known as the "FARC."[38]

One of several anti-government groups that had waged an ongoing insurgency against the Colombian government since the early 1960s, the FARC was the best organized and most active. Its successes forced the Colombian military and National Police to devote enormous resources to defeat it. Compounding the issue of insurgent movements were the drug cartels whose violence also compelled decisive government action. These requirements established a military culture receptive to asymmetric thinking and responses to the challenges faced. Arguably a leader in asymmetric thinking, Colombia nevertheless requested and received support from the AWG starting in 2010. Operational advisors assisted in analyzing a changing problem set. Instead of focusing on capabilities gaps as those were known and understood, the advisory examined the "seams" of the gaps—that nebulous area between clearly defined capabilities and deficiencies from which new threats emerged. Included in this were the conceptual lessons of the SLP and camouflage evaluation programs and how friendly actions could concurrently generate new problems or threats. Also included were lessons learned from the American experiences in Southwest Asia and elsewhere.[39]

On November 4, 2011, the Colombian military killed FARC military commander Guillermo Vargas, better known as Alfonso Cano. Credit for this action rightfully belongs to the Colombian Joint Special Operations Command. It is interesting, however, to note that after years of pursuing Cano, actual success followed the AWG-Colombian Army engagement.[40] A reasonable conclusion one can draw from this is the effort to assist the Colombian Joint Special Operations Command in thinking imparted problem-solving ideas.

Reassigning the AWG from the Army Staff to a Permanent Command

While the AWG demonstrated the soundness of its operational and organizational concept and its ability to perform its mission, the question of where to permanently place the organization within the Army remained open. With the transitions in senior Army leadership, not everyone who came into power believed in the need for the AWG as a separate unit and one of them possessed a great deal of clout within the Army—the Vice Chief of Staff, General Chiarelli. By September 2010, Chiarelli concluded the Army did not require the AWG as a stand-alone organization. This opinion led him to recommend the unit's deactivation to General Casey. Some AWG members and Special Forces officers, including Mis, believed Chiarelli disliked the AWG owing to its Special Operations origins and his incorrect assessment of the unit as a Special Operations organization. Instead, Chiarelli's attitudes towards the unit showed a far more sophisticated perspective. He believed in keeping the AWG's functions by transferring them to already existing commands. In theory, such an approach would retain the AWG's perceived essential elements—providing operational advisors and providing a broader understanding of threats across the Army—without the personnel, fiscal, administrative, and logistical burdens of a separate organization. Implicit in this was oversight by individuals with a more conventional perspective toward military operations. Chiarelli's assessment profoundly affected the AWG for several months by initiating a series of actions by the Army Staff, the AWG itself, and, most significantly, the Department of the Army Secretariat. For Jim Mis, the next few months of his command would be stormy ones.[41]

On September 20, 2010, General Casey ordered the AWG's deactivation. Over several days, the Army G-3/5/7 Operations staff developed a basic plan per Casey's decision and Chiarelli's guidance. The early briefing slides prepared to communicate this plan are important because they captured the four major issues associated with the unit, irrespective of its status as active or inactive. These issues were the unit's mission, its mission essential task list (METL), its construction budget, and its operational budget. The mission and METL's import lay in identifying the functions that the Army would have to migrate to other units. The construction budget was relevant since the Army had to decide if it needed to sustain those projects authorized but not yet started, and how it would either manage or repurpose completed projects, or those on the verge of completion. The operational budget was important because it included over $138,000,000 distributed among contracting, Operations and Maintenance—Army (OMA), and Overseas Contingency Operations (OCO) funds. There are strict restrictions on how the Army may spend its funds. To prevent improper expenditures, the Army identifies funds by their specific accounts. The Army could thus take AWG OMA or OCO funds and apply them against other authorized OMA or OCO requirements. As for contracting budgets, the Army's money managers took a close look at every dollar and its use. Indeed, it was Cody's decision to initially take OMA funds from other units to stand up the AWG that caused some of the early institutional resentment the AWG encountered.[42]

The METL identified seven discrete tasks, one of which (conduct command and control) was common to virtually all Army units, and six reflecting the AWG's unique operational and organizational concept and task organization. At the time, commanders developed their unit METL and did not prioritize tasks; thus, there is no significance in their listed order. The AWG METL included:

- Supporting force commanders by advising and assisting forces both in the United States and overseas to reduce vulnerabilities and close capability gaps.
- Deploying forces globally to assess threats and friendly interactions to identify offensive and defensive countermeasures to defeat threats.
- Identify, develop, and disseminate both material and non-material solutions for full spectrum (offensive, defensive, stability and civil support) operations.
- Assist in integrating material and non-material solutions for full spectrum operations, assist in Army innovation and adapting the operating force to be more agile, versatile, and lethal.
- Staff the unit through both the selection process and the modified personnel management system.[43]

If viewed considering the functions of the period's three major commands—Forces Command (FORSCOM), Training and Doctrine Command (TRADOC), and Army Materiel Command (AMC)—the majority of the METL lay within the TRADOC's primary responsibilities. The deployment function was a FORSCOM responsibility. Elements of the material solution tasks aligned with AMC's mission. Properly migrating these separate tasks while ensuring effective connections between executors would be an essential part of the deactivation plan. The senior officers who had experience with the AWG and who possessed first-hand knowledge of its contributions intuitively understood the fundamental problem with breaking the unit apart: the loss of the unity of effort that made the AWG a fully functional organization. As a unit, the AWG was more than just the sum of its parts and parceling its missions out amongst multiple commands would not provide the Army the same results.[44]

The deactivation timeline was aggressive. It called for completing the final plan by November 29, 2010. This meant identifying the different organizations that would theoretically assume the AWG's specific responsibilities and then coordinating with them to ensure they could accomplish the tasks required to ensure they continued uninterrupted. Most METL tasks displayed both a clear threat identification and training component. Did this mean the task should go to TRADOC and then the Intelligence Center of Excellence, or should it go to TRADOC and reside with the Combined Arms Center? Inevitably there are elements of friction in developing plans and that would be most evident in allocating AWG tasks. The plan also called for the unit's formal deactivation on July 1, 2011, with all tasks completed no later than September 30. The July 1 deactivation date coincided with the same day the contract with Wexford expired. This timeline ensured full capability through deactivation without incurring any costs associated with prematurely canceling Wexford's contract for operational advisors.[45]

For the AWG, it appeared the handwriting was on the wall. It would cause its colors and send them to a warehouse for safekeeping within a year. The unit would suspend recruiting while the Army would cease reassigning military personnel to the unit. Jim Mis would preside over the deactivation and then move on. Other organizations would take over the AWG's facilities.

Not every decision by the Army Chief of Staff, however, is a final one. Civilian control of the service means the Secretary of the Army, in this case, John M. McHugh, has the final say on many issues—and the activation or deactivation of units remains within the Secretary's purview. Two individuals, Gen. Martin Dempsey and Lieut. Gen. Daniel P. Bolger, not only recognized the AWG's value but advocated for its retention.[46]

Dempsey was a career Armor officer with a self-deprecating manner and a penchant for singing Irish ballads. When told *Time* magazine included him on its list of the world's 100 most influential people, he quipped he "wasn't even the most influential person in my house."[47] A 1974 graduate of West Point, his career reflected the Army's Cold War focus with repeated service in tank or armored cavalry units in Germany and the United States. His first combat tour was as the brigade executive officer of the 3rd Armored Division's 3rd Brigade during the Persian Gulf War. Later command assignments included the 2nd Battalion, 67th Armor; and the 3rd Armored Cavalry Regiment. He then commanded the 1st Armored Division; the Multi-National Security and Transition Command in Iraq; and briefly served as commanding general, of U.S. Central Command before assuming command of TRADOC in 2008. Dempsey was very much of the "Big 5" generation with its major components at the heart of most of his assignments. His strong conventional background did not, however, prevent him from appreciating the AWG and its contributions. Like Chiarelli, he recognized a fundamental problem—the unit's placement directly under an Army staff section.[48]

Bolger had replaced Lieut. Gen. Thurman as the Army G-3/5/7 in May 2010 with the former's assignment to command FORSCOM. Commissioned as an Infantry officer after graduating from the Citadel, Bolger served in both mechanized and air assault infantry divisions as well as on the faculty at West Point. He commanded the 1st Battalion, 327th Infantry, 101st Airborne Division (Air Assault), and then the 2nd Brigade of the 2nd Infantry Division in Korea. Additionally, he held the position of Chief of Staff of the 2nd Infantry Division; commanded the Joint Readiness Training Center at Fort Polk, Louisiana; and commanded the 1st Cavalry Division in Iraq. His professional experience was very much within the Army's mainstream. Holding a doctorate in history from the University of Chicago and the author of several books, Bolger possessed one of the Army's most incisive and insightful minds. He was a firm believer in the Army and its many strengths, but he was also unafraid to point out its shortcomings and he was unemotionally critical of poor decisions. His unquestioned intellectual honesty and encyclopedic knowledge of the service made him the ideal choice to replace Thurman. Bolger was also part of that cohort of division commanders whose personal experiences with the AWG made them believers in the unit, especially the operational advisors who accepted the risks of the common soldier.[49]

Through late 2010 and into early 2011, both Dempsey and Bolger made a case for moving the AWG from the direct supervision of the G-3/5/7 and its retention on the Army's rolls. Two arguments formed the core of the retention/reposition argument: the AWG's informing current combat efforts and its assistance in preparing the Army for future operations. Bolger's frequent meetings with both Secretary McHugh and Undersecretary of the Army Joseph Westphal provided periodic opportunities to reiterate the value offered by the AWG. Bolger ensured McHugh understood the AWG filled critical military skills gaps in the

Army's conventional forces—and did so at a low cost in both human resources and money. $138,000,000 is a large amount of money for most people, but to contextualize the amount, it equated to roughly the cost of five AH-64 Apache helicopters or nine M1A2 Abrams main battle tanks. Viewed through another lens, the AWG produced 37 different material or non-material solutions to asymmetric problems in fiscal year 2011 alone—and produced these in considerably less time than the standard TRADOC model. These results were in addition to the training support provided to units preparing for deployment. Through these actions, the unit created a cohort of advocates—division commanders who saw their soldiers benefit from the training and assistance that the unit provided.[50]

Secretary of the Army John McHugh came to office by way of Congress. A Republican serving a Democratic administration, McHugh grew up in the shadow of Fort Drum, New York, and had previously represented New York's 23rd Congressional District. His service in the House of Representatives included his co-chairing the Army Caucus. He also served 14 years as a member of the U.S. Military Academy Board of Visitors. These experiences gave him an understanding of the Army, and its leaders, and a desire to support tactical commanders. Dempsey's and Bolger's advocacy efforts prompted McHugh to visit the AWG himself so he could get a sense of the unit and what it meant in terms of Army resources. When informed by Mis of the AWG's cost, McHugh observed the amount represented "budget dust" when viewed as part of the total Army budget. Swayed by such arguments and the AWG's demonstrated value, McHugh directed its retention as an active unit.[51]

With the decision to retain the AWG as an active unit made, there remained the question of its final positioning. The original AWG concept called for making the unit subordinate to the 20th Support Command (20th SUPCOM). With the evolution of both the AWG and the 20th SUPCOM, however, this concept no longer made sense. The fundamental problem was the 20th SUPCOM's overarching mission and its misalignment with that of the AWG.[52]

Created in 2004, and subordinate to FORSCOM, the 20th SUPCOM is the Army's consequence management unit for weapons of mass effect. Known collectively as "CBRNE," these include chemical, biological, radiological, nuclear, and high-yield explosive devices. The potential use of such weapons by both state and non-state actors represents a unique problem in their ability to not only create significant casualties but also the capacity to render ineffective large portions of essential infrastructure. The Army had significant experience in working with chemical, biological, and radiological weapons, but this was within the context of a general war scenario in Europe or Korea. The migration of such weapons to non-state actors recast the paradigm for both employment and defense and drove the creation of the 20th SUPCOM.[53]

By 2009, the 20th SUPCOM's organization consisted of three Explosive Ordnance Disposal groups of three or four battalions, a separate ordnance company, a chemical brigade of four battalions, and an Analytical and Remediation Directorate including both a dedicated laboratory and aviation detachment. AWG's subordination to the 20th SUPCOM made sense if one considered the asymmetric threat as one limited to CBRNE weapons and their use. With an increasingly broad understanding of what constituted asymmetric threats, and the need to both advise units and create full-range DOTMLPF solutions to threats, it was clear that the AWG-20th SUPCOM relationship was fundamentally problematic. These problems would only increase as the 20th SUPCOM increasingly assumed a role in the homeland security mission.[54]

Assigning the AWG to the 20th SUPCOM accomplished the goal of moving it from the Army staff to a command. This same assignment, however, would likely degrade the effectiveness of both the 20th SUPCOM and the AWG as both organizations tried to execute and control operations that were mostly out of alignment. Equally important was the matter of unit funding. Subordination to the 20th SUPCOM would also likely mean decreased funding for the AWG as the 20th SUPCOM commander would allocate resources based on his priorities. This was hardly a small consideration. As difficult as the decisions were at the strategic level, they were even more so at the tactical level (the 20th SUPCOM was a tactical unit) where abstract priorities translated into specific tasks and requirements and where leaders more keenly felt the pressures of tight budgets. With assignment to the 20th SUPCOM creating as many—if not more—problems than it solved, the 20th SUPCOM ceased providing a viable assignment option.

In 2007, the AAWO once again examined the question of placing the AWG within the Army's force structure. AAWO considered seven courses of action (COA). The first, maintaining the *status quo*, was manifestly undesirable by keeping the unit directly subordinate to the Army staff—something initially intended as a short-term action necessary to create the organization. The second was the original plan—subordination to the 20th SUPCOM. The AAWO expressed an additional reservation to this idea: decreasing the unit's ability to "change the culture of the Army" by removing from the direct supervision of one of the Army's most senior leaders. COA 3 would place the unit under TRADOC, but the authors expressed concerns over TRADOC's ability to manage the AWG's wartime functions and how well TRADOC would resource the unit. COA 4 called for establishing an Asymmetric Warfare Joint Task Force under the supervision of US Joint Forces Command (a command subsequently deactivated) or directly under the Office of the Secretary of Defense. This action would remove the Army from any significant role in directing activities while leaving the Army in the position of providing personnel. The fifth COA would establish the US Army Special Operations Command as the higher headquarters with the noteworthy weaknesses residing in the command generally not controlling conventional units and concerns about how it would resource AWG requirements. COA 6 was fundamentally a shell game in placing the AWG under the administrative control of the Military District of Washington while having the AAWO exercise day-to-day supervision. Finally, COA 7 called for incorporating the AWG, the Rapid Equipping Force, the IED Defeat Task Force, and several other elements into a single command functioning as a Field Operating Agency to the Army G-3/5/7—a situation that General Thurman previously identified as problematic. This COA had at its heart the core problem requiring resolution—getting the AWG out from under the G-3/5/7. COA 7 represented the proverbial "throw away" course of action.[55]

With most of the AWG's METL falling in line with TRADOC's mission, this command was its logical home. While the 2007 AAWO analysis found several disadvantages in placing the unit under TRADOC, several changes had taken place by 2011 with the most significant found in General Dempsey's assignment as TRADOC's Commanding General. Dempsey appreciated the benefits of the AWG and sought its placement under his aegis. Dempsey relinquished his command in April 2011. Dempsey's successor, Gen. Robert Cone, shared his views on the AWG. With supportive TRADOC leadership, the most significant disadvantages to assignment under the command disappeared. Secretary McHugh accordingly directed the AWG move from the G-3/5/7 to TRADOC in June. With this decision, the G-3/5/7, TRADOC, and the AWG worked through the details of the command's coming change in status.[56]

Continuing Operations and Another Change of Command

One of Colonel Mis' strengths laid in his ability to prioritize his activities and workload. Despite the distractions of imminent deactivation and subsequent reprieve, Mis ensured his unit successfully completed its various missions. At the same time, he continued efforts to make the AWG better understood by the rest of the Army. In spite of its small size, the unit was busy with personnel seeing frequent deployments. Between 2009 and 2011, the number of military personnel deployed to theater or on temporary duty in other locations went from roughly 700 days per year to 1,400. This was in addition to the high deployment rate for its contracted operational advisors.[57]

One gains a sense of just how diverse the unit's efforts were during this period in reviewing the subjects of the projects successfully worked and products produced. These included tactical communications transmitters, counter-unmanned aerial vehicle techniques, the emerging (and lethal) insider threat, techniques to counter indirect fire, ambush awareness, non-lethal vehicle interdiction, inter-agency teaming, route clearance, and subterranean warfare. By the time Mis relinquished command, the AWG could claim credit for almost

60 solutions or products that found their way to the larger Army. Not included in this total were additional projects focusing on partners or allies and those meant to enhance the AWG's own capabilities—efforts keeping the unit at the forefront of a slowly evolving community of practice. [58]

As Mis prepared to relinquish command, he could look back on an impressive list of wide-ranging accomplishments and a sense of satisfaction in how his unit's performance continuously reinforced the case for its retention. While he initially had concerns about coming under TRADOC, Mis quickly found those dissipating as General Cone took the position "you work right here for me."[59] Rather than finding the unit subordinated to one of the command's centers of excellence (the Army's various schools), Mis found his organization retained access to key decision-makers. Such availability ensured the funding and support required to both maintain the unit's high standards and safeguard its reputation as an innovative, effective, and customer-focused organization capable of improving both a unit's pre-deployment training and battlefield performance. Innovation was particularly important as it was an element missing from much of a unit's pre-deployment training. [60]

July 14, 2011, saw the AWG undergo its third change of command as Col. Pat Mahaney took possession of the AWG's colors. Like Mis, Mahaney was a Special Forces officer. Originally from New York, he was commissioned as a Military Police officer after graduating from New York University. After serving an assignment with the 96th Civil Affairs Battalion (Airborne), he graduated from the Special Forces officer qualification course. As a Special Forces officer, his assignments included command of Company C, 1st Battalion, 7th Special Forces Group (Airborne), command of the 1st Battalion, 7th Special Forces Group (Airborne), and both the operations officer and deputy commander of the Combined Forces Special Operations Component Command-Afghanistan. Combat and operational assignments included service in Panama, Saudi Arabia, Iraq, Afghanistan, Colombia, Peru, Honduras, Ecuador, and El Salvador. Mahaney's background also included service in a Special Mission Unit and on the Joint Special Operations Command staff. He was also a graduate of the Spanish Army's Command and General Staff College, Italy's Institute for High Defense Studies, and completed a master's degree at Italy's *Universitá degli Studi di Perugia.* Moving into command, Mahaney did not have direct experience with the AWG but had knowledge of the intent behind its creation going back to his time on the Joint Special Operations Command staff while still a major.[61]

As Mis had done before him, Mahaney began his command with a critical look at the organization and its existing relationships—including the one with the Army staff. Mahaney recognized the strained relationship between his unit and General Chiarelli. As Mahaney saw the problem, Chiarelli perceived the AWG as operating with a set of priorities at odds with his own. As a result, every contact only reminded Chiarelli of the differences and thus provided him a source of irritation. For Mahaney, fixing this relationship was a high priority. One method employed by Mahaney was frequent face-to-face meetings with Maj. Gen. Joseph Anderson. As one of Bolger's deputies, Anderson's responsibilities included overseeing the AAWO, and thus the AWG. In Mahaney's meetings with Anderson, he delivered a recurring message: the AWG's focus lay in supporting the Army's priorities. This was not so much a case of doing things differently under a new commander as it was a new commander taking steps to ensure others possessed a clear understanding of what was going on—essentially a shift in messaging. As a result of these efforts, the relationship between the AWG and Chiarelli saw a change in trajectory. Mahaney believed a deteriorating relationship turned into one that was neutral and trending positively.[62]

An important early effort in Mahaney's command was subterranean warfare. These efforts continued work begun under Mis. One of two major problems that the Army faced—the other was the closely related problem of combat in dense urban areas—subterranean fighting presented a monumental list of tactical problems. With the 2011 publication of *Field Manual 3-0, Operations,* the Army aggregated its myriad tasks into six different warfighting functions. Consisting of mission command, movement and maneuver, intelligence, fires, sustainment, and protection, fighting underground stressed existing capabilities in all six areas. Controlling forces was difficult as radios did not perform in the same way they did above ground and space-based systems such as the global position system became useless. The narrow tunnels characterizing subterranean combat restricted both movement and maneuver. Intelligence capabilities decayed dramatically as most collection systems lost their utility with a reciprocal reliance on scouts and observers. Indirect fires could not materially support forces operating underground, and their use on surface targets could significantly imperil the underground fight. Additional protection issues revealed themselves, particularly when operating in the tunnels providing such infrastructure as heat, sewage, or chemical pipelines. Sustainment suffered from the competition for limited space and the challenges of moving commodities. Despite the challenges of subterranean fighting, however, a variety of state and non-state actors were developing increasingly sophisticated capabilities for such combat.[63]

Recognizing the increasing likelihood that Army units would find themselves operating in such an environment, the AWG began researching and validating effective tactics, techniques, and procedures that units could use underground. These activities included liaison with other government agencies and the military and security forces of allied and partner nations. Communicating the AWG's efforts in preparing the Army to win in such an environment demonstrated the unit's alignment with Army priorities. [64]

Assignment to TRADOC

As the AWG continued its day-to-day activities under its new commander, the Army staff completed preparations for transitioning the unit from the G-3/5/7 to TRADOC. On July 11, the Army finalized a memorandum of agreement between the AWG and the Rapid Equipping Force, outlining the changes in processes effective with the AWG's new placement. On July 29, TRADOC presented a briefing to Secretary McHugh articulating the command's concept for command and control of the AWG. These, and similar efforts among the staff, worked the details of the AWG's pending change in status. Late September saw the publication of three significant documents—two from the Army staff and one from the Secretary of the Army.[65]

General Order 2011-06, dated September 21, 2011, directed the AWG's disestablishment as a Field Operating Agency under the G-3/5/7 and its formal assignment as a subordinate command of TRADOC effective November 11, 2011. It directed the commanding general of TRADOC, then General Robert Cone, to "all authority, direction, and control" of the AWG and for the commanding officer, AWG to "report to the Commanding General, TRADOC (...) as he may prescribe."[66] This essentially closed what some considered a key communication channel—directly from the AWG to the Army's most senior leadership—with a corresponding loss in the leadership's understanding of the circumstances driving the AWG's activities.[67]

HQDA EXORD 304-11, "Army Asymmetric Warfare Group Assignment to The U.S. Army Training and Doctrine Command (TRADOC) provided the details of the AWG's new assignment. These required no changes to the unit's mission and "no disruption of support to current operations and missions."[68] The order also specified two future reporting requirements to the Secretary of the Army: a May 11, 2012, "integration status update," and May 11, 2013, fully staffed "AWG integration plan."[69] The former would inform the Secretary of issues and opportunities associated with the unit's new placement. The latter would provide an assessment of the unit's integration into TRADOC after 18 months. This reporting date provided time to identify issues, develop corrective actions, and then assess them.[70]

On September 23, Secretary McHugh signed a memorandum providing additional details on the May 11, 2012, reporting requirement. This report would include recommendations for the AWG's enduring mission and functions in the future operational environment, and a basis for manning and funding as the conflicts in Iraq and Afghanistan drew to a close It would include a potential enduring role as an "action arm" in maintaining contact with the combatant commands and the training base, and realignments in organization and structure to achieve its enduring mission. McHugh's objective was a plan laying out how the AWG would "remain agile and retain the capability to expand rapidly during periods of conflict and grow smaller as mission demands abate."[71] This aligned with some of the key elements of the original operational and organizational concept.

The publication of the general order, the execution order, and the Secretary of the Army memorandum assured the AWG's continued existence—at least over the near term. The only significant action remaining was the formal execution of the orders. The concurrent planning efforts of the TRADOC and Army staff and the AWG itself ensured a smooth uncoupling from the G-3/5/7. TRADOC published TASKORD IN 1279-212, Asymmetric Warfare Group Assignment, to TRADOC on October 6, 2011. The order's mission statement read "TRADOC receives the AWG as an assigned Table of Distribution and Allowances (TDA) organization NLT 11 NOV 2011 in order to enhance TRADOC's ability to execute its core functions and support the Generating and Operating Forces while ensuring continuity of current and planned AWG operations."[72] The order's intent statement was even briefer: "AWG transitions to TRADOC with no interruption of AWG support to current operations and missions" with this language mirroring that of the Department of the Army order.[73] This tasking order assigned specific responsibilities to each of TRADOC's three deputy commanding generals and coordinating staff sections. It also identified three Commander's Critical Information Requirements intended to provide the commanding general with immediate visibility of any major issues impairing the AWG's ability to accomplish its mission. Finally, and perhaps most importantly, it codified Cone's earlier statement to Mis by not assigning the AWG to any of TRADOC's subordinate organizations. FRAGO (Fragmentary Order) 1 to this order followed on October 14 and identified additional requirements for the Resource Management and Safety offices.[74]

For all the drama accompanying the decision to move the command to TRADOC, the unit did not celebrate the actual transfer itself on November 11, 2011, in any fashion. Corresponding to the Veterans Day holiday, Pat Mahaney was home in New York City and participating in the city's annual parade as other unit members celebrated the holiday either at home—or deployed in support of the Army's combat units.[75]

Conclusions

The AWG's transition under its second and third commanders from the supervision of the Army staff to TRADOC showed just how various individuals viewed the unit, its contributions, and its future. Under Generals Schoomaker and Cody, the unit's future was never in doubt. As their retirements brought new leaders into position, there was no guarantee that would remain the case. General Casey "saw value in the unit" as constituted while General Chiarelli believed other units could take parts of the AWG's various missions.[76] Generals Dempsey and Bolger viewed the AWG in the same favorable light as Schoomaker and Cody. More importantly, Secretary of the Army McHugh realized the many advantages gained by retaining the AWG and ensured its continued existence. What made Dempsey and Bolger such strong advocates for the unit were their personal experiences with it. Bolger successfully communicated his belief in the AWG's people and capabilities to Casey and Chiarelli. More significantly, he influenced Secretary McHugh. The impact of personality on change once again presented itself in Bolger's actions. The Secretary's September 23, 2011, memorandum, with its future reporting requirements, clarifies that McHugh's interest in the unit lay not in its past accomplishments, however laudatory, but in its *future*. Aware of the challenges posed by surprises in the field, and cognizant of the fact that the larger Army did not seem to have the ability to assess and report on emerging asymmetric threats, McHugh wanted an agency within the Army capable of doing that.[77]

Placing the unit under TRADOC instead of the 20th SUPCOM illustrates a second, if the subtler, point. Senior leaders were uncomfortable with the AWG under the G-3/5/7 for a variety of reasons, most of them related to questions of resources. The original plan to subordinate the unit to the 20th SUPCOM would have addressed most of those concerns but at a cost to both AWG and 20th SUPCOM capabilities. Aligning the unit under TRADOC reflected the belief that *capabilities* and not resource questions should determine the AWG's final placement.

Under Jim Mis and Pat Mahaney, the AWG successfully built on the foundation established by Bob Shaw and Greg Birch. Misinherited a functionally sound and tactically proficient unit and properly assessed its strengths and weaknesses. He recognized a significant issue—how the unit's original branding created barriers to the Army understanding what the unit was and what it did. In reorienting the AWG's image while retaining the unit's standards, Mis improved the AWG's ability to perform its mission by removing some of the mystery and making more transparent just how the AWG could support units both before and during their deployments. "Normal is a cycle of a washing machine" made sense to people who thrived in an environment of constant uncertainty and change—the kind of environment that was anathema to most conventional soldiers. "Think... Adapt... Anticipate" resonated with the larger Army. If the original motto captured the typical soldier's approach to combat the Army would not have needed the AWG. Units would have readily adapted to new battlefield realities without outside assistance—and without the high cost of lessons learned in Afghanistan and Iraq.

If "normal is a cycle on a washing machine" did not resonate with the larger Army, it still captured the AWG's daily activities. A review of the products produced and solutions provided under Mis and Mahaney shows just how diverse the unit's work was. The AWG was the only unit in the Army to simultaneously consider the challenges of air, ground, and underground operations. Between July 2009 and November 2011, the AWG produced over 70 products. Concurrent with this, the unit's operational advisors maintained training relationships with units in the United States and battlefield ones with them in the combat zone. The latter were highly valued by senior tactical leaders because they provided them with insights about the state of their unit's training and operations otherwise unavailable. These "informal reports" were as important as the handbooks, studies, graphic training aids, and material-based solutions the AWG provided to the operational Army.[78]

Secretary of the Army John McHugh once referred to the AWG's cost as "budget dust." In considering the totality of the AWG's contributions to the Army as a whole, the Secretary would have been equally correct to term it "gold dust." Small in size, cheap in cost, and disproportionate in its impact on the Army as a whole, the AWG's performance validated its organizing principles, the skill and dedication of its members, and the confidence of the leaders and soldiers who benefitted from its service and advocated for its retention.

Conclusions

"Not what to think, but how to think." [1]

Army Asymmetric Warfare Office Annual Review 2007

This chapter highlights the Asymmetric Warfare Group's history since it joined the Training and Doctrine Command in 2011. It also considers the forces compelling change within the Army. Understanding these forces is necessary to fully understand why the Army established the AWG, and how it did so. The "how" of the AWG is the most significant element of the AWG story since it provides a template for the future. The Army's history ultimately only proves one thing: the Army's peacetime establishment is usually not what the Army actually requires in times of war. The Army's battlefield history is its most exciting, but the history of how the Army changes is the most useful if one seeks to understand its structure at any point in time—and how it must change to meet real requirements.

The Asymmetric Warfare Group Since 2011

The ten years following the AWG joining the Training and Doctrine Command saw significant changes occur within the Army. One of the most significant involved a marked reduction in the American military presence in both Afghanistan and Iraq. With Iraq still incapable of maintaining its own security and Syria suffering a civil war, a resurrected Al Qaida created the Islamic State in Iraq and Syria (ISIS). Through early 2014, ISIS increased its control over portions of Iraq. By June, it had seized Mosul, a key city in Nineveh Governorate in northern Iraq. In response, the United States created the Combined Joint Task Force Inherent Resolve (CJTF-OIR) on October 17, 2014. The task force's mission was two-fold. First, it would defeat ISIS in designated areas of Iraq and Syria. Second, it would establish conditions for follow-on operations establishing greater regional stability. CJTF-OIR emphasized Special Operations Forces, aviation, long-range precision fires, and Iraqi and Syrian Democratic Forces. As a practical matter, this meant a relatively small number of Americans faced the threats posed by improvised explosive devices or snipers. The smaller presence coupled with the tactical and technical lessons learned since 2001, many facilitated by the AWG, meant a marked reduction in casualties.[2]

The same period also saw the Army reconsidering the questions of modernization. To that end, the Army established a new major command, Army Futures Command (AFC), in 2018. The command's mission lay in modernizing "the Army for the future" by integrating the "future operational environment, threat, and technologies to develop and deliver future force requirements, designing future force organizations, and delivering materiel capabilities."[3] To realize this, the command organized "cross-functional" teams addressing six modernization imperatives. These included future combat vehicles; long-range precision fires; the Army network; vertical lift; air and missile defense; and soldier lethality. While all these had implications for asymmetric warfare, the focus of this effort lay in ensuring that the Army could defeat any near-peer competitor. The Army's self-conceptualization as a new decade dawned returned to what it was before September 11, 2001: a force focused on using its soldiers, tanks, guns, and helicopters to defeat the enemy's tanks, guns, and helicopters.[4]

Also driving the modernization effort was the realization that the Army lacked sufficient command and control structure in Europe to face an increasingly aggressive Russian regime. The Cold War drawdown that precipitated a reduction of soldiers in Europe underwent a change beginning in April 2014 with the ATLANTIC RESOLVE exercise series. ATLANTIC RESOLVE deployed Army combat and support units to Europe for nine-month periods during which they trained with various NATO allies or European partners. By design, these rotations were also information operations and demonstrations to Russia that the United States would honor its commitments in Europe. To reinforce this point, Poland, Romania, Bulgaria, Latvia, Lithuania, Estonia, and Ukraine made their forces available to train with American combat units. To provide the missing command and control and improve allied integration, the Army reactivated V Corps and established a permanent forward command for the corps at Poznań, Poland, in November 2020. Again, the unmistakable focus laid in maintaining the ability to defeat a technologically sophisticated opponent.[5]

In the minds of much of the Army's leadership, the AWG did not align with the returning focus on modernization and the threats posed by hostile states fielding large, technologically sophisticated armies. Part of the problem was of the organization's own making: it excelled in accomplishing its mission in Afghanistan and Iraq. Consequently, many on the Army staff came to view the AWG as a no longer-needed counter-insurgency unit. As a result, Secretary of the Army Ryan McCarthy, Ryan and Army Chief of Staff James McConville decided to deactivate the command and transfer its functions across the Army. Some would move to the Maneuver Center of Excellence at Fort Benning, Georgia, while others would migrate to the newly created Army Futures Command. Army Execution Order 110-20 articulated the plan for the AWG's deactivation. The first milestone was October 1, 2020, when the command would become "non-mission capable." This meant the only new activities in which it engaged were those associated with its ultimate end: work-in-progress would continue but the AWG would not undertake any new projects. May 13, 2021, saw the casing of the AWG's colors. The unit's shoulder-sleeve insignia would remain only on the right sleeves of a few—and in the memories of those who benefitted from the AWG's efforts. Finally, on September 30, 2021, the AWG disappeared from the Army's rolls with the retirement of its unit identification code—the computer code that represented its bureaucratic heart.[6]

Through the AWG's inactivation, it remained a subordinate element of TRADOC. The group's headquarters still occupied a small complex of buildings near its original home in the Fort Meade confinement facility. Until February 13, 2021, it also maintained its state-of-the-art training facility at Fort A. P. Hill. This included a mix of permanent and temporary purpose-specific training sites. The former consisted of an urban training site including multi-story buildings and subterranean infrastructure. This site contained a subway station, complete with rails and discarded subway cars from the Washington Metropolitan Area Transportation Authority. This sophisticated facility enabled the testing and validation of tactics, techniques, and procedures for a variety of combat scenarios. It also facilitated the training of both AWG members as well as members of other Army and Marine Corps units, along with personnel from other government agencies. Also included was a more austere subterranean training complex consisting of tunnels and galleries of different sizes, shapes, supports, and linings. Like the urban training site, this facilitated the testing, validation, and training of the methods most effective in underground combat. Twenty-first-century experience shows such environments are increasingly the site of military or security operations. The realization that soldiers would likely fight underground spurred another AWG innovation—locally producible tunnel trainers. These modular structures allow soldiers to practice the unique skills underground combat requires. Several installations now have such training devices and units routinely use them to practice tactics and techniques developed by the AWG.[7]

The Army's capstone doctrine continued to evolve through the AWG's active service. While most of the Army remained equipped for Air-Land Battle, its new doctrinal concept became "multi-domain operations." Simply expressed, this meant dominating the battlefield on land, sea, air, space, cyberspace, and the electromagnetic spectrum. Concurrent with this, the Army continued expanding its ability to provide security assistance training to friendly nations, seeking to improve their ability to operate with the Army. Even as the AWG remained the lead organization for countering asymmetric threats, the Army invested heavily in a new approach to assisting partners—dedicated advisory units. Termed Security Force Assistance Brigades (SFAB), these units would "train, advise, assist, support and liaise with partner forces."[8] Together, the doctrine of "multi-domain operations" and the rise of dedicated advisory brigades demonstrated the Army was rapidly and decisively turning away from counterinsurgency.[9]

Until 2020, AWG personnel still came from across the Army, from civilian government service, and private contractors. Successful completion of the AWG's selection course remained a requirement for soldiers seeking an assignment. Conducting two selection courses annually, approximately one-third of those applying for service with the AWG completed the program and received an invitation to join the unit. Demographics changed as the group matured. As Colonel Shaw preferred, soldiers came not only from the Special Operations community but the larger Army. Increasingly, AWG soldiers were female. By 2015, the Army had lifted all restrictions on female soldiers and allowed them to attend every Army school. For the AWG, this meant that by 2018, 5 percent of the unit's military personnel were female. Interestingly, the selection rate for females was approximately 35 percent, while males achieved an approximate selection rate of 30 percent. Branch assignments and military occupational specialty included not only Special Forces, had now included such diverse specialties as Air Defense Artillery, Armor, Chemical, Civil Affairs, Medical, Military Intelligence, and Ordnance.[10]

Those with ties to the AWG did not welcome the deactivation decision. Col. Bob Shaw lamented the loss of a critical capability. For Shaw, deactivation meant losing a unit able to "identify/define the problem (...) create and test the tactical or technical solutions (...) and get the solutions to the units in contact with training and follow through into Army doctrine."[11] The AWG not only did this, it did so much more quickly than the standard methods of doctrine or material development. Col. Jim Mis was similarly disappointed. As he observed, "the AWG was formed in a time of crisis so that our Army (...) could rapidly learn, adapt, and innovate to save the lives of soldiers (...) help them win on the battlefield. The deactivation of the AWG leaves a gap that destines our Army to repeat the failures of the past with its Soldiers ultimately being the ones who will pay a high price."[12] Col. Pat Mahaney observed that "even with the current focus on Great Power Competition, many strategic leaders within the Army are overlooking the fact that our competitors are overwhelmingly most likely to continue to use asymmetric, irregular and hybrid approaches."[13]

The Army recognized the AWG's contributions to the fights in Afghanistan and Iraq in 2015 by awarding it the Army Superior Unit Award (ASUA) for its operations from November 1, 2011, through August 22, 2013. What makes this award particularly interesting are the details of the accompanying citation. These noted: "(T)he cited unit demonstrated the ability to innovate and accomplish a mission unique to their organization alone. The unit was second to none in its support of the Army and Joint Command's global and rapid solutions development. The unit enhanced Soldier survivability, and combat effectiveness, and enabled the defeat of current and emerging threats in support of Unified Land Operations."[14]

The citation's language was consistent with that required for an award of the Meritorious Unit Commendation. The AWG, however, was ineligible for such an award as its headquarters remained in the United States. In conferring the ASUA, the Army formally recognized the vision of Generals Schoomaker and Cody and the work of Colonel Shaw, Command Sgt. Maj. Birch, and their successors in creating and sustaining a highly specialized unit with a disproportionate impact for its size.[15]

Considering the Forces of Change

The AWG's history illustrates several complementary themes. The two most important are the distribution of power within the Army, specifically between transient senior leadership and the enduring bureaucracy, and the influence of self-conceptualization. Other themes include the inertia inherent (particularly evident in the Program Objective Memorandum, or POM) in the Army's organization and the implications of organizational leadership. In *On War,* Carl von Clausewitz described war "as a fascinating trinity (...) composed of violence, hatred, and enmity (passion), the play of chance and probability (...) an instrument of policy, which makes it subject to pure reason (politics)."[16] Clausewitz continued by noting the task was to "develop a theory that maintains a balance between these three tendencies (...) like an object suspended between three magnets."[17] The metaphor of "suspension between magnets" could describe change within the Army. In this case, the magnets are internal forces (contained within the service), controllable external forces (those over which the Army effectively exerts some influence or control), and uncontrollable external forces (those over which the Army exerts little or no influence or control) (Figure 2).

These three magnets, however, are not static. Instead, they are subject to multiple internal pressures determining their behaviors, and thus their influence. Foreign interaction and emerging threats represent the greatest of these shaping "uncontrollable external forces." Presidential directives, Department of Defense policy, and directives from the Joint Chiefs of Staff profoundly shaped "controllable external forces." Pressures affecting the "internal forces" magnet include self-conceptualization, bureaucratic processes, leadership, institutional norms, branch parochialism, tribal culture, and soldier acceptance. All these components affected the AWG's creation, growth, and ultimate transition to the Training and Doctrine Command. Looking at the AWG (both its elements and its duration) provides insight into what magnets and associated internal forces proved most influential as the Army adjusted to new battlefield realities in the wake of September 11, 2001.

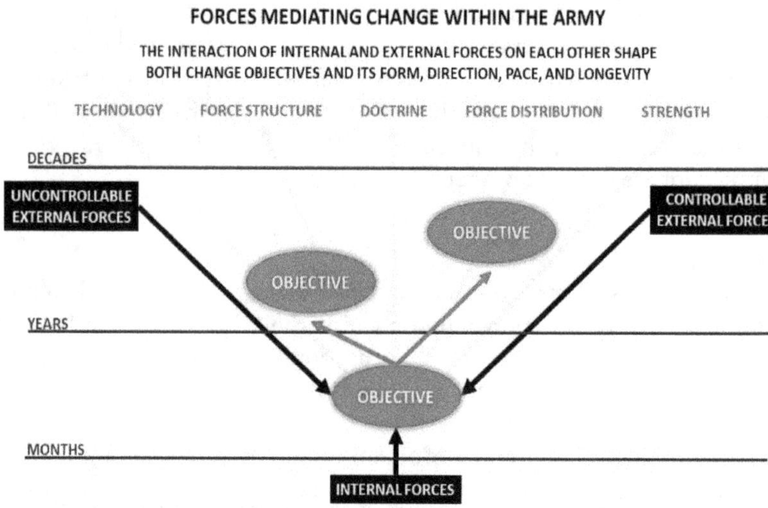

Figure 2. Forces mediating change within the Army

Implications—Uncontrollable External Forces

The most significant of these pressures was the emerging threat of new weapons and their proliferation and foreign interaction. While elements of both the Joint Staff and the Army recognized the threat posed by non-state actors, the implications associated with this threat never led to its occupying a place of primacy in either the Department of Defense or the service prior to 2001. Occupying greater attention were the Army's modernization efforts and its overseas operations. Seemingly never-ending peacekeeping or humanitarian assistance operations began with Kurdish Iraq and continued through Somalia, Haiti, Bosnia, Rwanda, Kosovo, and the Sinai. These operations proceeded against the backdrop of an ongoing presence in Kuwait as a check on military action by Iraq.[18]

In retrospect, the attack on the *U.S.S. Cole* and the bombings of the Khobar Towers in Dhahran, Saudi Arabia, and the American embassies in Kenya and Tanzania should have led to a reprioritization of threats by both the Department of Defense and the Army. As a practical matter, however, there was no political appetite for large-scale operations to eliminate the threats posed by non-state actors. This condition existed not only in the United States but in countries capable of assisting American efforts and in the failing/failed states harboring such actors. To the extent that the United States took action against those suspected of terrorist actions, it restricted the action to strikes by precision weapons that theoretically allowed engagement without the risk of actual commitment. A famous metaphor describes the traditional "ham and eggs" breakfast: the chicken is involved while the pig is committed. Until September 11, 2001, there was tepid national policy interest in playing the role of the chicken—and none in playing the role of the pig. Employing such a strategy meant there was no real need to create additional capabilities within the Army to resolve the complex problems associated with defeating an opponent whose structure or methods did not align with the service's existing capabilities.

Technological change is also important. The Army cannot control how potential adversaries respond to advances in military or associated technology. A variety of factors influence how different countries respond to technological advances. These include their own perceptions of the strategic environment, their capacity to invest in technological advances, and the self-conceptualization of their own military forces.

Implications—Controllable External Forces

The Army cannot exercise sole control over such external forces, but it does have the ability to influence them. The Army's position within the Department of Defense allows it to inform policies, the Joint Strategic Capabilities Plan, and Joint Chiefs of Staff directives. While not positioned to directly participate in crafting the National Security Strategy, presidential directives, or congressional action, the Army indirectly informs them. The Army's Office of Congressional Legislative Liaison (OCLL) provides direct engagement with senators and congresspersons and thus can communicate concerns on a variety of issues. Under the provisions of the 1947 National Security Act and its 1949 amendments, National Security Council membership includes the Secretary of Defense and the chairman of the Joint Chiefs of Staff and these provide the conduit for Army concerns. Presidential directives are the most difficult for the Army to influence, but again the Secretary of Defense and the chairman of the Joint Chiefs of Staff provide a means of expressing departmental positions.[19]

Implications—Internal Forces: Bureaucratic Processes

Where power resides within the Department of the Army is an issue almost as old as the service itself. The modern bureaucracy's historical roots extend back to the War Department's earliest days and the hiring of civilian clerks to fill a variety of administrative posts. By 1817, requirements reached the point where Secretary of War John C. Calhoun introduced the bureau system placing a combination of senior officers and civilians into offices possessing significant institutional power. The tensions between the bureaus and the commanding general (later the Chief of Staff) created the sort of institutional infighting familiar to many in today's Army. How much (or how little) power a senior leader will exert reflects directly on his ability to influence the bureaucracy effectively. General Shinseki's tenure as Army Chief of Staff witnessed the creation of the Stryker brigades and the introduction of the Rapid Equipping Force. Generals Schoomaker and Cody succeeded in bending existing structures and processes more dramatically to create the AWG. They then succeeded in sustaining the REF and AWG and reconstituting critical elements of the IED Defeat Task Force after the formal organization migrated to the Joint Staff as the Joint IED Defeat Organization (JIEDDO). In doing so, they provided an example of organizational behavior validating two of Benjamin Jensen's arguments in *Forging the Sword: Doctrinal Change in the U.S. Army*: the powers of bureaucracy and personality in mediating bureaucratic power. Their success in doing this lay in Schoomaker clearly expressing what he wanted as the outcome—and providing Cody the freedom to realize it. It also lay in Cody providing unambiguous guidance, placing individuals in positions to effectively engage the bureaucracy (including forcing significant changes in business practices when necessary), and maintaining an exceptionally high level of personal involvement in every aspect of the AWG's creation. This combination of personality, vision, and engagement ensured the outcome Schoomaker and Cody desired. Similarly, their replacement with individuals possessing a differing vision consigned the AWG to the Army's long list of inactive units.[20]

The AWG story also illustrates the powerful inertia produced by the bureaucracy and its processes. Many elements of the bureaucracy serve as brakes on progress and innovation, but the most powerful is the Program Objective Memorandum (POM). The POM process requires committing funding, the service's lifeblood, in five-year increments. While mechanisms exist to change the POM when circumstances warrant, the reality is the actual changes, as a percentage of the Army's budget, are relatively small—budget line allocations contain reductions representing small percentages with premature eliminations occurring infrequently. This approach is not necessarily a bad thing. By requiring the Army to think ahead, the service finds it must think as rationally as possible to glean the best possible returns on the funds allocated. The POM provides a necessary brake against momentary passions or unrealistic expectations associated with emerging technologies or concepts. For all its drawbacks, the POM still represents the best way to force the Army to think before it spends, particularly when the dollar amounts can reach the billions.

The POM historically casts a long shadow over the Army. Codifying the one essential resource—funding—multiple actors covet its provisions and seek to either attack or defend its contents. The AWG's unusual initial placement under the G-3/5/7 reflected this fact, as did its final placement under TRADOC. For Cody, it was essential that the AWG not fall prey to bureaucratic or organizational predation by actors believing they could better utilize the AWG's funding by applying it to other programs or units. General Cody ultimately achieved two victories in the bureaucracy with the creation of the AWG: creating the unit and preventing it from dying in the proverbial crib by others reallocating its funding.

In opposition to the POM, stood the Army Strategic Planning Board (ASPB), later renamed the Army Requirements and Resourcing Board (AR2B). Created by Cody, this board imparted the dynamism necessary for the Department of the Army to favorably affect operations in the field. The agility missing from the POM came from the ASPB/A2RB. Those actors whose institutional positions or programs suffered seldom appreciated such agility. It was essential, however, if the Army staff was to enable rather than disable units fighting a global war.

If "global war" sounds hyperbolic, consider the disposition of Army forces during this period, which involved combat operations in Afghanistan, the Horn of Africa, and Iraq. In addition, soldiers engaged in peacekeeping operations in Bosnia, Kosovo, and the Sinai; peace enforcement on the Korean peninsula; and training to allies conducting counterinsurgency operations in the Philippines, Thailand, and Colombia. Routine approaches were woefully insufficient in providing the units, equipment, and training required to execute these missions.

The POM and ASPB/AR2B, however, represent only two implements in the bureaucracy's toolbox. The DOTMLPF construct provides another. The DOTMLPF elements fall across the Army's staff and major commands and each element represents multiple bureaucratic components. Doctrine, training, leader development, and organization are inherent responsibilities of both TRADOC and the Army staff. Material, facilities, and personnel have legs in TRADOC, the Army Materiel Command and, since 2018, the Army Futures Command.

Additionally, they also have elements falling under not only the G-3/5/7 but also the Army's G-1 (Personnel) and G-4 (Logistics). Depending on the function considered, the G-2 (Intelligence), G-6 (Communications), Chief of Engineers, Provost Marshal General, Surgeon General, Judge Advocate General, the Director of the Army National Guard, the Chief of the Army Reserve, or even the Chief of Chaplains may play a role. The Army's various service component commands and direct reporting units (those reporting directly to the Army Chief of Staff) are also involved. The former includes such diverse organizations as the Army Special Operations Command, the Army Space and Missile Defense Command, the Army Cyber Command, and the Surface Deployment and Distribution Command. The Intelligence and Security Command, the Network Enterprise Command, the Army Medical Command, and the Criminal Investigation Division Command are examples of the latter. Finally, the Army's Forces Command plays a role. While the most common portrayal of all these elements is a line-and-block chart showing vertical relationships, a functional portrayal depicts horizontal relationships as well. It would thus resemble a spider's web: touching it in one location transmits vibrations across the broader web. All these entities possess bureaucratic structures, and these locations are where the web's vibrations most often resonate: a proposed doctrinal change from the Intelligence Center of Excellence at Fort Huachuca, Arizona, could involve TRADOC, AFC, AMC, the G1, the G2, the director of the Army National Guard and the chief of the Army Reserve.

In the years since its creation, the AWG achieved a degree of institutional and organizational maturity. This growth placed it entirely within the institutional Army's bureaucratic structure. Unlike its earliest years, it no longer received its funding by pulling money from other organizations or programs. Fully embedded in the Army, the AWG possessed an identified funding line and went through the POM process like the rest of the service. No longer a direct reporting agency to an Army staff element, it relied on TRADOC and its bureaucratic muscle when institutional battles required such action.

There is another facet of the Army's bureaucracy evident in the AWG's creation—the flexibility to reach outside the Army when circumstances warrant solving problems beyond its organic capacity. The institutional Army could not develop the operational and organizational concept providing the road map for building and employing the AWG. One solution to the problem resided in developing the required skills within the Army. This would require reassigning the few officers possessing the required experience from other critical assignments. The second was in exploiting outside expertise. In opting for the second solution, the use of contractors, the Army's bureaucracy saved a tremendous amount of time, capitalized on hard-won experience, and ensured the unit retained the flexibility required to remain effective. One of the paradoxes of the Army's bureaucratic structures is that while they fundamentally induce inertia, they sometimes act as an accelerating force. Fundamentally rigid, the structures nevertheless possess enough suppleness to change in the presence of active and engaged leadership—something both Schoomaker and Cody provided—and exigent circumstances.

"Buy, try, decide" also had significant bureaucratic implications. In granting the AWG the ability to consider and acquire items, the Army tacitly recognized the problems inherent in its standard materiel acquisition model. Reduced to its core elements, this model consisted of centers or schools associated with each of the Army's subordinate branches. These centers or schools had a staff of combat developers who identified future requirements and who then initiated the process for procurement. This process included creating the request for proposal, determining a provider, testing the provided item, determining if the provided item met service requirements through a long test and evaluation process, and then acquiring the item. As this process played out, the center or school prepared both the doctrine for employment and the training plan. Concurrent with this, the Army determined the units and locations receiving the item. When completed, the equipment, doctrine, training, and fielding plans addressed all the DOTMLPF elements and provided the Army with the item desired along with the capability to employ it fully.

In granting the AWG "buy, try, decide" authority, the Army, in effect, transferred some of a center or school's responsibilities to a unit not subordinate to that center or school. Equally important, it provided the unit with the ability to operate across branch lines. Stated another way, the AWG focused on the Army while the centers or schools focused on specific branches. Linking "buy, try, decide," to the AWG to REF had an additional effect: a marked compression of time. Where the standard model could sometimes play out over many years, the AWG could go from idea to test to initial procurement in months. Granting the AWG this authority created the agility necessary to counter the diverse threats soldiers faced in the Global War on Terrorism. The AWG prevented the Army from recreating the failures of the Vietnam War's TELACS and XM-2 Personnel Detector Set (introduction). The unit also largely eliminated the need for units to subvert the Army's bureaucracy to find solutions to their problems as was the case in prior conflicts. One can argue that "Hillbilly armor" was the Iraq War equivalent of the Vietnam War's artillery firing platforms, but its development and use pre-dated (and helped spur) the AWG's creation.

Implications—Internal Forces: Self-Conceptualization

The influence of self-conceptualization is also clearly apparent in the AWG story. The events of September 11, 2001, came at a critical time in the Army's history. For several years, three currents dominated the Army. The first was recognizing the need to re-develop a force optimized for a then non-existent threat as the Soviet Union faded into history. The second was recognizing the military potential of the cyber domain. Together, these provided the focus for long-term modernization. The third was the realities of a force locked into peacekeeping and peace enforcement missions in the Balkans and elsewhere.

The Army's major equipment items—the "Big 5"—reflected perceptions of Soviet-led military power and the requirements of battle in central Europe. These systems—and the soldiers who employed them—demonstrated the ability to dominate the battlefield in southwest Asia in 1991. That same combat experience also identified some significant limitations with these systems. Compounding this was the effect of the post-Cold War drawdown of forces. The Army's 1990 order of battle (the forces then available) consisted of 18 combat divisions on active duty in the Regular Army, with an additional 10 divisions active in the Army National Guard. By 1996, this number decreased to ten Regular Army Divisions and eight Army National Guard divisions. Concurrently, the Army eliminated a corps headquarters and the associated combat, combat support, and combat service support units. The Army's fundamental mission had not changed, but the resources available were both markedly reduced and not necessarily optimized for a different political and military environment.[21]

The post-Cold War period saw advances in cybernetics, which ushered in a perceived revolution in military affairs. Theoreticians, both in and out of the Army, envisioned a force built not around weapons systems, but information networks allowing a much higher level of situational awareness at all echelons. Such awareness would compress time by giving decision-makers information much more quickly. Presenting information graphically would provide necessary context and reduce the potential for processing mistakes. The unambiguous, simultaneous portrayal of friendly forces, enemy forces, terrain, and weather facilitated rapid decision-making and rapid responses by the soldiers in the tanks, fighting vehicles, helicopters, and other systems maneuvering on the battlefield. A force with these capabilities (or so went the reasoning) would soundly defeat a force lacking such capabilities either by pitting its strengths against the enemy's weakest points or by allowing it to overwhelm the enemy's center of gravity. Implicit in this was the belief of a future opponent possessing a modern, mechanized force. The future Army would face better tanks than the Soviet T-72s and BMP infantry fighting vehicles used by Iraq in 1991—but it would still battle such conventional opposition.[22]

Operations in the Balkans marked both Gen. Eric Shinseki and the larger Army by compelling levels of cooperation between the service's three components (Regular, National Guard, and Reserve) previously unimagined. Starting with the 1995 deployment of the 1st Armored Division as the American element of the Bosnia-Hercegovina implementation force, the region has continuously hosted an Army presence. Throughout this commitment, the mission has changed several times, with each change forcefully driving home some painful lessons for the Army. The most significant lay in strategic deployability, the need to maintain Army National Guard and Army Reserve readiness, and the difficulties inherent in moving large, mechanized forces in the absence of suitable infrastructure.[23]

The 1991 Gulf War precipitated deployments ranging from 8,000 to 12,000 miles from locations in the United States and Europe to well-developed port facilities in eastern Saudi Arabia. Not only could these ports accommodate large, deep-draft vessels, but they were also adjacent to fully developed airfields and connected by a modern road network. Adjacent to these ports were large areas of vacant land suitable for staging units and stockpiling supplies. Such facilities were largely invulnerable to any serious attack by Iraqi forces.

In contrast, the 1995 deployment from Germany to Bosnia required traveling 1,000 miles while the 1999 deployment to Kosovo stretched to roughly 1,200 miles. Both countries were landlocked, which necessitated deployment by air and rail. While Bosnia's Tuzla airfield could support operations, Kosovo lacked any such facility. As a result, Tirana, Albania's, Rinas airfield became the focal point of air movement. Rinas, however, was not usable as a tactical facility without considerable improvements. The combined effects of restricted terrain, poor weather, and substandard infrastructure meant the Army had to overcome myriad challenges to get forces into the area and then conduct operations. Shinseki was one of many both in and outside the Army who realized that future deployments would more likely look like the Balkans than the Persian Gulf and believed the Army needed to make fundamental changes to ensure its deployability.[24]

The long-term nature of Balkans deployments also forced a change in how the Regular Army viewed its reserve components, and how they viewed themselves. The inability of Army National Guard roundout brigades to meet readiness standards in 1991 led to ill feelings between the Guard and the Army. The Balkans, however, provided an opportunity to re-cast the relationship. The smaller Army soon found it could not sustain operations in Bosnia-Hercegovina without significant changes in force structure. The reduced number of units in Europe, coupled with other commitments in the Persian Gulf and Asia forced the Army to task the Army National Guard with responsibility for the Bosnian stabilization force mission starting in 2000 with the 49th Armored Division (Texas) taking the mission from the 10th Mountain Division. One more Regular Army division headquarters would control troops in Bosnia, but it would deploy with significant Army National Guard augmentation. Before the Global War on Terrorism, the Army National Guard was transitioning from a strategic reserve to an operational one. Concurrent with these actions was the mobilization of various Army Reserve elements to both support Balkan operations and the resettlement of refugees fleeing the region. While the primary purpose of mobilization was to meet Army requirements, an additional benefit came in both the bridge-building between the components necessary to complete these missions and the "re-greening" inherent in bringing reserve component soldiers to active duty for extended periods. "Re-greening" meant making the mental and psychological transition from civilian to soldier. These were the first steps in transitioning the reserve components to the operational role required by the changes the twenty-first century brought to American security policy and strategy. [25]

Self-conceptualization, however, reflected more than the Army's internal perspectives on its place in national strategy: external forces also shaped it. These included legislative pressures, funding, public opinion, the broader strategic environment, and technological change. Legislative pressures included the various constituencies engaging Congress to shape its decisions. These groups ranged from the Association of the United States Army (seeking support for the Army's vision) to corporations or consortiums (seeking procurement of their products) to members of think tanks (seeking to validate or refute the Army's guiding vision). Closely tied to this was funding—both the amount provided and the limitations on how the Army could spend it. At the same time, additional guidance came from the President, Congress, the Secretary of Defense, and the Chairman of the Joint Chiefs of Staff. All these elements could shape the Army's self-conceptualization by compelling changes in priorities or by eliminating programs associated with its vision. The Crusader field artillery system, Comanche advanced attack helicopter, and the Future Combat System all experienced this fate.[26]

Another force that influenced the Army's self-conceptualization, however, was public opinion. Ironically, the Army's stunning performance against the Iraqi Army in 1991 made many people wonder why the country still required either a large Army or an investment in a new generation of weapons. Having defeated the best systems produced by its long-term opponent, the Soviet Union, and with no formidable perceived threat on the near horizon, many wondered about the actual size of the coming "peace dividend." A smaller Army meant money for other purposes that had gone underfunded over the long Cold War. Paradoxically, the Army's recent successes created circumstances limiting the funding necessary for upgrades. [27]

The strategic environment played an important—perhaps decisive—role in the Army's self-conceptualization during the immediate post-Cold War period. While the service observed many indications that the next security challenges would originate from non-state actors rather than nation-state that did not synchronize with the Army's overarching vision. This vision rightfully showed a long-term perspective. In practical terms, this meant the Army pursued programs optimized to fight the large mechanized formations of foreign states and did not heavily invest in the capabilities most useful against asymmetric actors.

Finally, technological change profoundly shapes the Army's self-conceptualization. The "Big 5" represented a massive, and continuing, investment in a broad range of technological advances. As the Army employed multi-domain doctrine and concurrently sought replacements for the "Big 5," the focus remained on what that paradigm originally provided: greatly improved capabilities against similar forces.

Implications—Internal Forces: Senior Leadership

The Army is not through changing: change remains the one constant within the service. What the story of the AWG tells us is that *effective* change only takes place when the Army's most senior leadership takes bold action on two fronts. The first is in recrafting the service's self-conceptualization to reflect new perspectives the leadership introduces. This assessment is essentially Rosen's argument in *Winning the Next War.* In creating the AWG, one observes how leaders who can recognize a new problem set, visualize possible solutions, and then effectively describe those remedies can establish the conditions necessary to solve those problems. The key to Gen. Schoomaker's and Gen. Cody's visualization lay in their unusual backgrounds: both had extensive Special Operations experience, but both also had served in divisions as generals. Schoomaker's service as an Assistant Division Commander with the 1st Cavalry Division provided the opportunity to draw comparisons between the Special Operations and conventional Army. It also allowed him to recognize how one could significantly enhance the other. Cody's experiences with the 101st Airborne Division (Air Assault) imbued him with similar insights and a deeper understanding of the techniques (and benefits) of rapid procurement.

The second is in engaging the bureaucracy at multiple levels and points to create the environment needed for change. This point also reinforces Jensen's principal argument in *Forging the Sword.* If senior leadership succeeds in these actions, the desired change is more likely to not only occur but also survive beyond the initiating leader's tenure of office. Cody succeeded in these efforts by both bending the bureaucracy to meet his needs and positioning individuals sharing his perspectives to ensure the resistance inherent in creating new organizations not fully understood by the institutional Army did not smother his objectives even after he retired.

Change *can* create its own inertia enabling success—and survival. If the proposed change involves creating new units, then one last element appears—the initial leadership of those units. If these leaders establish firm footings for their organizations, then the change is not only successful, it is enduring.

Implications—Internal Forces: Unit Leadership

The AWG's creation and growth also demonstrate just how much difference organizational leadership makes. The tasks of senior leadership include visualizing, describing, and directing. Visualization means a leader possesses a clear mental picture of what he or she seeks from the organization. Most often described as the end state, this picture includes the actions or activities that will lead the unit where the leader desires. Describing is the way the leader communicates his or her vision to the unit. Most often, leaders achieve this through direct engagement with subordinates. This is perhaps the most crucial action as it provides the transmission linking vision and outcome. The most successful leaders are those who possess the ability to describe their objectives. This provides an understanding of not only the mission's "what," but also the "why" and "how." Subordinates, armed with such an understanding, can act independently. Units led by such leaders are generally the most effective as subordinates are not only able to behave proactively but can link their actions with the unit's overall objectives or goals.[28]

The AWG represented a unique unit, and its initial personnel cohort was highly unusual in terms of individual and collective skill sets. This mix of high-performing individuals used to independent action meant directing the unit could pose problems. Over-controlling the unit could retard individual performance and work against the objectives assigned by Generals Schoomaker and Cody. Under-controlling could also pose problems by allowing the over-investment of time, effort, or other resources in efforts less likely to create desired outcomes.

Col. Robert Shaw and Command Sgt. Maj. Alfred G. Birch possessed different responsibilities but successfully aligned their respective efforts to build the AWG effectively and efficiently. As the commander, Shaw's primary responsibility was ensuring the unit accomplished its assigned mission, and this was not possible without his developing and promulgating a clear vision for the unit. The O&O provided the necessary aid. Ultimately, it was up to Shaw to visualize the organization able to perform the tasks articulated in the O&O and then build a unit capable of achieving them. As the Command Sergeant Major, Birch both reinforced Shaw's efforts and concurrently ensured the unit's uniformed members possessed the martial skills required to operate effectively on a very complex battlefield. Together, Shaw and Birch established the framework for the unit's enduring culture. This culture prioritized forward thinking by unit members and close coordination between deployed and non-deployed elements. It also emphasized the understanding that members supported the units with which they embedded and did not grade them. Operational advisors were not like the observer-controllers at the Army's combat training centers who observed activities and then provided feedback for after-action reviews. Neither were they "inspectors" or "evaluators" tasked with reporting observations to a higher headquarters. Instead, operational advisors provided commanders with information about tactical trends and what they meant in terms of training requirements. AWG members understood their own success came from the commands they supported enjoying tactical success. These cultural elements characterized the AWG through its deactivation and played a major role in developing the reputation the unit enjoyed.

Subsequent commanders, Cols. James Mis and Pat Mahaney, each built on the efforts of their predecessors. Mis and Mahaney saw the need to adjust unit activities and did so in a manner that sustained AWG's reputation as an innovative, capable, and user-friendly organization. The changes reflected the different approaches of different commanders. While both had similar personal and professional backgrounds, they nevertheless had unique perspectives on the unit, its mission, and the best way to ensure mission success. The shift away from "normal is a cycle on a washing machine" to "Think. Adapt. Anticipate." marked a change in perspective from Bob Shaw's. Just as new commanders made changes, so too did SGTs. Maj. Ray Devens and Mike Akridge. Using the work of their predecessor as a foundation, they adjusted procedures and requirements in accordance with the priorities of their commander. This provided the continuity of vision and clarity of purpose required to accomplish the AWG's mission. Establishing the AWG required one approach, but sustaining it took others.

Implications—Internal Forces: Institutional Norms, Branch Parochialism, Tribal Culture, and Soldier Acceptance

These four components display the Army's organization from the highest to the lowest levels and play a role in change. Considering them from the lowest to the highest levels, however, most easily facilitates understanding their relationships.

The individual soldier represents the lowest level. New soldiers enter the Army from an amazingly wide range of backgrounds. Some come from foreign countries with almost no previous exposure to American society. Others represent an unbroken family tradition of service stretching back generations. Turning such a diverse group of people into a cohesive team with shared values is an important part of a soldier's basic combat training. The process of instilling the Army's professed values of loyalty, duty, respect, selfless service, honor, integrity, and personal courage is also the first step in equipping the soldier to accept or reject change. The overwhelming majority of recruits know little of the soldier's codes, profession, or tools. As the new soldier acquires this knowledge, he or she often develops strong opinions about what is good, bad, right, or wrong. As the soldier matures in service and acquires greater professional knowledge, those opinions frequently become stronger. Efforts at change lacking soldier acceptance quickly become difficult to sustain. One of the AWG's great successes was getting the individual soldier to accept the advice of operational advisors.[29]

The Army's tribal culture stems from the amalgam of branch and unit identity. The Army consists of 22 basic branches. Each branch has its own culture rooted in its primary mission. Each unit also has a unique culture reflecting its branch assignment, its mission, and its soldiers. Battalions and brigades permanently stationed overseas possess a culture different from those stationed in the United States. A unit's mission also profoundly shapes its culture. The 3rd Infantry, 25th Infantry, 82nd Airborne, and 101st Airborne are all infantry divisions—and all have markedly different cultures reflecting their missions. The 3rd is a mechanized division; the 25th consists of airborne, light, and Stryker infantry units; the 82nd is a parachute infantry division; and the 101st is an air assault division. This means each of these divisions has the same basic infantry soldier as the centerpiece of its infantry battalions, but these soldiers operate in very different ways on the battlefield. Most paratroopers look upon mechanized infantry soldiers with disdain since they move with the protection of armored vehicles. To the uninitiated, the Army's various badges, tabs, and patches may seem curious. For the soldier, these loudly proclaim qualifications and tribal identity. The latter quality is important and is the basis for the *esprit de corps* found in high-functioning units of all types. Since tribal identity both cuts across branch lines and expresses itself on the battlefield, it is a powerful force within the Army. The story of the black beret vividly illustrates this power as it relates to change. Army Chief of Staff Gen. Eric Shinseki's 2001 decision to make a black beret, formerly reserved for the 75th Infantry Regiment (Ranger), the Army's standard headwear led to protests across the Ranger community. Shinseki saw the beret as symbolic—the beret proved the Army was on the road to the objective force. The Rangers saw the beret decision as insulting. Strong protests from the Ranger community led Shinseki ultimately led Shinseki to approve a tan beret (tan) for Rangers. To allow both parties to save face, the Army invented a narrative linking the Ranger beret's color to both past accomplishments and "the leather caps worn by the original Rangers of American heritage and lore." [30] The story is farcical in many respects but illustrates the power of tribal culture and soldier acceptance.

Branch parochialism has influenced change from the Army's beginnings. Each branch has a specific function and there is little debate over obvious core functions. For instance, infantry squad doctrine is the responsibility of the Infantry School. When things become more complex, parochialism presents itself. Crafting doctrine for a heavy brigade combat team consisting of multiple battalions and all its supporting units it becomes the responsibility of the Combined Arms Center (CAC) rather than the Infantry or Armor school. The CAC incorporates officers from virtually every branch, and their contributions to the doctrine writing efforts reflect their professional experiences, including those within their branch. Branch parochialism is both desirable and a curse. While the doctrine is where branch parochialism most frequently asserts itself, the other DOTMLPF considerations provide additional opportunities. Every major technological advance in protection, firepower, mobility, cybernetics, communications, or logistics finds champions within at least one branch—and sometimes several. This change, particularly as it relates to tactical capabilities, strongly displays branch perspectives.

Institutional norms represent the proverbial Gordian Knot. They are a synthesis of individual soldier, tribal and branch attitudes, and standards. They also reflect the desires imposed by American society, the Department of Defense, and the Department of the Army. A recurring challenge remains defining these norms, and then bringing the Army into compliance with them. Most often, these reflect changing social norms. A recent example of this is the Army's failure to end sexual harassment or assault. The Army recognized this as a problem as early as 1982 and developed training programs to combat it. Despite this, the problem persisted. In 2005, the Defense Department issued Directive 6495.01, "Sexual Assault and Response (SAPR) Program." Updated in 2013, the directive required the Army to establish a comprehensive sexual harassment and assault prevention program. This included not only training soldiers in acceptable standards but providing a broad range of support services for victims. In 2020, a series of incidents at Fort Hood, Texas, resulted in a high-profile investigation that found sexual harassment and assault remained a major concern. Establishing new institutional norms required two actions. First, the Army had to promulgate the new norm (itself a very bureaucratic action). Second, it had to adjust soldier attitudes and tribal and branch culture to create the desired outcome.[31]

Implications—Contracted Expertise

Contracted expertise is not one of the pressures constituting internal forces. It does, however, touch on the elements of leadership, self-conceptualization, institutional norms, tribal culture, and soldier acceptance and thus warrants final consideration. Contractors exercised a significant role in developing the Improvised Explosive Device Task Force (IED TF) which became the Army's counter-IED organization and gave rise to the AWG. The Wexford Group both developed the basic concept for the IED Defeat Task Force and provided the initial cohort of contracted personnel for the AWG. The Burdeshaw Group put together the Operational and Organizational Concept for the AWG. Both Wexford and Burdeshaw leveraged the expertise of retired or former military personnel whose service provided their basic skills. The Army paid for these skills twice: first in developing them while contractors served on active duty and then in the form of contracted service as civilians. Despite the cost, contracted employees, especially by Wexford, represented a bargain. The demands of operations in Afghanistan, Iraq, and elsewhere meant the full employment of the seven Special Forces groups in both the regular and reserve force structure. The only way Special Forces could provide similar support to the conventional force would be by abandoning operations in one or more locations. Such a course of action entailed tremendous risk by reducing the Special Operations presence in the combat zones.[32]

Wexford employees provided an additional advantage: the distinctions of rank largely disappeared from relationships. Contractors did not have military rank, but their tactical and training expertise gave them access to all levels of a unit's chain of command. Contracted Operational Advisors successfully embedded with units, supported pre-deployment training, and accompanied units during combat operations. During combat operations, advisors provided on-the-spot advice to both small soldiers and leaders. Such activity entailed considerable personal risk. Contracted Operational Advisors found themselves in hundreds of firefights and IED attacks during their tours and suffered casualties with four wounded and one killed in action. Retired Special Forces Master Sergeant Robert "Pitt" Pittman received a posthumous Secretary of Defense Medal of Valor, the civilian equivalent of the Distinguished Service Cross in recognition of his battlefield performance during a company-sized engagement in Afghanistan. Killed in action near OBJECTIVE BAKERSFIELD (approximately ten kilometers northwest of Kandahar), the tactical objective of the 1st Battalion, 320th Field Artillery, Pittman was advising the battalion commander during a difficult fight. Multiple IED detonations and the death and injury of several soldiers occurred during this battle. The battalion was not fighting as an artillery unit, rather, it deployed as a provisional infantry battalion (another example of the Army's lack of readiness for the fights it faced). Its commander, Lt. Col. David Flynn, noted two aspects of Pittman's performance in a letter that included the AWG as one of the addressees. He credited the decisions he made to Pittman's sage advice. Flynn also reported that Pittman always volunteered for the most dangerous missions with the troops most needing his expertise.[33]

Occupying such places on the battlefield typified the use of contracted operational advisors. These civilians preferred to be with the soldier at the lowest level conducting the most dangerous operations. The fact that such individuals had spent years in training and conducting combat operations meant they possessed a tremendous advantage over the soldiers serving their first combat tours. Embedding in a small unit environment, where requirements were incredibly fluid, placed operational advisors where commanders could best leverage their tactical strengths. Their ability to pass on their skills facilitated tremendous growth in an infantry squad or platoon's overall capabilities while concurrently exposing them to problems not yet appreciated by senior leaders.[34]

The Final Analysis

Despite an impressive record of service and sacrifice and the regard of many within the Army, the AWG simply could not compete with new priorities. By the early 2020s, the Army had largely turned its back on insurgencies. The 65 campaign streamers on the Army flag notwithstanding, the service appears to believe such conflicts do not represent a priority. Counterinsurgency is akin to the proverbial "crazy uncle" about whom families avoid speaking. The campaigns of World War II ended with decisive, if costly, victories and provided the Army with a cultural touchstone. The Army's new "Army Green Service Uniform" (actually brown and taupe in color) vividly expresses the service's current attitude. Officially, it honors the generation of soldiers who won World

War II. Unofficially, it recognizes the fact that the Army has not achieved a truly decisive outcome since that war. The victory in Kuwait in 1991 only set the stage for IRAQI FREEDOM. Toppling Saddam Hussein's regime was an impressive feat, but subsequent strategic mismanagement coupled with insufficient forces on the ground meant IRAQI FREEDOM soon turned into a long, expensive, and inconclusive fight. Operations in Afghanistan tell a similar story culminating in an embarrassing 2021 withdrawal. The counterinsurgencies of the twentieth and twenty-first centuries fail to provide happy endings when the enemy refuses to submit and the nation declines to commit all the resources required to change its mind. Only in the Philippines did the Army achieve a clear-cut victory. Little wonder the Army would rather focus on the types of wars it does well.[35]

By 2021, the "Big 5" had served two generations of soldiers. All five systems benefitted from ongoing investments that improved capabilities, but prudence called for fielding replacements. Consistent with its self-conceptualization, the Army's modernization emphasis shifted to highly sophisticated systems. To replace tanks and infantry fighting vehicles the Army now seeks "optionally-manned vehicles" incorporating artificial intelligence and advanced robotics linked to an intelligence network providing real-time information on the enemy and the environment. To replace its current helicopter fleet, the Army wants aircraft incorporating stealth and tilt-rotor technologies allowing greater speed and range. Such weapons are more suited to defeating a large, mechanized opponent and small groups of insurgents. The Army's evolution continues, but like the dinosaurs, the AWG's evolutionary line proved unsustainable. The AWG reflected the vision of Schoomaker and Cody—how they saw a critical problem facing the Army and then successfully bent the bureaucracy to solve it. Today, the Army's senior leadership sees the critical problem residing in fighting the type of war posited by the revolution in military affairs that drove the thinking behind Shinseki's objective force. With an institutional structure optimized to support such a force, there is far less need to manipulate bureaucratic processes.

In removing the AWG from the Army's rolls, the service is unwittingly accepting two fallacies. The first is that it will not again face the sort of insurgencies confronted in Afghanistan and Iraq. The Army may not want to face such fights again, but adversaries have learned from observing these conflicts. They have seen the effects of such wars both on the battlefield and on the American home front. To think opponents will ignore the lessons of these fights is to willingly blind one's self to options an enemy may exercise. The second fallacy is that future enemies will only employ capabilities in ways that play to the Army's strengths. This means adversaries will not operate in the "gray zone" described by David Barno and Nora Bensahel in *Adaptation Under Fire: How Militaries Adapt in Wartime.* It also means an enemy will ignore opportunities because we cannot or will not recognize them. Finally, it means opponents will neither exploit emerging technologies nor employ large units in ways fundamentally different from the Army. Such thinking ignores the fact that humans control armies, and different humans think about things differently. People who think differently are the types of people who provide unpleasant surprises in wartime. While an insurgency spawned the AWG, the organization identified a wide array of asymmetric methods available in a more conventional conflict. In eliminating the AWG, the Army eliminates the one unit focusing on asymmetric warfare in all the domains in which the Army operates.

The Asymmetric Warfare Group left two enduring legacies. The first is the soldiers who benefitted from the unit's operations. Most significantly, this cohort includes those whose thinking about basic tactics changed. It also includes those who benefitted from the tactics, techniques, and equipment generated by AWG efforts. While the Army has largely closed the door on insurgencies, it recognizes the need to fight in and under cities and the AWG fundamentally changed the way the Army considers these challenges. Similarly, the range of products the AWG produced that did not focus on counterinsurgency demonstrated its ability to inform the Army on a broad range of asymmetric warfare issues. Unfortunately, the circulation of these products all too often did not include the Army staff. Consequently, this staff was unable to assess or appreciate the AWG's true value in a changing world.

The second is the AWG provided a case study on how to effect change in an Army at war. Changing a large institution is difficult, particularly when its fundamental structure tends towards stasis. Despite this, Schoomaker, Cody, Lovelace, Shaw, Mis, Mahaney, Birch, Devens, Akridge, and the AWG's other personnel proved the difficulties are, in fact, surmountable. Equally important, they proved the benefits outweighed the difficulties encountered along the way. Unfortunately, the AWG's end proved a tragic corollary. If the change does not align with perceptions of the next war, it will not endure. This is unfortunate as predictions of the next conflict are rarely accurate. In deactivating the AWG, the Army improperly identified the AWG as a "one-trick pony" suited only for counterinsurgency and consequently accepted tremendous risk. This risk resides in losing its ability to think critically about the problems determined actors can present by being smart enough to challenge the U.S. Army's weaknesses rather than its strengths.

Only time will tell if McCarthy and McConville's decision to deactivate the AWG was the correct one. Some doubt the decision was correct. Pat Mahaney succinctly argued against it and noted the "AWG will need to be re-created in some form, because the complex, emerging challenges that it works on are not going away."[36] One might like to be optimistic. Military history, however, is largely a chronicle of failures rooted in unrealistic optimism. The latest proof of this observation resides in Russia's failure to rapidly realize its objectives in Ukraine in early 2022.

Endnotes

Introduction

[1] Jim McDonough, "Change Can Be Hard, Irrelevance is Easy," https://ivmf.syra zcuse.edu/2017/10/06/change-can-hard-irrelevance-easy/#:~:text=Retired%20U.S.%20Army%20General%20Eric%20Shinseki%20in%20December,change,%20you're%20going%20to%20dislike%20irrelevance%20even%20more" (accessed, 4 January 2020).

[2] John S. Brown, *Kevlar Legions: The Transformation of the U.S. Army, 1989-2005* (Washington, D.C.: Department of the Army, 2011), 27; Christopher G. Pernin, Elliot Axelband, et al., *Lessons from the Army's Future Combat Systems Program* (Santa Monica: The Rand Corporation, 2012), 1-3; U.S. Government, Office of the Secretary of Defense, Director of Operational Test and Evaluation, "FY 2002 Annual Report," https://www.dote.mil/Portals/97/pub/reports/FY2002/other/2002DOTEAnnualRpt.pdf?ver=2019-11-07-180204-860 (accessed, 4 January 2020).

[3] U.S. Government, Department of Defense, Joint Chiefs of Staff, *Joint Publication 1-02, Department of Defense Dictionary of Military and Associated Terms, 8 November* 2010 (Arlington: United States Department of Defense, 2010, As Amended Through 15 February 2016), 17.

[4] Lida Mayo, *United States Army in World War II—The Technical Services: The Ordnance Department: On Beachhead and Battlefront* (Washington, D.C.: Center of Military History, 1968), 253-254; John G. Westover, ed., *U.S. Army in Action Series: Combat Support in Korea* (Washington, D.C.: Center of Military History, 1955 [1987 reprint]), 111-114; Reid Kirby, "Operation Snoopy: The Chemical Corps' People Sniffer, "*CML: Army Chemical Review,* January-June 2007 (Ft. Leonard Wood: U.S. Army Maneuver Support Center, 2007), 20-22; Tom Mangold and John Penycate, *The Tunnels of Cu Chi: The Untold Story of Vietnam* (New York: Random House, 1985), 110; U.S. Government, Department of Defense, *Joint Publication 1-02, Department of Defense Dictionary of Military and Associated Terms as of January 2019* (Arlington, VA: The Joint Staff, 2019), 2.

[5] U.S. Government, Department of Defense, Department of the Army, *FM 100-15, Corps Operations* (Washington D.C.: Department of the Army, 1989), 3-0, 5-12, B3; U.S. Government, Department of Defense, Department of the Army, *FM 71-100, Division* Operations (Washington, D.C.: Department of the Army, 1990), 1-7, 5-6; U.S. Government, Department of Defense, Department of the Army, *FM 17-15 (Test), Tank Platoon: Division 86* (Washington D.C.: Department of the Army, 1984), 2-50.

[6] Department of the Army, *Army Regulation 350-1: Army Training and Leader Development* (Washington, D.C.: Department of the Army, 2014), 2.

Chapter 1

[1] Andrew J. Bacevich, *The Pentomic Era: The U.S. Army Between Korea and Vietnam* (Washington, D.C.: National Defense University Press, 1986), 4.

[2] John F. Dulles "The Evolution of Foreign Policy," Evolution of Foreign Policy Text Of Speech By John Foster Dulles, Secretary Of State, Before The Council On Foreign Relations, New York, N. Y., January 12, 1954 (Department Of State, Press Release No. 8, 1954): John Foster Dulles: Free Download, Borrow, and Streaming: Internet Archive (accessed, 20 October 2023).
[3] T. R. Fehrenbach, *This Kind of War: A Study in Unpreparedness* (Washington, D.C.: Brassey's, 1986), x; Maxwell D. Taylor, "Transcript of Remarks Made by General Maxwell D. Taylor, Chief of Staff, United States Army at the Conference of Civilian Aides to the Secretary of the Army, 6 April 1956, Army Heritage and Education Center, papers of General (RET) Maxwell D. Taylor, Box 1; Address by Maxwell D. Taylor to the U.S. Army Infantry Conference, Ft. Benning, Georgia, December 5, 1958, Army Heritage and Education Center, papers of General (RET) Maxwell D. Taylor, Box 1; Address by Maxwell D. Taylor at the Tenth Annual Reunion of the 101st Airborne Division Association, September 3, 1955 Army Heritage and Education Center, papers of General (RET) Maxwell D. Taylor, Box 2; Robert W. Coakley, "The Army as a Flexible Instrument of National Policy" in Gordon B. Turner and Richard Challener, eds., *National Security in the Nuclear Age: Basic Facts and Theories* (New York: Praeger, 1960), 258; Bacevich, *The Pentomic Era,* 20-21.
[4] Janice E. McKenney, *The Organizational History of Field Artillery, 1775-2003* (Washington, D.C.: Center of Military History, 2007), 169, 190-92; Donald A. Carter, *The U.S. Army Before Vietnam, 1953-1965* (Washington, D.C.: Center of Military History, 2015), 25.
[5] John B. Wilson, *Maneuver and Firepower: The Evolution of Divisions and Separate Brigades,* (Washington, D.C.: Department of the Army, 1998), 279-84; Department of the Army., *Artillery Trends: Reference Data Issue* (Ft. Sill: U.S. Army Artillery and Missile School, 1966), 10; McKenney, *Organizational History,* 214-15, 243; Address by Maxwell D. Taylor before Annual Convention of the Military Order of the World Wars, Annapolis, Maryland, October 19, 1956. Hereafter Taylor, "Address to Order of the World Wars;" Army Heritage and Education Center, papers of General (RET) Maxwell D. Taylor, Box 2; Department of the Army, *Field Manual 100-5, Operations* (Washington, D.C.: Department of the Army, 1962), 1; Department of the Army, *Field Manual 100-5, Operations (with Change 3)* (Washington, D.C.: Department of the Army, 1958), 1-44.
[6] Brian M. Linn, *The Echo of Battle: The Army's Way of War* (Cambridge: Harvard University Press, 2007), 99; Bowling W. Smith and William C. Gaines, "Coast Artillery Organization: A Brief Overview," http://cdsg.org/wp-content/uploads/pdfs/FORTS/C ACunits/CACorg20 11.pdf; Shelby Stanton, *World War II Order of Battle* (New York: Galahad Press, 1984), 37-38; Maxwell D. Taylor "Statement on the United States Army's Role in Continental Air Defense," May 22, 1959, Army Heritage and Education Center, Papers of General (RET) Maxwell D. Taylor, Box 2; Address by Maxwell D. Taylor to the U.S. Army Infantry Conference, Ft. Benning, Georgia," 5 December 1958; Taylor, Maxwell, "Address to Order of the World Wars," Annapolis, Maryland, October 19, 1956, Army Heritage and Education Center, papers of General (RET) Maxwell D. Taylor, Box 1; Address by Maxwell D. Taylor, to the Committee of One Hundred Surf Club, Miami, Florida, "January 10, 1956, Army Heritage and Education Center, papers of General (RET) Maxwell D. Taylor, Box 1; Bacevich, *The Pentomic Era,* 25.
[7] Lewis Sorley, *Thunderbolt, From the Battle of the Bulge to Vietnam and Beyond: General Creighton Abrams and the Army of His Times* (New York: Simon and Schuster, 1992), 139.
[8] Combat Studies Institute, *CSI Report No. 14, 60 Years of Reorganizing for Combat: A Historical Trend Analysis* (Ft. Leavenworth, KS: Combat Studies Institute, 1999), 22; Taylor, *Swords and Plowshares,* 153; Wilson, *Maneuver and Firepower,* 281-82; Roger J. Spiller, *"Not War, But Like War": The American Intervention in Lebanon* (Ft. Leavenworth, KS: Combat

Studies Institute, 1981), 26, 37; Gary W. Wade, *Rapid Deployment Logistics: Lebanon, 1958,* (Ft. Leavenworth, KS: Combat Studies Institute, 1984), 85-88.

[9] *CSI No. 14,* 23; Association of 3rd Armored Division Veterans, "3AD Unit Pages," http://www.3ad.org/unit-web-sites/ (accessed, 22 October 2018); California State Military Museum; 40th Armored Division Order of Battle, 1968, (accessed, 28 October 2018); Wilson, *Maneuver and Firepower,* 296.

[10] Wilson, *Maneuver and Firepower,* 308; J. Ronald Fox. *Defense Acquisition Reform, 1960-2009: An Elusive Goal,* (Washington, D.C.: Center of Military History, 2011), 35-37; James E. Hewes, *Special Studies: From Root to McNamara, Army Organization and Administration* (Washington, D.C.: Center of Military History, 1975), 311; U.S. Government, Department of Defense, Department of the Army, U.S. Army War College, *How the Army Runs: A Senior Leader Reference Handbook, 2011-2012,* (Carlisle: U.S. Army War College, 2011), 131.

[11] Hewes, *Special Studies, 316;* Department of the Army, *Reader's Guide: The United States Army in World War II,* (Washington, D.C.: Center of Military History, 1992), 99-101.

[12] Hewes, *Special Studies,* 316, 325-35, 404-08; *Department of Defense Reorganization Order 10,* dated January 10, 1962; *Department of the Army General Order 8,* dated February 15, 1962; John L. Romjue, *The Army of Excellence: The Development of the 1980's Army* (Fort Monroe, VA: US Army Training and Doctrine Command, Office of the Command Historian and the Center of Military History, 1997), 6.

[13] Ingo Trauschweizer, *The Cold War U.S. Army: Building Deterrence for Limited War* (Lawrence: University Press of Kansas, 2008), 180-83; State of California National Guard, *Military Support of Law Enforcement During Civil Disturbances: A Report Concerning the California National Guard's Part in Suppressing the Los Angeles Riot, August 1965,* 6-9, www.militarymuseum.org ; (accessed, 23 October 2018); National Guard Association of the United States, "National Guard Association of the United States Report to Secretary of the Army Resor, July 28, 1967" (Washington, D.C.: National Guard Association of the United States, 1967), National Guard Association of the United States archives, Washington, D.C.; Paul J. Scheips, *The Role of Federal Military Forces in Domestic Disorders, 1945-1992,* (Washington, D.C.: Center of Military History, 2012); 30-31; U.S. Government, Department of Defense, Department of the Army, "Airborne Sensor Operator Checklist, 1968" (Ft. Huachuca, AZ: USACSS, 1968), October 2018); Richard P. Winert, *A History of Army* Aviation (Fort Monroe, VA: Office of the Command Historian, U.S. Army Training and Doctrine Command, 1991), 175-76. James W. Williams, *A History of Army Aviation From Its Beginnings to the War on Terror* (New York: iUniverse, 2005), 120-21, 132-33.

[14] Mark A. Lawrence, *The Vietnam War: An International History,* (Oxford: Oxford University Press, 2008), 56; Francis J. Kelly, USA, *Vietnam Studies: U.S. Army Special Forces: 1961-1971,* (Washington, D.C.: Center of Military History, 1973), v, 19-49, 133-48, 161, 166-68, 180; Gabriel Kolko. *Anatomy of a War: Vietnam, the United States, and the Modern Historical Experience,* (New York: The New Press, 1985), 111; Fredrik Logevall. *Choosing War: The Lost Chance for Peace and the Escalation of the War in Vietnam,* (University of California Press, 1999), xiii; Jeffrey J. Clarke, *The United States Army in Vietnam: Advice and Support: The Final Years, 1965-1973,* (Washington, D.C.: Center of Military History, 1988), 53-61, 195, 198, 201-05; Charles C. Pritchett, "End-of-Tour Interview, Captain Joseph W. Kinzer, Advisory Team 163, Liaison Officer, 3d Airborne Brigade, Airborne Division, 11 June 1968," https://history.army.mil/documents/vietnam/vnit/vnit0101.htm (accessed, 13 November 2018); Ron Thayer, e-mail to Author, October 29, 2018; Department of the Army, *The Development and Training of the South Vietnamese*

Army, 1950-1972 (Washington, D.C.: Center of Military History, 1975), 38; Department of the Army. "Lineage and Honors, 5th Special Forces Group, 1st Special Forces Regiment, https://history.army.mil/html/forcestruc/lineages/branches/sf/005.htm; (accessed, 21 October 2018); William G. Bell, *Department of the Army Historical Summary, Fiscal Year 1970* (Washington, D.C.: Center of Military History, 1973), 16; "Lineage and Honors, 7th Special Forces Group, 1st Special Forces Regiment, https://history.army.mil/html/forces truc/lineages/branches/sf/007sfgp1sf.htm (accessed, 21 October 2018); "Lineage and Honors, 10th Special Forces Group, 1st Special Forces Regiment, (accessed, October 21, 2018), https://history.army.mil/html/forcestruc/lineages/branches/sf/010sfgp1sf.htm (accessed, 21 October 2018); "Lineage and Honors, 19th Special Forces Group, 1st Special Forces Regiment, https://history.army.mil/html /forcestruc/lineages/branches/sf/019sfgp 1sf.htm; (accessed, 21 October 2018); United States Special Operations Command, "Special Operations Forces Medal of Honor Recipients," https://www.socom.mil/hall-of-heroes/med al-of-honor (accessed, 26 October 2018).

[15] Williams, *A History of Army Aviation*, 96; 1st Cavalry Division Association, "Vietnam War, 1965-1972," http://www.first-team.us/tableaux/chapt_08/ (accessed, 24 October 2018); United States Army Aviation Center of Excellence, "1st Aviation Brigade," http://www.ru c ker.army.mil/usaace/1ab/ (accessed, 24 October 2018).

[16] Richard Crockatt, *The Fifty Years War: The United States and The Soviet Union in World Politics, 1941-1991*, (New York: Routledge, 1995), 145; Francis J. Gavin, "The Myth of Flexible Response: United States Strategy in Europe during the 1960's," *The International History Review*, 23, (December 2001): 847-75, https://www.belfercenter.org/sites/default/files/l egacy/files/CMC50/FrancisGavinTheMythOfFlexibleResponseUnitedStatesStrategyIn EuropeDuringThe1960sInternationalHistoryReview.pdf (accessed, 23 October 2018); Michael D. Doubler, *I Am the Guard: A History of the Army National Guard, 1636*-2000 (Washington, D.C.: National Guard Bureau, 2001), 220-21; Federal Research Service, Library of Congress. *Historical Attempts to Reorganize the Reserve Components, A Report Prepared by the Federal Research Division, Library of Congress under an Interagency Agreement with the Commission on the National Guard and Reserve* (Washington, D.C.: Library of Congress, 2007), 8-9, 13.

[17] Graham A. Cosmas, *The United States Army in Vietnam: MACV, The Joint Command in the Years of Withdrawal, 1968-1973* (Washington, D.C.: Center of Military History, 2006), 61, 84; William G. Bell, *Department of the Army Historical Summary, 1970* (Washington, D.C.: Department of the Army, 1973), 27; United States Army War College, Study on Military Professionalism, (Carlisle: United States Army War College, 1970).

[18] Department of the Army, *Field Manual 100-2-3, The Soviet Army: Troops, Organization and Equipment* (Washington, D.C.: Department of the Army, 1991), 5-22, 5-44, 5-59, 5-102. George W. Gawrych, *Leavenworth Papers Number 21, The 1973 Arab-Israeli War: The Albatross of Decisive Victory*, (Ft. Leavenworth: Combat Studies Institute, 1996), 81; U.S. Government, Department of Defense, Department of the Army, "Horseblanket on Transformation," Papers of GEN George W. Casey, Jr., Chief of Staff, United States Army, Box 5, Annual Reviews and Historical Interviews, National Defense University, Ft. Lesley J. McNair, Washington, D.C., 20319. Hereafter "HQDA Transformation Horseblanket."

[19] U.S. Army War College, Professionalism Study, iii.

[20] Sorley, *Thunderbolt*, 346.

[21] U.S. Army War College, Professionalism Study, 13-15; Selective Service System, "The Vietnam Lotteries", https://www.sss.gov/About/History-And-Records/lotter1; (accessed, 13 June 2018); Sorley, *Thunderbolt*, 341, 346.

[22] Jean R. Moenk and Brooks E. Kleber, *Operation Steadfast Historical Summary: A History of the Reorganization of the Continental Army Command, 1972-1973* (Ft. McPherson, GA and Ft. Monroe, VA: Office of the Command Historian, U.S. Army Forces Command and U.S. Army Training and Doctrine Command, 1974), 2, 12, 19, 23, 26, 34, 146; Department of the Army, Memorandum from Brigadier General Jones to Chief of Staff, CONARC, March 15, 1972, OPERATION STEADFAST files, Box 3, archives of the U.S. Army Training and Doctrine Command, Fort Eustis, VA. William G. Belle and Carl E. Cocke, eds., *Department of the Army Historical Summary, Fiscal Year 1972*, (Washington, D.C.: Center of Military History, 1977), 45; U.S. Government, Department of the Defense, Department of the Army by LTC Palmiei, Talking Paper, Subject: TRADOC and CDC, 28 November 1972, OPERATION STEADFAST files, Box 1, archives of the U.S. Army Training and Doctrine Command, Ft. Eustis, VA; U.S. Government, Department of Defense, Department of the Army by Maury S. Crallel, Message DTG 241907MAR72Z, Subject: Installation Organization Charts, OPERATION STEADFAST files, Box 1, archives of the U.S. Army Training and Doctrine Command, Ft. Eustis, VA. Department of the Army, *Report of the Special Review Panel on Department of the Army Organization, Volume I, Executive Summary, 1 March 1971*, I-2. STEADFAST BRIEFINGS file, archives of the U.S. Army Training and Doctrine Command, Ft. Eustis, VA; U.S. Government, Department of Defense, Department of the Army, REORG ONE DETAILED PLAN BRIEFING FOR CDC COMMANDERS' CONFERENCE (25 Jul 72), STEADFAST BRIEFINGS file, archives of the U.S. Army Training and Doctrine Command, Ft. Eustis, VA; Department of the Army, Operation Steadfast Detailed Plan, Volume B, 20 July 1972, B-1-1, OPERATION STEADFAST files, Box 2, archives of the U.S. Army Training and Doctrine Command, Ft. Eustis, VA; Anne Chapman, Carol J. Lilly, John Romjue and Susan Canedy, *Prepare the Army for War: A Historical Overview of the Army Training and Doctrine Command, 1973-1998* (Ft. Monroe, VA: U.S. Army Training and Doctrine Command, 1998), 5, 90; Department of the Army, Continental Army Command, General Orders No. 21, 23 February 1973, STEADFAST Collection, Box 10, archives of the U.S. Army Training and Doctrine Command, Ft. Eustis, VA; Department of the Army, Continental Army Command, General Orders No. 59, 19 March 1973, STEADFAST Collection, Box 10, archives of the U.S. Army Training and Doctrine Command, Ft. Eustis, VA; U.S. Government, Department of Defense, Department of the Army, Office of the Chief of Staff, Memorandum, Subject: Guidance for Implementation, CONUS Reorganization 1973, STEADFAST Collection, Box 5, archives of the U.S. Army Training and Doctrine Command, Ft. Eustis, VA.

[23] HQDA Transformation Horseblanket; Richard W. Stewart and Edgar Raines, *Operation Urgent Fury: The Invasion of Grenada*, (Washington, D.C.: Center of Military History, 2008), 25; R. Cody Phillips, *Operation Just Cause: The Incursion into Panama* (Washington, D.C: Center of Military History, 2004.), 20, 26, 34; Frank N. Schubert, Teresa Krause eds., *The Whirlwind War: The United States Army in Operations Desert Shield and Desert* Storm (Washington, D.C.: Center of Military History, 1995), 246-250, 252, 254-256; Hereafter Whirlwind War.

[24] HQDA Transformation Horseblanket.

[25] Department of the Army, *Field Manual 100-5: Operations* (Washington, D.C.: Department of the Army, 1976), 1-4.

[26] Paul H. Herbert, *Deciding What Has to be Done: General William E. DePuy and the 1976 Edition of FM 100-5, Operations* (Fort Leavenworth: Combat Studies Institute, 1988), 38-42; Frank N. Schubert and Theresa L. Kraus, eds., *Whirlwind War,* 27; *FM 100-5 (1976),* 1-1-1-4.
[27] John L Romjue, *From Active Defense to Airland Battle: The Development of Army Doctrine, 1973-1982* (Fort Monroe, VA: Historical Office, United States Army Training and Doctrine Command and Center of Military History, 1984), 8, 10; Edith M. Boldan, ed., *Department of the Army Historical Summary, Fiscal Year 1979* (Washington, D.C.: Center of Military History, 1982),15-16.
[28] Wilson, *Maneuver and Firepower,* 379.
[29] *Fort Campbell Courier,* June 22, 1979; Schubert and Kraus, *Whirlwind War,* 31; David C. Trybula, *"Big Five" Lessons for Today and Tomorrow* (Alexandria, VA: Institute for Defense Analysis, 2012), 29; Cheryl Morai-Young, ed., *Department of the Army Historical Summary, Fiscal Year 1983* (Washington, D.C.: Center of Military History, 1990), 9, 12.
[30] Each combatant—soldier or vehicle—in the maneuver area carried MILES. Each weapon's laser fired a code corresponding to the weapon it replicated. The codes corresponded to the weapons capabilities. Riflemen or machine guns could kill riflemen while tanks could kill anything.
[31] Anne W. Chapman, *The National Training Center Matures, 1985-1983* (Fort Eustis, VA: US Army Training and Doctrine Command Military History Office, 1997), 5; Dale R. Steinhauer, *A Short History of the Center for Army Lessons Learned, 1985-2000* (Ft. Leavenworth: Combined Arms Center, 2000), 9.
[32] Department of the Army, *Field Manual 100-5, Operations* (Washington, D.C.: Department of the Army, 1982), 1-1.
[33] U.S. Naval War College, *Maritime Advanced Warfighting School,* https://usnwc.edu/college-of-naval-command-and-staff/Additional-Academic-Opportunities/Maritime-Advanced-Warfighting-School (accessed, 22 June 2018).
[34] HQDA Transformation Horseblanket.
[35] Romjue, *Army of Excellence,* 10, 67-82, 94, 99-107.
[36] The attempted hostage rescue, codenamed OPERATION EAGLE CLAW, effectively ended with a collision between two aircraft which killed eight Americans. This operation led to a rethinking of tasking, employing, and controlling Special Operations Forces (SOF) and the creation of the U.S. Special Operations Command (USSOCOM).
[37] Department of the Army, *FM 100-5, Operations,* (Washington, D.C.: Department of the Army, 1982), 15-1.
[38] Department of the Army, *FM 100-5, Operations* (Washington, D.C.: Department of the Army, 1986), 2, 6, 10, 15, 26-29, 33, 161-164.
[39] Mary L. Haynes, *Department of the Army Historical Summary, Fiscal Year 1987* (Washington, D.C.: Center of Military History, 1995), 69-75.
[40] US Special Operations Command, "US Special Operations Command," in *USSOCOM Fact Book: Introduction to US Special Operations Command* (Tampa: US Special Operations Command, 2006), https://www.socom.mil/FactBook/2006%20Fact%20Sheet.pdf; (accessed, 26 July 2018); Haynes, *Historical Summary,* 32; Department of the Army, *Field Manual 31-20: Doctrine for Special Forces Operations* (Washington, D.C.: Department of the Army, 1990), 3-1—3-2.
[41] Department of the Army, *Field Manual 25-100, Training the Force,* (Washington, D.C.: Department of the Army, 1988), i.
[42] Ibid., 1-4, Ibid., 2-1.

[43] U.S. Department of State, "Highlights in the History of U.S. Relations with Russia, 1780-June 2006," https://www.state.gov/p/eur/ci/rs/200years/c30273.htm#fall_communism (accessed, 29 January 2018); U.S. Department of State, "U.S. Relations with Kuwait," https://www.state.gov/r/pa/ei/bgn/35876.htm; (accessed, 29 January 2018); *Washington Post,* January 29, 1992; Wallace J. Theis, "Twenty-first Century Army," *Parameters* 21 (July 1991): 62-76, https://www.press.armywarcollege.edu/cgi/viewcontent.cgi?article=1612 &context=parameters (accessed, 8 January 2021).
[44] Ibid., 49.
[45] Rickey E. Smith, interview with Author, September 26,2018, Arlington, Virginia; Gordon Rudd, *Humanitarian Intervention: Assisting the Iraqi Kurds in Operation Provide Comfort, 1991* (Washington, D.C.: Center of Military History, 2004), 63; L. Martin Kaplan, *Department of the Army Historical Summary, Fiscal Year 1994* (Washington, D.C.: Center of Military History, 2000), 64; Dwight Oland and David W. Hogan, *Department of the Army Historical Summary, Fiscal Year 1992* (Washington, D.C.: Center of Military History, 2001), 49; Richard W. Stewart, *The United States Army in Somalia, 1992-1999* (Washington, D.C.: Center of Military History, 2002), 16-25, 49; Department of the Army, *FM 100-5, Operations* (Washington, D.C.: Headquarters, Department of the Army, 1993), ii, 1-4, 2-1, 4-1- 4-6, 5-1-5-5.
[46] Paul E. Menoher, "Force XXI: Redesigning the Army Through Warfighting Experiment," *Military Intelligence Professional Bulletin,* April-June 1996, 6-8.
[47] Jerry R. Bolzak, ed., *The Collected Works of the Thirty-Second Chief of Staff of the United States Army, Gordon R. Sullivan, General, United States Army, June 1991-June 1995,* (Washington, D.C.: Center of Military History, undated), 103-104; Oland and Hogan, *Historical Summary, 1992,* 72; U.S. Army Training and Doctrine Command, *TRADOC Pamphlet 525-5, Force XXI Operations* (Ft. Monroe, VA: U.S. Army Training and Doctrine Command, 1994), 1-3; John S. Brown, *Kevlar Legions: The Transformation of the United States Army, 1989-2005* (Washington, D.C.: The Center of Military History, 2011), 5.
[48] Cheryl Morai-Young, ed., *Department of the Army Historical Summary, Fiscal Year 1983* (Washington, D.C.: Center of Military History, 1990), 83.
[49] General Gordon R. Sullivan, USA to General Officers of the United States Army, November 10, 1994, "Force XXI Campaign Plan—Execution and Synchronization" in Bolzak, *The Collected Works,* 436-37; General Gordon R. Sullivan, USA Chief of Staff Message, March 7, 1994 "Building the Force for the 21st Century—Force XXI" in Bolzak, *The Collected Works,* 318-20; Patrick Cathcart, "TACFIRE Deployment and Training," *Field Artillery Journal,* January-March 1981, 10-13; Forest G. Clark, "TACFIRE and the Maneuver Commander," *Field Artillery Journal,* January-March 1981, 14-16; Department of the Army, "Lineage and Honors, Headquarters and Headquarters Battalion, 4th Infantry Division (Ivy Division), https://history.army.mil/html/forcestruc/lineages/branches/div/004id.htm; TRADOC Pamphlet 525-5, *Force XXI Operations* (Ft. Monroe: U.S. Army Training and Doctrine Command, 1994), 4-1, 4-2; Major General John W. Hendrix, USA, "Commandant's Note: Infantry Force XXI—Facing the Challenge" *Infantry* 85 (January-February 1995): 1-3. Lon Maggart, "Your Mind is Your Primary Weapon," *Armor: The Professional Bulletin of the Armor Branch,* 105 (July-August 1996): 5, https://www.benning.army.mil/armor/eARMOR/content/issues/1996/JUL_AUG/ArmorJulyAugust1996web.pdf (accessed, 10 October 2018); Timothy D. Cherry, "Future Command and Control Systems: IVIS and B^2C^2 Only Scratch the Surface," *Armor: The Professional Bulletin of the Armor Branch* 103 (November-December 1994): 16-18; U.S. Government, Department of Defense, Department of the Army,

U.S. Army Training and Doctrine Command, "Briefing: Force XXI Heavy Division "F" Series TOEs—As of 3 MAR 97" (Ft. Monroe: U.S. Army Training and Doctrine Command, 1997).

[50] William G. Bell, *Commanding Generals and Chiefs of Staff, 1775-2013* (Washington, D.C.: Center of Military History, 2013), 162; R. Phillip Cody, *Bosnia-Herzegovina: The U.S. Army Role in Peace Enforcement Operations, 1995-2004* (Washington, D.C.: Center of Military History, 2015), 7; North Atlantic Treaty Organization (NATO), *Peace Support Operations in Bosnia and Herzegovina,* https://www.nato.int/cps/en/natohq/topics_52122.htm?selectedLocale=en (accessed, 3 December 2018); North Atlantic Treaty Organization (NATO), *NATO Kosovo Force, Enduring Commitment: About Us,* https://jfcnaples.nato.int/kfor/about-us/ (accessed, 3 December 2018); HQDA Transformation Horseblanket; Dennis J. Reimer and James J. Carafano (ed.), *Soldiers are our Credentials: The Collected Works and Selected Writings of the Thirty-Third Chief of Staff, United States Army* (Washington, D.C.: Center of Military History, 2000), 29-35, 120, 212-13;Dennis J. Reimer, e-mail to Author, October 6, 2018.

[51] Chief, National Guard Bureau, *Annual Review of the Chief, National Guard Bureau, 1995* (Washington, D.C.: National Guard Bureau, 1995), 35; Chief, National Guard Bureau. *Annual Review of the Chief, National Guard Bureau, 1993* (Washington, D.C.: National Guard Bureau, 1993), 37-38; Richard G. Greene, Jr. "Assessing the Army National Guard's Enhanced Brigade Concept: Searching for Readiness And Relevance" (master's thesis, U.S. Army Command and General Staff College, 2003), 13; Reimer to Author, October 6, 2018.

[52] Reimer to Author, October 6, 2018.

[53] Reimer, *Soldiers,* 78, 207-08; Reimer to Author, October 6, 2018.

[54] Center of Military History, "Army Chiefs of Staff," https://history.army.mil/faq/FAQ-CSA.htm; (accessed, 17 January 2019); GlobalSecurity.org.,"Task Force Falcon/IFOR/SFOR Deployments," https://www.globalsecurity.org/military/ops/sfor-d eploy.htm; (accessed, 1 October 2018); Center of Military History, Transcript: "1999 Fletcher Conference, Wednesday, November 3, 1999 9:00 to 10:30 a.m., Panel 4, Serving the Nation in the 21st Century," https://www.history.army.mil/documents/Fletcher/Fletcher-99/F99-P4-T.htm (accessed, 29 October 2018); U.S. Government, Government Accountability Office, *Report to the Chairman, Committee on Armed Services, House of Representatives: Kosovo Air Operations—Army Resolving Lessons Learned Regarding the Apache Helicopter,* (Washington, D.C.: Government Accountability Office, 2001), 12-14; Stewart, *American Military History 2:* 447-48; R. Cody Phillips, *Operation Joint Guardian: The U.S. Army in Kosovo* (Washington, D.C.: Center of Military History, 2007), 16, "NATO's War Party," *Newsweek,* May 2, 1999, 44; Christopher N. Koontz, *Department of the Army Historical Summary, Fiscal Year 2001* (Washington, D.C.: Department of the Army, 2011), 1.

[55] Department of the Army, Briefing: "Army Transformation, 17 October 2000," 5, personal papers of Donald F. Schenk, Warren, Michigan; Department of the Army, *Operational and Organizational Concept for Future Combat Systems, Chapter 1: The Army's Transformation Strategy, 2 November 1999 Draft* (Washington, D.C.: Department of the Army, 1999), 6-9; Department of the Army, Briefing: "The Army Vision," (Washington, D.C.: Office of the Secretary of the Army and the Chief of Staff, 2000), 6, papers of Donald F. Schenk, Warren, Michigan.

[56] L. Martin Kaplan, *Department of the Army Historical Summary, Fiscal Year 1994* (Washington, D.C.: Center of Military History, 2000), 63; Department of the Army, *FM 3-0, Operations* (Washington, D.C.: Department of the Army, 2001), 1-15, 2-1.

[57] Donald F. Schenk, interview with Author, August 27, 2018; 1999 Fletcher Conference; Program Manager, Future Combat System, "White Paper: To Provide an Overview of the Future Combat Systems (FCS) Program, 30 September 2004" (Warren, Michigan: Program Manager, Future Combat Systems, 2004); 1-15.
[58] U.S. Government, Department of Defense, Department of the Army by Marion Van Fossen, Briefing: FCS Industry Day Brief, 11 January 2000, (Warren, Michigan: Program Manager, Future Combat System, 2000), 4.
[59] U.S. Army Training and Doctrine Command, Briefing: *The Operational Environment—Where We've Come From,* September 8, 2007, 1-4, Papers of Rickey E. Smith.
[60] Van Fossen, Industry Day Brief, 11 January 2000.
[61] Christopher G. Pernin and Elliot Axelbend et al, *Lessons from the Army's Future Combat Systems Program,* (Santa Monica, CA: RAND Corporation, 2012), 28; William M. Donnelly, *Department of the Army Historical Summary, Fiscal Year 2002* (Washington, D.C.: Center of Military History, 2012), 44; Kathleen J. Nawyn, *Department of the Army Historical Summary, Fiscal Year 2003* (Washington, D.C.: Center of Military History), 63; Schenk interview.
[62] Schenk interview.
[63] Donnelly, *Historical Summary 2002,* Schenk interview.
[64] Mark J. Reardon and Jeffrey A. Charlston, *From Transformation to Combat: The First Stryker Brigade at War* (Washington, D.C.: Center of Military History, 2007), 3, 5, 9; Donnelly, *Historical Summary 2002,* 24, 27; Nawyn, *Historical Summary 2003,* 24; Alan Vick, David Orlesky et al, *The Stryker Brigade Combat Team: Rethinking Strategic Responsiveness and Assessing Deployment Options* (Santa Monica, CA: RAND Corporation, 2002), 7.
[65] Schenk interview; Government Accountability Office, *Military Transformation: Army's Evaluation of Stryker and M113-A3 Infantry Carrier Vehicles Provided Sufficient Data for Statutorily Mandated Comparison* (Washington, D.C.: General Accountability Office, 2003), 1-3; John Hendren, "Army Holds its Ground in Battle with Rumsfeld," *The Los Angeles Times,* November 29, 2002, https://www.globalsecurity.org/org/news/2002/021129-sbct.htm Koontz, *Historical Summary 2001,* 23; Nawyn, *Historical Summary 2003,* 31.
[66] National Guard Bureau, "About the Army National Guard, History Since September 11, 2001,", http://www.nationalguard.mil/About-the-Guard/Army-National-Guard/ (accessed, 9 January 2021); Marc Acton, "Civil Support Teams - New Tactics For New Threats," http://www.nationalguard.mil/News/Article-View/Article/576636/civil-support-teams-new-tactics-for-new-threats/ (accessed, 9 January 2021); GlobalSecurity.org, "Weapons of Mass Destruction Civil Support Teams, https://www.globalsecurity.org/military/agency/army/wmd-cst.htm (accessed, 9 January 2021); Nawyn, *Historical Summary 2003,* 49-50, 58; U.S. Army Reserve, "Office of Army Reserve History, Post 9/11," https://www.usar.ar my.mil/OurHistory/SinceSept11/ (accessed, 9 January 2021).

Chapter 2

[1] Bevin Alexander, "We Cannot Win the War in Iraq," https://www.bevinalexander.com/commentary/we-cannot-win-war-iraq.htm (accessed, 8 August 2020).
[2] Exec. Order 10670, 21 C.F.R. 4069 (1956), https://www.archives.gov/federal-register/codification/executive-order/10670.html (accessed, 14 March 2020); Walter Kretchik, Robert F. Baumann and John T. *Fishel, Invasion, Intervention, "Intervasion:" A Concise History of the U.S. Army in Operation Uphold* Democracy (Fort Leavenworth, Kansas: Combat Studies Institute, 1997), 93-160; B.C. Mossman, *American Military History,* Morris J.

McGregor, ed. (Washington, D.C.: Center of Military History, 1989), 537-38, 544; R. Cody Phillips, *Bosnia-Herzegovina: The U.S. Army's Role in Peace Enforcement Operations, 1995-2004* (Washington, D.C.: Center of Military History, 2005), 5-41; James A. Sawicki, *Cavalry Regiments of the U.S. Army* (Dumfries, Virginia: Wyvern Publications, 1985), 152-71. Hereafter Sawicki, *Cavalry Regiments*; James A. Sawicki, *Infantry Regiments of the U.S. Army* (Dumfries, Virginia: Wyvern Publications, 1981), 43-44, 50-58, 60-66, 69, 77-80, 83, 87-88, 94-98, 268-69, 314-15; Robert K. Sawyer, *Military Advisors in Korea: KMAG in Peace and* War (Washington, D.C.: Office of the Chief of Military History, 1961), v, 40, 96-100; Thomas E. Skidmore and Peter H. Smith, *Modern Latin America* (Oxford: Oxford University Press, 1984), 307, 314; Richard W. Stewart, *The United States Army in Somalia, 1992*-1994 (Washington, D.C.: Center of Military History, 2002), 5-26; U.S. Army Center of Military History, "Listing of the Campaign Streamers Displayed on the U.S. Army Flag," https://history.army.mil/html/reference/campaigns.html (accessed, 13 March 2020); U.S. Army Special Operations Aviation Command, "Special Operations Aviation History," https://www.soc.mil/USASOAC/USASOAChome page.html (accessed, 21 April 2020).

3 Minnesota Historical Society, "The U.S.-Dakota War of 1862," http://www.mnhs.org/fortsnelling/learn/us-dakota-war (accessed, 16 March 2020); John Pope to H. H. Sibley, September 28, 1862, in U.S. War Department, *The War of the Rebellion: A Compilation of the Official Records of the Union and Confederate Armies,* 130 vols. (Washington, D.C.: Government Printing Office ,1880–1901), series 1st ser., vol. 13, pt. 1, 686. Hereafter cited as *OR.*

4 Dee Brown, *Bury My Heart at Wounded Knee: An Indian History of the American West* (New York: Holt, Reinhart, and Winston, 1970), 86-87.

5 Charles J. Brill, *Conquest of the Southern Plains: Uncensored Narrative of the Battle of the Washita and Custer's Southern Campaign* (Oklahoma City, OK: Golden Saga Publishers, 1938), 20, 155-56; Harry A. Kersey and Michael Petersen, "Was I a Member of Congress…": Zachary Taylor's Letter to John J. Crittenden, January 12, 1838, concerning the Second Seminole War," *Florida Historical Quarterly* 75, no. 4 (1997): 447-61, www.jstor.org/stable/30149063; (accessed, 14 March 2020); John K. Mahon, *History of the Second Seminole War, 1835-1842* (Gainesville: University Press of Florida, 1967), 27; CMH Campaigns; Allan R. Millett, Peter Maslowski, and William B. Feis, *For the Common Defense: A Military History of the United States from 1607 to 2012* (New York: Free Press, 2012), 127-28; Richard W. Stewart, ed. *The United States Army and the Forging of a Nation: 1775-1916* (Washington, D.C.: Center of Military History, 2005), 161-63; U.S. Government, Department of the Interior, *Letter of the Secretary of the Interior, Communicating, In Compliance with the Resolution of the Senate of the 14th Ultimo, Information Relating to the Late Battle on the Washita River, January 6, 1869,* 17; https://shareok.org/bitstream/ handle/11244/40073/Senate-40-3-Executive-13-Serial-1360.pdf?sequence=1&isAllowed=y, (accessed, 16 March 2020); Russell F. Weigley, *The American Way of War: A History of United States Military Strategy and Policy* (New York: MacMillan Publishing, 1973), 67-68; O. James Hollister, *Colorado Volunteers in New Mexico: 1862,* ed. Richard Harwell (Chicago: R.R. Donnelly, 1962), 97-100, 118, 142, 286; John M. Chivington, "To the People of Colorado. Synopsis of the Sand Creek Investigation, Denver, Colorado, June 1865," https://www-americanwest-amdigital-co-uk.libproxy.temple.edu/Documents/Details/Graff_698; (accessed, 13 December 2020); *OR,* ser. 1, vol. 41, pt. 1, 949; Brown, *Wounded Knee,* 91.

6 William B. Gardner, *American Military History,* Morris J. McGregor, ed. (Washington, D.C.: Center of Military History, 1989), 314.

[7] W.H. Hutchinson, foreword, *With General Crook in the Indian* Wars (Palo Alto, CA: Lewis Osbourne, 1968), 9-10 ; Brown, 358-362, 428-431; Brian W. Dippie, *The Vanishing American: White Attitudes and U.S. Indian Policy* (Middletown, CT: Wesleyan University Press, 1982), 130-131; Paul Magid, *The Gray Fox: George Crook and the Indian* Wars (Norman: University of Oklahoma Press, 2015), xiii, xv, 46-47, 78-79, 322-23; Joe Starita, "The Case of Standing Bear: Establishing Personhood under the Law," *Court Review* 45 (March 2008): 4–11, http://search.ebscohost.com.libproxy.temple.edu/lo gin.aspx?direct=true &db=lgs&AN=502599232&site=ehost-live&scope=site (accessed, 19 August 2020); Gardner, *American Military History,* 314; Gerald Vizenour, *Bear Island: The War at Sugar Point* (Minneapolis, Minnesota: 2006), xii; Janne Lathi, *Wars of Empire: Apaches, The United States, and the Southwest Borderlands* (Norman: University of Oklahoma Press, 2017), 167, 177, 206-07, 219.

[8] Andrew J. Birtle, *U.S. Army Counterinsurgency and Contingency Operations Doctrine, 1860-1941* (Washington, D.C.: Center of Military History, 1998), 104, 148, 153-68, 169-71; Harry G. Gole, "Don't Kill the Messenger: Vietnam War Reporting in Context," *Paramaters* 26 (Winter 1996-1997): 148-53, https://search-proquest-com.libproxy.temple.edu/docv iew/1306228507?accountid=14270&rfr_id=info%3Axri%2Fsid%3Aprimo (accessed, 20 April 2020); John H. Hay, *Vietnam Studies: Tactical and Material Innovations* (Washington, D.C.: Center of Military History, 1974), 3-9. Hereafter Hay, *Tactical Innovations*; Stanley R. Larson and James L. Collins, Jr., *Vietnam Studies: Allied Participation in Vietnam* (Washington, D.C.: Center of Military History, 1975), v, 25-137. Hereafter Larson and Collins; Allan R. Millet, "The General Staff and Cuban Intervention of 1906," *Military* Affairs 31 (Fall 1967): 113-19; Umio Otsuko, "Coalition Coordination During the Boxer Rebellion: How Twenty-Seven "Councils of Senior Naval Commanders" Contributed to the Conduct of Operations," *Naval War College Review* 71 (Autumn, 2018): 111-30, http://libproxy.te mple.edu/login?url=https://search-proquest-com.libproxy.temple.edu/docview/21092792 59?accountid=14270 (accessed, 1 June 2020); George S. Prugh, *Vietnam Studies: Law at War, 1964*-1973 (Washington, D.C.: Center of Military History, 1975), 61-78, 113-17, 123-26; Lars Schoultz, "Blessings of Liberty: The United States and the Promotion of Democracy in Cuba," *Journal of Latin American Studies* 34 (2002): 397–425, https://www-cambridge-org.libproxy.temple.edu/core/journals/journal-of-latin-american-studies/article/bless ings-of-liberty-the-united-states-and-the-promotion-of-democracy-in-cuba/F5FC FEA1CFE7F80584FFED393B982 D29 (accessed, 10 January 2021); "The Press: Farewell to the Follies," *Time,* February 12, 1973, http://content.time.com/time/magazine/article/0 ,9171,90 3831,00.html (accessed, 20 April 2020).

[9] Center of Military History, "Civil War Campaigns," https://history.army.mil/html/refere nce/army_flag/cw.html (accessed, 21 August 2020); Ted Ballard, *Staff Ride Guide: Battle of* Antietam (Washington, D.C.: Center of Military History, 2008), 75-76; Birtle, *Counterinsurgency and Contingency Operations,* 23; James McPherson, *Battle Cry of Freedom: The Civil War* Era (New York: Oxford University Press, 1988), 307, 784; Daniel E. Sutherland, *A Savage Conflict: The Decisive Role of Guerrillas in the American Civil War* (Chapel Hill, North Carolina: University of North Carolina Press, 2009), 19, 32, 61, 106, 132, 146. Hereafter Sutherland, *A Savage Conflict.*

[10] United States Statutes at Large 2 (1789-1848), 9th Cong., 1st Session, *Articles of War,* http://freepages.rootsweb.com/~pa91/military/cfawar.html (accessed, 18 March 2020).

[11] George W. Bacon, *The Life and Administration of Abraham Lincoln: Presenting his Early History, Political Career, Speeches, Messages, Proclamations, Letters, etc., with a General*

View of his Policy as President of the United States... Also the European Press on his Death (London: S. Low, Son, and Marston, 1865), 52, 58, 60, 71; Sutherland, *A Savage Conflict,* xi.

[12] Roberto Naranjo, "Partisan Warfare in the American Civil War," https://ehistory.osu.edu/articles/partisan-warfare-american-civil-war (accessed, 19 March 2020); Donald E. Sutherland, *American Civil War Guerrillas: Changing the Rules of* Warfare (Santa Barbara, CA: Praeger, 2013), 17-20, 38-43.

[13] John Gardner Bell, *Commanding Generals and Chiefs of Staff 1775-2005: Portraits and Biographical Sketches of the Army's Senior Officer* (Washington, D.C.: Center of Military History, 2005), 82; California State Military Museum, "Californians and the Military: Major General Henry Wager Halleck," http://www.militarymuseum.org/Halleck.html (accessed, 19 March 2020).

[14] Francis Lieber, *Guerilla Parties, Considered with Reference to the Laws and Usages of War, Written at the Request of Major-General Henry W. Halleck, General-in-Chief of the Army of the United States* (New York: Baker and Godwin Printers, 1862), 1 , https://heinonline-org.libproxy.temple.edu/HOL/Page?handle=hein.beal/guerp0001&id=3&collection=beal&index=(accessed, 19 March 2020).

[15] Ibid., 7, 15.

[16] Ibid., 1, 7, 15.

[17] U.S. War Department, General Orders Number 100 (Washington, DC: Adjutant General's Office, 1863) https://avalon.law.yale.edu/19th_century/lieber.asp#sec4 (accessed, 18 March 2020).

[18] Ibid.

[19] Mark L. Bradley, *The Army and Reconstruction,* 1865-1877 (Washington, D.C.: Center of Military History, 2015), 8.

[20] Warren D. Crandall, *History of the Ram Fleet and the Mississippi Marine Brigade in the War for the Union on the Mississippi and its Tributaries: The Story of the Ellets and their Men* (St. Louis, Missouri: Press of Buschart Brothers, 1907), 252; John D. Milligan, "Charles Ellet and His Steam Ram," *Civil War History* 9 (1963): 121-32, https://muse-jhu-edu.libproxy.temple.edu/article/415678/pdf (accessed, 22 March 2020); Robert R. Mackey, *The Uncivil War: Irregular Warfare in the Upper South, 1861*-1865 (Norman: University of Oklahoma Press, 2004) 55-58; H. W. Halleck to A. W. Ellet, November 11, 1862, in *OR* 3rd ser., vol. 2, 761-62.

[21] Sutherland, *A Savage Conflict,*178-79.

[22] Mackey, *The Uncivil War,* 66-68; J.M. Schofield to Brig. Gen. Prentiss, January 11, 1862, in *OR* 1st ser., vol. 8, 493-94; Sutherland, *A Savage Conflict,* 209.

[23] Birtle, *Counterinsurgency and Contingency Operations,* 41.

[24] Mackey, *The Uncivil War,* 58-60; U.S. Government, War Department, "General Orders No. 140, September 24, 1862, *OR,* 3rd ser., vol. 3, 586.

[25] "Text of the First Reconstruction Act, March 2, 1867," https://www.owleyes.org/text/reconstruction-acts-of-1867-and-1868/read/text-of-the-first-reconstruction-act#root-6 (accessed, 31 August 2020).

[26] Bradley, *Army and Reconstruction,* 15-16, 26, 38, 47, 50, 58-59, 62, 71; William A. Dobak, *Freedom by the Sword: The U.S. Colored Troops, 1862-1867* (Washington, D.C.: Center of Military History, 2011), 458-59, 464, 468-69, 471, 473; Eric Foner, *A Short History of Reconstruction* (New York: Harper Perennial Modern Classics, 1988), 184-87; Allen W. Trelease, *White Terror: The Ku Klux Klan Conspiracy and Southern Reconstruction* (New York: Harper & Row, Publishers, 1971), 36, 103, 135, 149-73, 180-83, 280; Otis A. Singletary,

Negro Militia and Reconstruction (Westport, CT: Greenwood Press, 1984), 13, 35-39; Margaret M. Storey, "The Military and Reconstruction, 1862–77," in *A Companion to American Military History*, J.C. Bradford, ed. (Hoboken, NJ: Wiley, 2009), 643-44, 646; Millet, Maslowski and Feis, *For the Common Defense*, 231; Sawicki, *Cavalry Regiments*, 152, 154, 156, 164, 167, 171; Albert G. Brackett, "The United States Cavalry, Past and Present," *United Service; A Quarterly Review of Military and Naval Affairs* 4 (August 1890), http://libproxy.temple.edu/login?url=https://www-proquest-com.libproxy.temple.edu/docview/126324331?accountid=14270 (accessed, 31 August 2020); George F. Price, *Across the Continent With the Fifth Cavalry* (New York: Antiquarian Press, 1959), 127-28; 6th Cavalry Museum, "A Brief History," https://www.6thcavalrymuseum.org/regimental-history (accessed, 31 August 2020); 42nd Cong., 2nd Session, Testimony Taken by the Joint Select Committee to Enquire into the Condition of Affairs in Late Insurrectionary States: Alabama, Vol. 3 (1872), Select Comm. 41-42, Part 10 at 1629, at 1717, at 1958, at 1975; 42nd Cong., 2nd Session, Testimony Taken by the Joint Select Committee to Enquire into the Condition of Affairs in Late Insurrectionary States: Mississippi, (1872), Select Comm. 41-42, Part 12 at 797, at 723, at 727, at 787-794, at 840, at 868; 42nd Cong., 2nd Session, Testimony Taken by the Joint Select Committee to Enquire into the Condition of Affairs in Late Insurrectionary States: North Carolina, (1872), Select Comm. 41-42, at 6, at 16, at 64, at 171, at 175, at 258, at 278-79; 42nd Cong., 2nd Session, Testimony Taken by the Joint Select Committee to Enquire into the Condition of Affairs in Late Insurrectionary States: South Carolina, (1871), Select Comm. 41-42, Part 5, at 1499, at 1509, at 1602; 42-1 (1871); 42nd Cong., 1st Session, Select Committee of the Senate to Investigate Alleged Outrages in the Southern States Ordered Printed March 10, 1871, S. Rep. 1-41, at XL, at XLI-II, at LXVII-VIII, at LXXVII, at 11, at 14, at 19, at 31, at 38-39, at 47, at 178, at 264, at 320, at 364, at 402 (1871); Andrew F. Lang, *In the Wake of the War: Military Occupation, Emancipation and Civil War* America (Baton Rouge, LA: Louisiana State University Press, 2017), 185, 189-80, 199, 204, 212-14, 217-18, 220, 225-26; Kenneth M. Stampp, *The Era of Reconstruction: 1865-1877* (New York: Alfred A. Knopf, 1966), 199-00; Mark W. Summers, *The Ordeal of the Reunion: A New History of Reconstruction* (Chapel Hill: The University of North Carolina Press, 2014), 42-43, 264-65; Gregory P. Downs, *After Appomattox: Military Occupation and the Ends of War* (Cambridge, MA: Harvard University Press, 2015), 216-17, 257-64; Douglas R. Egerton, *The Wars of Reconstruction: The Brief, Violent History of America's Most Progressive Era* (New York: Bloomsbury, 2014), 298.

[27] Dapo Akande and Lawrence Hill-Cawthorne, "The Lieber Code and the Regulation of Civil War in International Law" *Columbia Journal of Transnational Law* 53 (June 2015): 638–51, http://search.ebscohost.com.libproxy.temple.edu/login.aspx?direct=true&db=lgs&AN=108927711&site=ehost-live&scope=site (accessed, 22 August 2020); Rotem Gilhadi, *International Review of the Red Cross*, 94 (885) (Occupation): 81-116, https://www-cambridge-org.libproxy.temple.edu/core/journals/international-review-of-the-red-cross/article/different-sense-of-humanity-occupation-in-francis-liebers-code/8D0FC69EA6516A8909BE7211D874D7CB (accessed, 27 August 2020); Gideon M. Hart, "Military Commissions and the Lieber Code: Toward a New Understanding of the Jurisdictional Foundations of Military Commissions," *Military Law Review* 203 (March 2010): 1–77, http://search.ebscohost.com.libproxy.temple.edu/login.aspx?direct=true&db=lgs&AN=502156302&site=ehost-live&scope=site (accessed, 22 August 2020).

[28] Naval History and Heritage Command, "Battle of Manila Bay, 1 May 1898," https://www.history.navy.mil/research/library/online-reading-room/title-list-alphabetically/b/ba

ttle-of-manila-bay-1-may-1898.html (accessed, 10 June 2020); John M. Gates, *Schoolbooks and Krags: The United States Army in the Philippines, 1898*-1902 (Westport, CT: Greenwood Press, 1973), 15; Stewart, *The Forging of a Nation*, 346-47; U.S. Government, Department of State by Rounsville Wileman, Counsel-General, Consulate of the United States, Hong Kong, No. 63, July 18, 1898 in John R. M. Taylor, *The Philippine Insurrection Against the United States: A Compilation of Notes and Introduction with an Introduction by Renato Constanatino* (Pasay City, Philippines: Eugenio Lopez Foundation, 1971), 474.

[29] A.B. Feuer, *America at War: The Philippines, 1898*-1913 (Westport, CT: Greenwood Publishing Group, 2002), xx, 198; Brian M. Linn, *The Philippine War, 1899-1902* (Lawrence, KS: The University Press of Kansas, 2000), 9, 15, 19, 21, 24; Wileman, July 18, 1898; Birtle, *Counterinsurgency and Contingency Operations*, 114-19; Stewart, *The Forging of a Nation*, 346-47, 354; Millet, Maslowski, and Fies, *For the Common Defense*, 267-68, 273; Brian M. Linn, *Guardians of Empire: The U.S. Army and the Pacific, 1902-1940* (Chapel Hill: The University of North Carolina Press, 1997), 12-13, 18-19; Dwight D. Murphey, "Relevant Today: Lessons from the Spanish-American War." *The Journal of Social, Political, and Economic Studies* 38 (Spring, 2013): 80-98, http://libproxy.temple.edu/login?url=https://search-proquest-com.libproxy.temple.edu/docview/1381518510?accountid=14270 (accessed, 16 June 2020); Lazaro Segovia, *The Full Story of Aguinaldo's Capture*, trans. Frank de Thoma (Manila: MCS Enterprises, 1969), 143; *Washington Post*, March 29, 1901; Birtle, *Counterinsurgency and Contingency Operations*, 153, 158; Ronald E. Dolan, *Philippines: A Country Study* (Washington, D.C.: Federal Research Division, Library of Congress, 1993), 27, 75-87; Brian Linn, *The U.S. Army and Counterinsurgency in the Philippine War* (Lawrence, KS: University Press of Kansas, 2000), 23, 26, 46, 50, 73, 84, 111, 124; Alleyne Ireland, "The United States in the Philippines," *The Atlantic Monthly*, November 1904, 577, http://libproxy.temple.edu/login?url=https://search-proquest-com.libproxy.templ e.edu/docview/203601151?account zid=14270 (accessed, 16 June 2020); Filipinas Heritage Library, "The Moro to the Spanish Colonizers," https://www.filipinaslibrary.org.ph /articles/the-moro-to-the-spanish-colonizers/ (accessed, 13 December 2020); United States Government, War Department, U.S. Government, War Department, Department of the Pacific and 8th Army Corps, "General Order No. 16, September 10, 1898," *General Orders and Circulars issued from Headquarters Department of the Pacific and Eighth Army Corps, and Office of the U.S. Military Governor in the Philippine Islands, 1899* (Washington, D.C.: War Department, 1902); U.S. Government, United States Congress, *Federal and State Constitutions Colonial Charters, and Other Organic Laws of the States, Territories, and Colonies Now or Heretofore Forming the United States of America* by Thorpe F. Newton, ed. (Washington, D.C.: Government Printing Office, 1909), 3158-89; U.S. Government, War Department, Bureau of Insular Affairs, *What Has Been Done in the Philippines: A Record of Practical Accomplishments Under Civil Government* (Washington, D.C.: Government Printing Office, 1904), 7, 9, 15, 17, 22-23, 25; Gates, *Schoolbooks and Krags*, 270, 277.

[30] Linn, *The Philippine War*, 142, 145, 202, 271; W. N. Holden, "The Role of Geography in Counterinsurgency Warfare: The Philippine American War, 1899–1902. *GeoJournal* 85 (April 2020): 423-37, https://doi-org.libproxy.temple.edu/10.1007/s10708-019-09971-7 (accessed, 9 June 2020); *New York Times*, July 12, 1899; Secretary of State John Hay, https://www.nps.gov/prsf/learn/historyculture/spanish-american-war-a-splendid-little-war.htm (accessed, 21 June 2020); William H. Allen, "The Election of 1900." *The Annals of the American Academy of Political and Social Science* 17 (January 1901): 53–73, https://doi-org.libproxy.temple.edu/10.1177/000271620101700103 (accessed, 17 June 2020); Birtle,

Counterinsurgency and Contingency Operations, 112; Millet, Maslowski, and Fies, *For the Common Defense,* 268, 278, 280; Samuel Elliot Morrison, *The Oxford History of the American People* (New York: Oxford University Press, 1965), 801; *New York Times,* October 18, 1899; *New York Times,* May 15, 1899; Samuel E. Morrison, Frederick Merk and Frank Freidel, *Dissent in Three* Wars (Cambridge, MA: Harvard University Press, 2013), 78-79, 85-86, 88-90; Robert Lechie, *The Wars of* America (New York: Harper and Row, 1981), 570-72; *New York Times,* October 20, 1899; *Washington Post,* April 20, 1900; *Washington Post,* May 25, 1900; Erving Winslow, The Anti-Imperialist Faith," *The North American Review,* December 1, 1902, 811, https://search-proquest-com.libproxy.temple.edu/docview/1296786946/fulltextPDF/541BCE074F7A48F5PQ/1?accountid=14270 (accessed, 17 June 2020); Erving Winslow, Secretary of the New England Anti-Imperialist League, "The Anti-Imperialist Position," *The North American Review,* October 1, 1900, 460, http://libproxy.temple.edu/login?url=https://search-proquest-com.libproxy.temple.edu/docview/1296776931?accountid=14270 (accessed, 16 June 2020); *New York Times,* October 18, 1899; Richard E. Welch, Jr., *Response to Imperialism: The United States and the Philippine-American War, 1899-1902* (Chapel Hill: The University of North Carolina Press, 1979), 43-57.

31 *Washington Post,* Feb 17, 1899; Taylor, *The Philippine Insurrection,* 63, 66, 76; *Washington Post,* February 06, 1900; F.D. Millet, "The Military Position in the Philippines" *The Speaker: The Liberal Review* 1 (October 7, 1899): 11-12, http://libproxy. temple.edu/login?url=https://search-proquest-com.libproxy.temple.edu/docview/6433348?accountid=14270 (accessed, 16 June 2020); George Dewey to Secretary of the Navy in *Correspondence Relating to the War with Spain, Including the Insurrection in the Philippine Islands and the China Relief Expedition, April 15, 1898 to July 30, 1902,* Graham A. Cosmas, ed. (Washington, D.C.: Center of Military History, 1993), 893-94; *New York Times,* August 29, 1899; Taylor, *The Philippine Insurrection,* 192; *Washington Post,* August 5, 1900; *Washington Post,* February 11, 1899; *Washington Post,* February 8, 1899; Linn, *Guardians of Empire,*10-12; Mindanao (PH) State University, The Philippine-American War Documents, "Pedro Paterno's Proclamation of War, http://www.msc.edu.ph/centennial/pa990602.html (accessed, 10 June 2020); Linn, *The Philippine War,* 10, 17, 121, 136-37, 149, 185-87, 190-97, 209, 266; *New York Times* May 29, 1899; *Washington Post,* April 20, 1899; U.S. Government, 55th Cong., Session III, Ch. 352, "An Act for Increasing the Efficiency of the Army of the United States, and for Other Purposes, https://www.loc.gov/law/help/statutes-at-large/55th-congress/session-3/c55s3ch352.pdf (accessed, 18 June 2020); *Washington Post*, December 2, 1899; *Washington Post,* November 2, 1899; *Washington Post,* September 10, 1899; *Washington Post,* July 6, 1899; *Washington Post,* July 4, 1899; *Washington Post*, July 11, 1899; Millet, Maslowski, and Fies, *For the Common Defense,* 274; Taylor, *The Philippine Insurrection,* 38, 41-45, 71-76, 253, 299, 373, 378, 393-400.

32 Taylor, 419-21, 427-28; Linn, *The Philippine War,* 171, 187; Gates, *Schoolbooks and Krags,* 272.

33 William McKinley, "Benevolent Assimilation Proclamation" of December 21, 1898, http://www.msc.edu.ph/centennial/benevolent.html (accessed, 22 June 2020).

34 Ibid.

35 Department of the Pacific, *General Orders 1899,* "General Order No. 3, August 9, 1898."

36 *United States Army-Navy Journal and Gazette,* December 9, 1899.

37 Gates, *Schoolbooks and Krags,* 35-36; 137-41, 145, 274-75; Mary C. Gillett, *The Army Medical Department, 1865-1917* (Washington, D.C.: Center of Military History, 1995), 295-97, 301-06; Linn, *The Philippine* War, 213-14; Arthur MacArthur, telegram to War Department, in *Correspondence,* 1276; Rebecca R. Raines, *Getting the Message Through: A Branch History of*

the U.S. Army Signal Corps (Washington D.C.. Department of the Army, 1996), 101; Edmund Morris, *Theodore* Rex (New York: Random House, 2001), 102; Birtle, *Counterinsurgency and Contingency Operations,* 111; Geoffrey Hunt, *Colorado's Volunteer Infantry in the Philippine Wars, 1898*-1899 (Albuquerque: University of New Mexico Press, 2006), 196; Lechie, *Wars of America,* 570; Linn, *The Philippine War,* 284; *New York Times,* March 7, 1900; *Washington Post,* June 1, 1900.

[38] Birtle, *Counterinsurgency and Contingency Operations,* 114, 153-55, 158, 165; Linn, *Guardians of Empire, 13;* Linn, *The Philippine War,* 250, 286; Lechie, *Wars of America,* 551; Glenn A. May, "Why the United States Won the Philippine-American War, 1899-1902," *Pacific Historical Review* 52 (November 1983): 353-78, http://libproxy.temple.edu/login?url=https://search-proquest-com.libproxy.temple.edu/docview/1290903146?accountid=14270 (accessed, 24 June 2020); Robert Ferrell, *Reminiscences of Conrad S. Babcock: The Old U.S. Army and the New, 1898-1918* (Columbia: University of Missouri Press, 2012), 38-40.

[39] Adna Chaffee to Adjutant General, in *Correspondence,* 1295-96, Adna Chaffee to Adjutant General, in *Correspondence,* 1298, "Adna Chaffe to Adjutant General, December 27, 1901," 1306; Linn, *The Philippine War,* 233, 285; *New York Times,* October 27, 1900; May, 358; Arthur MacArthur to Adjutant General, in *Correspondence,* 1169; Arthur MacArthur to Adjutant General, in *Correspondence,* 1221-22; Elwell S. Otis to Adjutant General, in *Correspondence,* 865, Elwell S. Otis to Adjutant General, in *Correspondence,* 1100; "Philippine-American War," Stanley Karnow, *In Our Image: America's Empire in the Philippines* (New York: Random House, 1989), 194; *Washington Post,* June 11, 1901.

[40] Birtle, *Counterinsurgency and Contingency Operations,* 114; James R. Arnold, ed. *Americans at War: Eyewitness Accounts from the American Revolution to the 21st Century, Vol. II* (Santa Barbara, CA: ABC-CLIO, 2018), 487-88; Adna Chaffee to Adjutant General, in *Correspondence,* 1294-98.

[41] Linn, *The Philippine War,* 212; The Lieber Code; Will Smiley, "Lawless Wars of Empire? the International Law of War in the Philippines, 1898–1903," *Law and History Review* 36, no. 3 (August 2018): 511-50, http://libproxy.temple.edu/login?url=https://search-proquest-com.libproxy.temple.edu/docview/2084811024?accountid=14270 (accessed, 27 June 2020); *Department of the Pacific General Orders,1899,* General Orders No. 8; John M. Gates, "War-Related Deaths in the Philippines, 1898-1902." *Pacific Historical Review* 53 (1984): 367, http://libproxy.temple.edu/login?url=https://search-proquest-com.libproxy.temple.edu/docview/1290882199?accountid=14270 (accessed, 28 June 2020); Birtle, *Counterinsurgency and Contingency Operations,* 134; Fritz, David L. "Before the "Howling Wilderness": The Military Career of Jacob Hurd Smith, 1862-1902." *Military Affairs* 43 (1979): 186-90, https://www-jstororg.libproxy.temple.edu/stable/1986751?sid=primo&origin=crossref&seq=2#metadata_info_tab_contents (accessed, 27 June 2020); Linn, *The Philippine War,* 321; *Washington Post,* July 01, 1902; *Washington Post,* November 17, 1902; *Hearing Before the Committee on the Philippines, United States Senate, May 5, 1902,* S. Doc. No. 331-57, Part 3 at 2069. Hereafter S. Doc No. 331-57 (1902); U.S. Government, War Department, Headquarters of the Army, Adjutant-General's Office, General Orders No. 12, February 6, 1902, https://congressional-proquest-com.libproxy.temple.edu/congressional/result/pqpresultpage.gispdfhitspanel.pdflink/$2fapp-bin$2fgis-serialset$2fe$2f9$2f1$2fd$2f4244_sdoc331_from_1_to_1124.pdf/entitlementkeys=1234%7Capp-gis%7Cserialset%7C4244_s.doc.331 (accessed, 1 July 2020); U.S. Government, War Department, Headquarters, Department of Northern Luzon, General Orders No. 13, March 12, 1901, https://congr

essional-proquest-com.libproxy.temple.edu/congressional/result/pqpresultpage.gispdfhitspanel.pdflink/$2fapp-bin$2fgis-serialset$2fe$2f9$2f1$2fd$2f4244_sdoc331_from_1_to_1124.pdf/entitlementkeys=1234%7Cappgis%7Cserialset%7C4244_s.doc.331 (accessed, 1 July 2020); U.S. Government, War Department, Headquarters, Division of the Philippines, General Orders No. 63, August 8, 1902, https://congressional-proquest-com.libproxy.temple.edu/congressional/result/pqpresultpage.gispdfhitspanel.pdflink/$2fapp-bin$2fgis-serialset$2fe$2f9$2f1$2fd$2f4244_sdoc331_from_1_to_1124.pdf/entitlementkeys=1234%7Capp-gis%7Cserialset%7C4244_s.doc.331 (accessed, 1 July 2020); S. Doc. No. 331-57 at 2233-34; Adjutant General to Chaffee, in *Correspondence,* 1328-29; Adna Chaffee to Adjutant General, in *Correspondence,* 1329; *New York Times,* July 26, 1902; *New York Times,* June 13, 1902; *New York Times,* May 4, 1902; S. Doc. No. 331-57, Part 3 at 1169-72, 2061-62, 2252-53, 2237-39, 2243-44, 2236-37, 2061-62; *Washington Post,* May 06, 1902; *Washington Post,* May 4, 1902; *Washington Post,* April 29, 1903.

42 Herbert A. White, "The Pacification of Batangas," *The International Quarterly* 7, March 1, 1903, 431-44 http://libproxy.temple.edu/login?url=https://search-proquest-com.libproxy.temple.edu/docview/1296533081?accountid=14270 (accessed, 30 June 2020); *New York Times,* May 30, 1902; *Washington Post,* May 21, 1902; *Washington Post,* March 20, 1901; *Washington Post,* May 8, 1902; *Washington Post,* January 29, 1902; *Washington Post,* December 16, 1901; Gates, *Schoolbooks and Krags,* 271.

43 Elwell Otis to Adjutant General, in *Correspondence,* 1058.

44 Adna Chaffe to Adjutant General, in *Correspondence,* 1327.

45 U.S. Government, War Department, Head-Quarters of the Army, General Orders No. 66, in *Correspondence,* 1352-53; Donald Smythe, *Guerrilla Warrior: The Early Life of John J. Pershing* (New York: Charles Scribner's Sons, 1973), 90-91, 100-01; Linn, *Guardians of Empire,* 41; Elwell Otis to Adjutant General, in *Correspondence,* 1322; Adna Chaffee to Adjutant General, in *Correspondence,* 1336; Joseph P. McCallus, *Forgotten Under A Tropical Sun: War Stories by American Veterans in the Philippines, 1898-1913* (Ashland, OH: The Kent State University Press, 2017), 168, 189.

46 Smythe, *Guerilla Warrior,* 80.

47 G.O. 66 in *Correspondence,* 1352-53; Charles Byler, "Pacifying the Moros: American Military Government in the Southern Philippines, 1899-1913," *Military Review* 85 (May 2005): 41-45, http://libproxy.temple.edu/login?url=https://search-proquest-com.libproxy.temple.edu/docview/225312103?accountid=1427 (accessed, 1 July 2020); Frederick Palmer, *John J. Pershing, General of the Armies: A* Biography (Westport, CT: Greenwood Publishers, 1970), 49; Smythe, *Guerrilla Warrior,* 74, 79-84, 92, 108.

48 Elwell S. Otis to Adjutant General, in *Correspondence,* 931; Elwell S. Otis to Adjutant General, received March 19, 1900, 7:50 a.m., in *Correspondence,* 938; Elwell S. Otis to Adjutant General, received March 19, 1900, 9:52 a.m., in *Correspondence,* 938; Elwell S. Otis to Adjutant General, in *Correspondence,* 945; Elwell S. Otis to Adjutant General, in *Correspondence,* 948-49; Elwell S. Otis to Adjutant General, in *Correspondence,* 985; Elwell S. Otis to Adjutant General, in *Correspondence,* 988; Elwell S. Otis to Adjutant General, in *Correspondence,* 990; Elwell S. Otis to Adjutant General, in *Correspondence,* 993-94; Elwell S. Otis to Adjutant General, in *Correspondence,* 1001; Elwell S. Otis to Adjutant General, in *Correspondence,* 1042; Elwell S. Otis to Adjutant General, *Correspondence,* 1158; Elwell S. Otis to Adjutant General, in *Correspondence,* 1111; Elwell S. Otis to Adjutant General, in *Correspondence,* 1136; Elwell S. Otis to Adjutant General, in *Correspondence,* 1146; Elwell S. Otis to Adjutant General, in *Correspondence,* 1158; Arthur MacArthur to Adjutant

General, in *Correspondence,* 1214; Linn, *The Philippine War,* 132, 313; A. P. Niblack, "Operations of the Navy and Marine Corps in the Philippine Archipelago," *Proceedings* 30 (October 1904), https://www.usni.org/magazines/proceedings/1904/october/operations-navy-and-marine-corps-philippine-archepelago-1898 (accessed, 19 November 2020).

[49] Linn, *The Philippine War,* 202-03; *New York Times,* April 29, 1899; *Washington Post,* April 25, 1901; *Washington Post,* November 28, 1900; *Washington Post,* May 19, 1900; Adna Chaffee to Adjutant General, in *Correspondence,* 1349-50; Arthur MacArthur to Adjutant General, in *Correspondence,* 1250; *New York Times,* September 29, 1900; *New York Times,* May 29, 1899; *Washington Post,* September 29, 1900; Elwell S. Otis to Adjutant General, in *Correspondence,* 1091; Elwell S. Otis to Adjutant General, in *Correspondence,* 1120."

[50] Gregory J. W. Urwin, *The United States Infantry: An Illustrated History, 1775*-1918 (London: Blanford Press, 1988), 112, 121; U.S. Government, War Department, "Ordenance Report Number 15: Report of the Board of Officers Appointed in Pursuance of the Act of Congress June 6, 1872 for the Purpose of Selecting a Breach-System Muskets and Carbines for the Military Service, Together With Their Report on the Subject of Trowel-Bayonets" (Washington, D.C.: Government Printing Office, 1873), 12; Daniel R. Beaver, "The U.S. War Department in the Gaslight Era: Stephen Vincent Benét at the Ordnance Department, 1870-91," *Journal of Military History* 68 (January 2004): 105–32, http://web.b.ebscohost.com.libproxy.temple.edu/ehost/pdfviewer/pdfviewer?vid=1&sid=d5e35e5b-c3e0-44f6-96a9-4abd7804d06b%40pdc-v-sessmgr03 (accessed, 24 August 2020); Patrick McSherry, "Model 1896 Krag-Jorgensen," (accessed July 1, 2020), http://www.spanamwar.com/krag.htm; Paul J. Daugherty and Herbert C. Eidt, "Wound Ballistics: Minie Ball vs. Full-Metal Jacketed Bullets: A Comparison of Civil War and Spanish-American War Firearms," *Military Medicine* 174 (April 2009): 403-09, http://web.a.ebscohost.com.libproxy.temple.edu/ehost/pdfviewer/pdfviewer?vid=1&sid=1bed7899-cf10-4f18-bc4e-607240a1dca8%40sessionmg (accessed, 30 June 2020); *New York Times,* September 21, 1894; Linn, *The Philippine War,* 55, 57, 110, 126.

[51] Paul Scarlata, "United States .38 Cal. Military Revolvers Part I: Long Denigrated Rut Still Serving Proudly. Like Other Nations, The USA Moved Its Sidearms Down in Caliber at the end of the 19th Century. Results Were Mixed, To Say The Least," *Firearms News,* April 10, 2017, 44, https://link-gale-com.libproxy.temple.edu/apps/doc/A491782110/ITOF?u=temple_main&sid=ITOF&xid=a00bacd1 (accessed, 24 August 2020); Stewart, *The Forging of a Nation,* 367; "The U.S. Military and the M1911," *Shotgun News,* May 10, 2014, 15, https://link-gale-com.libproxy.temple.edu/apps/doc/A369365692/ITOF?u=temple_main&sid=ITOF&xid=f8b9e6a4 (accessed, 24 August 2020); Department of the Army, *TM 9-1005-211-35 Pistol, Caliber .45, Automatic, M1911A1* (Washington, D.C.: Department of the Army, 1968), 3.

[52] Holden, "The Role of Geography," 431; Alfred McCoy "Policing the Imperial Periphery: The Philippine-American War and the Origins of U.S. Global Surveillance." *Surveillance & Society* 13 (July 2014): 4-26, http://libproxy.temple.edu/login?url=https://search-proquest-com.libproxy.temple.edu/docview/1662059002?accountid=14270 (accessed, 2 July 2020); Ralph H. Van Deman and Ralph E. Weber, *The Final Memoranda: Major General Ralph H. Van Deman, USA (Ret), 1865-1952, The Father of U.S. Military Intelligence* (Wilmington, DE: SR Books, 1988), xi, 7, 123; Birtle, *Counterinsurgency and Contingency,* 118; Linn, *The U.S. Army and Counterinsurgency in the Philippine War,* 155; Gates, *Schoolbooks and Krags,* 271.

[53] Edward M. Coffman *The Embattled Past: Reflections on Military History* (Lexington, KY: University Press of Kentucky, 2014), 69; Birtle, *Counterinsurgency and Contingency,* 116-18;

Linn, *The Philippine War,* 128; Victor Holman, "Negro Indians, Macabebes, and Civilian Irregulars: Models for the Future Employment of Indigenous Forces" (master's thesis, U.S. Army Command and General Staff College, 1995), 33, https://archive.org/details/DTIC_ADA299732/page/n41/mode/2up (accessed, 2 July 2020).

54 Edward M. Coffman *The Embattled Past: Reflections on Military History* (Lexington, KY: University Press of Kentucky, 2014), 69; Birtle, *Counterinsurgency and Contingency,* 116-18; Linn, *The Philippine War,* 128; Holman, "Models for Future Employment," 28-30, 32-33; Robert M. Cassidy, "The Long Small War: Indigenous Forces for Counterinsurgency," *Parameters* 36 (Summer 2006):47-62, https://link-gale-com.libproxy.temple.edu/apps/doc/A148856123/AONE?u=temple_main&sid=AONE&xid=b11b32ca (accessed, 2 July 2020); *New York Times,* March 29, 1901; U.S. Government, War Department, Annual Reports of the War Department for the Fiscal Year Ended June 30, 1901, by J. Franklin Bell, "Report of Brig. Gen. J. F. Bell, Commanding First District, Department of Northern Luzon, April 30, 1901," 26, https://congressional-proquest-com.libproxy.temple.edu/congressional/result/pqpresultpage.gispdfhitspanel.pdflink/$2fapp-bin$2fgis-serialset$2f6$2f4$2f0$2f6$2f4273_hdoc2-5_0001_from_1_to_50.pdf/entitlementkeys=1234%7Capp-gis%7Cserialset%7C4273_h.doc.2_5 (accessed, 2 July 2020); *Washington Post,* March 29, 1901; Arthur MacArthur to Adjutant General, in *Correspondence,* 1265.

55 56th Cong., 2nd Sess. CHS. 191-92, 1901, 748, https://www.loc.gov/law/help/statutes-at-large/56th-congress/session-2/c56s2ch192.pdf (accessed, 27 July 2020).

56 Ibid., 748.

57 56th Cong., 2nd Sess., H. Rep. 2010: Efficiency of the Military Establishment, https://congressional-proquest-com.libproxy.temple.edu/congressional/result/congresultpage:pdfevent?rsId=172F803BE19&pdf=/app-bin/gis-serialset/8/0/a/6/4212_hrp2010_from_1_to_11.Pdf&uri=/app-gis/serialset/4212_h.rp.2010 (accessed, 27 July 2020); 56th Cong., 2nd Session, CHS. 191,192, 1901, 748, https://www.loc.gov/law/help/statutes-at-large/56th-congress/session-2/c56s2ch192.pdf (accessed, 27 July 2020).

58 U.S. Government, 56th Cong. 2nd Session, H. Rep. Doc. No. 2, "Annual Reports of the War Department for the Fiscal Year Ended June 30, 1900," 72-73; U.S. Government, 57th Cong., 1st Session, H. Rep. Doc. No. 2, "Annual Reports of the War Department for the Fiscal Year Ended June 30, 1901: Report of the Philippine Commission," 386, https://congressional-proquest-com.libproxy.temple.edu/congressional/result/congresultpage:pdDfevent?rsId=1727AA360BE&pdf=/app-bin/gis-serialset/9/1/3/1_to_786.pdf&uri=/app-gis/serialset/4277_h.doc.2_9 (accessed, 3 July 2020); U.S. Government, 58th Cong., 2nd Session, H. Rep. Doc. No. 2, "Annual Reports of the War Department 5/4277_hdoc2-9_from_ for the Fiscal Year Ended June 30, 1903: Volume VI, Report of the Philippine Commission," 27-33, https://congressional-proquest-com.libproxy.temple.edu/congressional/result/congressional/congdocumentview?accountid=14270&groupid=96626&parmId=1727AE19660&rsId=1727AE10F75#1100.800048828125 (accessed, 3 July 2020); James W. Hurst, *Pancho Villa and Blackjack Pershing* (Westport, CT: Praeger, 2008), 29, 36; H. Rep. Doc. No. 2, "Annual Report," 275.

59 U.S. Government, War Department, *Official Register for* 1900 (Washington, D.C.: Adjutant General's Office, 1899), 5-7, 9, 60-61, 64, 70, 76, 79, 117, 125, 144, 159, 163, 179, 183, 187, 190, 202; Linn, *The Philippine War,* 93, 218.

60 Birtle, *Counterinsurgency and Contingency,* 201; David Craven, "Lineages of the Mexican Revolution (1910–1940)," *Third Text* 28 (July 2014): 223-34, https://www.tandfonline-com.libproxy.temple.edu/doi/full/10.1080/09528822.2014.89 9789 (accessed, 11 January 2021);

Virginia Gueda, "The Process of Mexican Independence," *The American Historical Review* 105 (2000): 116-30, https://www-jstor-org.libproxy.temple.edu/stable/2652439?seq=14#metadata_info_tab_contents (accessed, 25 March 2020); Frank McLynn, *Villa and Zapata: A History of the Mexican Revolution* (New York: Basic Books, 2000), 309, 321-25; Skidmore and Smith, *Modern Latin America,* 228; *Washington Post,* March 12, 1916.

[61] James W. Harper, "The El Paso-Juárez Conference of 1916," *Arizona and the West* 20 (Autumn, 1978): 231-44, 2020; www.jstor.org/stable/40168729 (accessed, 25 March 2020); McLynn, *Villa and Zapata,* 325; Lars Schoultz, *Beneath the United* States (Cambridge, MA: Harvard University Press, 1998), 248-51; Robert W. Tucker, "Woodrow Wilson's "New Diplomacy," *World Policy Journal* 21 (Summer 2004): 92-107, http://libproxy.temple.edu/login?url=https://search-proquest-com.libproxy.temple.edu/docview/232586825?accountid=14270 (accessed, 25 March 2020); United States Government, Department of State, *Papers Relating to the Foreign Relations of the United States, The Lansing Papers, 1914-1920, Vol. I,* Document 647, https://history.state.gov/historicaldocuments/frus1914-20v01/d647 (accessed, 21 April 2020); United States Government, Department of State, *Papers Relating to the Foreign Relations of the United States with the Address of the President to Congress December 5, 1916* (Washington, D.C.: U.S. Government Printing Office, 1925), 530-31, http://digicoll.library.wisc.edu/cgi-bin/FRUS/FRUS-idx?type=turn&entity=FRUS.FRUS1916.p0607&id=FRUS.FRUS1916&isize=M, (accessed, 25 March 2020).

[62] Julie I. Prieto, *The Mexican Expedition: 1916-1917* (Washington, D.C.: Center of Military History, 2016), 25; Smythe, *Guerrilla Warrior,* 254; Ellen Wellsome, *The General and the Jaguar: Pershing's Hunt for Pancho Villa* (New York: Little, Brown & Co., 2003), 171, 186; William H. Donaldson, *Biographical Register Of The Officers and Graduates Of The U. S. Military Academy At West Point, New York Since Its Establishment In 1802: Supplement, Volume VII 1920-1930* (Chicago, IL: R. R. Donnelley & Sons Company, Lakeside Press, 1930), 143, 148, 246-47, 257, 318, 417, 430, 439-40, 569, 678; United States Air Force, "Major General Benjamin Delahauf Foulois," https://www.af.mil/About-Us/Biographies/Display/Article/107091/major-general-benjamin-delahauf-foulois/ (accessed, 10 August 2020); Association of Graduates, *Seventy-Second Annual Report of the Association of Graduates of the United States Military Academy at West* Point (Newburgh, NY: Moore Printing, 1941), 157-58, 189-90, 227, 283, 306; Association of Graduates, *Seventy-First Annual Report of the Association of Graduates of the United States Military Academy at West* Point (Newburgh, NY: Moore Printing, 1940), 133-36, 151, 233-34; Association of Graduates, *Seventieth Annual Report of the Association of Graduates of the United States Military Academy at West* Point (Newburgh, NY: Moore Printing, 1939), 118, 227, 244; Association of Graduates, *Sixty-Ninth Annual Report of the Association of Graduates of the United States Military Academy at West* Point (Newburgh, NY: Moore Printing, 1938), 98, 231-32, 237-38; Association of Graduates, *Sixty-Eighth Annual Report of the Association of Graduates of the United States Military Academy at West* Point (Newburgh, NY: Moore Printing, 1937), 98-99, 181-83, 196-97; Association of Graduates, *Sixty-Seventh Annual Report of the Association of Graduates of the United States Military Academy at West* Point (Newburgh, NY: Moore Printing, 1936), 181-82; Association of Graduates, *Sixty-Sixth Annual Report of the Association of Graduates of the United States Military Academy at West* Point (Newburgh, NY: Moore Printing, 1935), 110, 154, 159-60, Association of Graduates, *Sixty-Fifth Annual Report of the Association of Graduates of the United States Military Academy at West Point* (Newburgh, NY: Moore Printing, 1934), 96-97; Association of Graduates, *Sixty-Fourth Annual Report of the Association of Graduates of the United*

States Military Academy at West Point (Newburgh, NY: Moore Printing, 1933), 174-75, 191, 211-13, 237-38; Association of Graduates, *Sixty-Third Annual Report of the Association of Graduates of the United States Military Academy at West* Point (Newburgh, NY: Moore Printing, 1932), 194-95; Association of Graduates, *Sixty-Second Annual Report of the Association of Graduates of the United States Military Academy at West* Point (Crawfordsville, IN: Lakeside Press, 1931), 172-73, 185-87, 241-42, 263-64, 417-18; Association of Graduates, *Sixty-First Annual Report of the Association of Graduates of the United States Military Academy at West* Point (Newburgh, NY: Moore Printing, 1930), 215-16; Association of Graduates, *Sixtieth Annual Report of the Association of Graduates of the United States Military Academy at West* Point (Newburgh, NY: Moore Printing, 1929), 75, 95, 245, 253; U.S. Government, War Department, *Annual Report of the Secretary of War to the President, 1917, Vol. 1* (Washington, D.C.: U.S. Government Printing Office, 1917), 481, https://babel.hathitrust.org/cgi/pt?id=uc1.b2980528&view=1up&seq=491 (accessed, 1 April 2020); 81; Hurst, *Villa and Pershing*, x, 28.

63 Detroit Historical Society, "Encyclopedia of Detroit: Ford Highland Park Plant," https://detroithistorical.org/learn/encyclopedia-of-detroit/ford-highland-park-plant (accessed, 8 April 2020); Benjamin King, Richard Biggs, and Eric Criner, *Spearhead of Logistics: A History of the United States Army Transportation Corps* (Fort Eustis, Virginia and Washington, D.C.: U.S. Army Transportation Center and the Center of Military History, 2001), 90; Quentin R. Skrabec, Jr. "Highland Park Ford Assembly Line (1910)," in *The 100 Most Significant Events in American Business: An Encyclopedia*, 136-38. Santa Barbara, CA: Greenwood, 2012), 126; United States Government, National Atmospheric and Space Administration, by Frank J. Zeleznick and Bonnie J. McBride, *NASA Reference Publication 1094: Modeling the Internal Combustion Engine* (Cleveland, Ohio: Lewis Research Center, 1985), 1.

64 *New York Times*, November 13, 1921.

65 King, Biggs and Criner, *Spearhead of Logistics*, 90; United States Government, War Department, 1st Aero Squadron, Signal Corps, War Diary: Period from March 12 to April 23, 1916," 1, https://texashistory.unt.edu/ark:/67531/metapth188057/ (accessed, 22 April 2020).

66 Sara E. Cothren and Alexander Barnes, "Long Distance Logistics: The Mexican Expedition," *Army Sustainment* 47 (May/June 2015): 45-49, https://search-proquest-com.libproxy.temple.edu/docview/1682660741/fulltextPDF/51F6BAA52A4D4CA1PQ/1?accountid=14270 (accessed, 25 March 2020); *Washington Post*, April 16, 1916; U.S. Army Transportation Museum, "War on Wheels: The Mexican Punitive Expedition-1916," https://web.archive.org/web/20041017154729/http://www.transchool.eustis.army.mil/Museum/Mexwar.htm. (accessed, 23 April 2020); Marc Blackburn, e-mail to Author, September 1, 2020.

67 Gregory Mason, "The Man Hunt in Mexico: The First of a Series of Articles on the Mexican Expedition," *Outlook*, April 19, 1916, http://libproxy.temple.edu/login?url=https://search-proquest-com.libproxy.temple.edu/docview/136649020?accountid=14270 (accessed, 26 March 2020); Marc Sammis, Curator, US Army Transportation Museum, e-mail to Author, May 8, 2020; USATC, "War on Wheels;" U.S. Army Transportation Museum, "The Liberty Truck," https://web.archive.org/web/20041017 211748/, http://www.transchool.eustis.army.mil/Museum/LIBTrucks.htm (accessed, 23 April 2020); Hurst, Villa and Pershing, 44, 131-34; McLynn, Villa and Zapata, 325.

68 Roger G. Miller, *A Preliminary to War: The First Aero Squadron and the Mexican Punitive Expedition, 1916* (Washington, D.C.: U.S. Air Force History and Museum Programs, 2003), 8; Joseph C. Scott, "The Infernal Balloon: Union Aeronautics During the American Civil War, *Army History: The Professional Bulletin of Army History* PB 20-14-4 (Fall 2014): 6-24.

69 Benjamin D. Foulois with C. V. Glines, *From the Wright Brothers to the Astronauts: The Memoirs of Major General Benjamin D. Foulois* (New York: McGraw-Hill Book Company, 1968), 119; Department of the Air Force, "Major General Benjamin Delahauf Foulois," https://www.af.mil//About-Us/Biographies/Display/Article/107091/major-general-benjamin-delahauf-foulois/ (accessed, 25 January 2021).
70 John C. Frederiksen, *The U.S. Air Force: A Chronology* (Santa Barbara, CA: Greenwood Publishing, 2011), 13; Miller, *The First Aero Squadron* 8; U.S. Air Force, Beale Air Force Base, "First Reconnaissance Squadron," https://www.beale.af.mil/Library/Fact-Sheets/Display/Article/279940/1st-reconnaissance-squadron/ (accessed, 27 March 2020); Wright Brothers.org, "Just the Facts: 1912-1913 Wright Model C," http://www.wright-brothers.org/Information_Desk/Just_the_Facts/Airplanes/Model_C.htm (accessed, 27 March 2020).
71 Nancy Weiz Aldrich, "1st Aero Squadron," https://firstaerosquadron.com/articles/the-1st-aero-squadron-a-history (accessed, 20 January 2017); Rick Leisenring, Curtiss Aircraft Museum, e-mail to Author, May 18, 2020; 1st Aero Squadron War Diary, 2-4; Hurst, *Villa and Pershing,* 122-24; measurements and distances obtained from googlearthpro.com using the program's distance measuring and terrain profile tools, March 28, 2020.
72 Aldrich, "1st Aero Squadron"; Miller, *The First Aero Squadron,* 41; Raines, *Getting the Message Through, 148*; 1st Aero Squadron War Diary, 2-4; *New York Times,* April 23, 1916.
73 *New York Times,* August 30, 1916.
74 Miller, *The First Aero Squadron,* 31; 1st Aero Squadron War Diary, 5-6; John J. Pershing, John T. Greenwood, ed; *My Life Before the World War, 1860-1917* (Lexington, Kentucky: University Press of Kentucky, 2013), 347.
75 Discover Southeast Arizona, "The Heliograph in the Apache Wars," http://www.discoverseaz.com/History/Heliograph.html (accessed, 22 April 2020); Raines, *Getting the Message Through,* 4, 66, 108.
76 Birtle, *Counterinsurgency and Contingency,* 168-69; David Hatch, "The Punitive Expedition: Military Reform and Communications Intelligence," *Cryptolgia* 31 (January 2007): 38-45, https://www-tandfonline-com.libproxy.temple.edu/doi/full/10.1080/01611190600964264 (accessed, 22 April 2020); Prieto, *The Mexican Expedition,* 52; Raines, *Getting the Message Through,* 126, 141; Hurst, *Villa and Pershing,* xi.
77 Incorporating Radio-Electronics.com, "Spark Gap Transmitter History," https://www.electronics-notes.com/articles/history/spark-gap-transmitters/history.php (accessed, 23 April 2020); OneTubeRadio.com, "Army Signal Corps Field Buzzer, 1916" http://onetuberadio.com/2014/09/21/army-signal-corps-field-buzzer-1916/ (accessed, 23 April 2020); Prieto, *The Mexican Expedition,* 52; Raines, *Getting the Message Through,* 150.
78 Department of the Army, *Technical Manual 11-666, Antennas and Radio Wave* Propagation (Washington, D.C.: Headquarters, Department of the Army, 1953), 1-14, 17, 23-29; 1st Aero Squadron War Diary, 6-11; Idc.com, "Radio Wave Propagation," http://www.idc-online.com/technical_references/pdfs/electronic_engineering/Radio_Waves_Propagation.pdf (accessed, 23 April 2020); Prieto, *The Mexican Expedition,* 54; Raines, 138, 149-50.
79 Clarence C. Clendenen, *Blood on the Border: The United States Army and the Mexican Irregulars* (New York: McMillan, 1969), 251, 333; Raines, *Getting the Message Through,* 150.
80 Harper, "El Paso-Juarez Conference," 231-44; Hatch, Hurst, *Villa and Pershing,* 56.
81 Roger D. McGrath, "The Patton You Didn't Know," *Chronicles,* February 2013, https://link-gale-com.libproxy.temple.edu/apps/doc/A328282473/ITOF?u=temple_main&sid=ITOF&xid=7c568f2d (accessed, 25 August 2020); Wellsome, *The General and the Jaguar,*

188-93, 261; *New York Times*, May 16, 1916; *New York Times*, May 23, 1916; Prieto, *The Mexican Expedition*, 32, 34, 37, 40, 42-43; *New York Times*, April 2, 1916; *New York Times*, May 7, 1916; Hurst, *Villa and Pershing*, 84-85.

[82] Prieto, *The Mexican Expedition*, 46; *New York Times*, July 7, 1916; *New York Times*, June 30, 1916; *New York Times*, June 29, 1916; *New York Times*, June 26, 1916; Hurst, *Villa and Pershing*, 76-77; McLynn, *Villa and Zapata*, 329-30: Wellsome, *The General and the Jaguar*, 274-86; Lute Pease, cartoon "Still Coming," https://www.loc.gov/pictures/item/2016683947/ (accessed, 29 August 2020); William A. Rogers, cartoon "The Punitive Expedition," https://www.loc.gov/pictures/resource/cph.3c30777/ (accessed, 29 August 2020); Clifford K. Berryman, cartoon "Put 'Em Back in the Box: How Long is this to Keep Up?," https://www.loc.gov/pictures/item/2016678298/ (accessed, 29 August 2020).

[83] U.S. Government, Department of State, *Foreign Relations of the United States, 1952–1954, East Asia and the Pacific, Volume XII, Part 1* (Washington, D.C.: Department of State, 1984), 980-81.

[84] CMH, "Campaigns;" Graham A. Cosmas, *MACV The Joint Command in the Years of Escalation, 1962–1967* (Washington, D.C.: Center of Military History, 2006), xi, 21-29, 206, 212, 331, 343-44University of Quebec at Montreal, "Military Assistance Advisory Group, Indochina," http://indochine.uqam.ca/en/historical-dictionary/890-military-assistance-advisory-group-maag-indochina.html (accessed, 25 April 2020).

[85] Larry H. Addington, *The Patterns of War Since the Eighteenth* Century (Indiana University Press, 1994), 296; Collins, 16-18; David R. Palmer, *Summons of the Trumpet: A History of the Vietnam War from a Military Man's* Viewpoint (New York: Ballantine Books, 1978), 80-102; Stanley Larsen and Joseph L. Collins, Jr., *Vietnam Studies: Allied Participation in Vietnam* (Washington, D.C.: Center of Military History, 1975), 23, 25, 30, 39-45, 53-54, 63, 89-101, 104-10, 128-30; James L. Collins, Jr., *The Development and Training of the South Vietnamese Army* (Washington, D.C.: Center of Military History, 1975), 151; Byung-Kook Kim and Eric Vogel, *The Park Chung Hee Era: The Transformation of South Korea* (Cambridge, MA: Harvard University Press, 2011), 410-13; Fredrik Logevall, *Choosing War: The Lost Change for Peace and the Escalation of War in* Vietnam (Berkeley: University of California Press, 1999), 178-81, 372-73, 378-79.

[86] *New York Times*, July 18, 1899; Palmer, *Summons of the Trumpet*, 150; Daniel Smyth, "Avoiding Bloodshed? U.S. Journalists and Censorship in Wartime," *War and Society* 32 (November 2013), https://www-tandfonline-com.libproxy.temple.edu/doi/citedby/10.1179/0729247312Z.00000000017?scroll=top&needAccess=true (accessed, 27 April 2020); Andrew F. Krepenevich, Jr., *The Army and* Vietnam (Baltimore, MD: The Johns Hopkins University Press, 1986), 78-80, Spencer C. Tucker, *Vietnam* (Lexington, KY: The University Press of Kentucky, 1999), 125; Odd A. Westad, *The Global Cold War* (Cambridge, MA: Cambridge University Press, 2005), 192-93; William C. Westmoreland, *A Soldier Reports* (New York: Dell, 1976), 335-41.

[87] Collins, *South Vietnamese Army*, 8-13, 25-26; Mark A. Lawrence, *The Vietnam War: A Concise International History* (Oxford: Oxford University Press, 2008), 75; Palmer, *Summons of the Trumpet*, 16-17, 41-42; David M. Toczek, "They Did Everything But Learn From It: The Battle of Ap Bac, 1963" (master's thesis, Texas Tech, 1999), 82-109, https://apps.dtic.mil/sti/tr/pdf/ADA341526.pdf

[88] Palmer, *Summons of the Trumpet*, 37-51; Rod Paschall, "Prescient At The Creation: Ridiculed After the 1963 Battle of Ap Bac as being too Old, Out of Touch and Mired in Outdated Military Thinking, Was General Paul Harkins Actually the Most Accurate American Prophet of the

Vietnam War?, *Vietnam* 23, December 2010, 46, https://go-gale-com.libproxy.temple.edu/ps/i.do?id=GALE%7CA238979160&v=2.1&u=temple_main&it=r&p=AONE&sw=w (accessed, 29 April 2020); Toczek, "Everything But Learn," 82-109.

[89] *New York Times,* January 7, 1963.

[90] *New York Times,* January 5, 1963; *Washington Post-Times Herald,* January 4, 1963.

[91] *New York Times,* January 7, 1963; William M. Hammond, *Public Affairs: The Military and the Media, 1962-1968* (Washington, D.C.: Center of Military History, 1990), 11-17; Don North, "Holding the Line at Credibility Gap," *Vietnam* 24 June 2011, 32-37, http://libproxy.temple.edu/login?url=https://search-proquest-com.libproxy.temple.edu/docview/865927453?accountid=14270 (accessed, 28 April 2020); Toczek, "Everything But Learn" 111; Westmoreland, *A Soldier Reports* 61, 82, 360.

[92] Robert J. Cima, ed. *Vietnam: A Country Study* (Washington, D.C.: Department of the Army, 1989), xv-xix, 8-10, 12, 17; Stanley Karnow, *Southeast* Asia (New York: Time, 1962), 66; Gabriel Kolko, *Anatomy of a War: Vietnam, the United States, and the Modern Historical* Experience (New York: The New Press, 1985), 208, 216; Logevall, 10-11; Mark Moyar, "Political Monks: The Militant Buddhist Movement During the Vietnam War," *Modern Asian Studies* 38 (2004): 749–84, https://www-cambridge-org.libproxy.temple.edu/core/journals/modern-asian-studies/article/political-monks-the-militant-buddhist-movement-during-the-vietnam-war/C001C40FD025C46671FD24FEA700C4DC (accessed, 30 April 2020); Mark Moyar, "Villager Attitudes During the Final Decade of the Vietnam War," Texas Tech University 1996 Vietnam Symposium, "After the Cold War, Reassessing Vietnam," https://www.vietnam.ttu.edu/events/1996_Symposium/96papers/moyar.php (accessed, 30 April 2020); Palmer, *Summons of the Trumpet,* 12-13; U.S. Government, Department of State, United States Embassy Vietnam (South), *Major Religious and Armed Groups 1954,* http://libproxy.temple.edu/login?url=https://search-proquest-com.libproxy.temple.edu/docview/1679150229?accountid=14270 (accessed, 30 April 2020).

[93] Addington, *Patterns of War,* 293; *New York Times,* July 7, 1970; Kolko, *Anatomy of a War,* 214, 216; Michael E. Latham, *The Right Kind of Revolution: Modernization, Development and U.S. Foreign Policy from the Cold War to the* Present (Cornell, NY: Cornell University Press, 2011), 2-4, 133, 142; Logevall, *Choosing War,* 3, 6, 43, 63-64, 292-93, 310, 320-22, 362; Tom Mangold and John Penycate, *The Tunnels of Cu Chi: The Untold Story of* Vietnam (New York: Random House, 1985), 38; Jeffrey H. Michaels, "Helpless or Deliberate Bystander: American Policy towards South Vietnam's Military Coups, 1954–1975," *Small Wars & Insurgencies* 25 (2014): 560-83, https://www.tandfonline.com/doi/abs/10.1080/09592318.2014.913546rds (accessed, 30 April 2020); Moyar, "Villager Attitudes;" Lindsey A. O'Rourke, *Covert Regime Change: America's Secret Cold War* (Ithaca, New York: Cornell University Press, 2018), 158, 174-78, 182-85; Palmer, 32, 56-58, 61-62, 106; Shelby Stanton, *Vietnam Order of Battle: U.S. Army and Allied Ground Forces in* Vietnam (New York: New Books, 1981), 49, 126-28, 131-32, 136-60; Nicholas Turse, *Kill Anything That* Moves (New York: Metropolitan, 2013), 173, 177-81; U.S. Government, National Security Council Memorandum # 362, Responsibility for U. S. Role in Pacification (Revolutionary Development), 5/9/1967, "NSAM 362 [Responsibility for U.S. Role in Pacification]," National Security Action Memorandums, NSF, Box 9, LBJ Presidential Library, https://www.discoverlbj.org/item/nsf-nsam362 (accessed, 30 April 2020); U.S. Government, National Security Council, Memorandum # 288, Implementation of South Vietnam Programs, 3/17/1964, "NSAM 288 Implementation of South Vietnam Programs," National Security Action Memorandums, NSF, Box 3, LBJ Presidential Library, https://www.discoverlbj.org/item/nsf-nsam288 (accessed,

30 April 2020); U.S. Government, Department of Defense, "Memorandum for the President of the United States, March 16, 1964, Folder, "[NSC Meeting on] U. S. Policy toward Vietnam, 3/17/1964, Volume 1, Tab 5," National Security Council Meetings Files, NSF, Box 1, LBJ Presidential Library, https://www.discoverlbj.org/item/nsf-nscm-b1-f06 (accessed, 10 May 2020); U.S. Government, Folder, "[November 24, 1963 - 3:00 p.m. Meeting with Ambassador Lodge and Advisors on Vietnam - McCone Memo]," Meeting Notes Files, Box 1, LBJ Presidential Library, https://www.discoverlbj.org/item/mtgnotes-b1-f03 (accessed, 30 April 2020); U.S. Government, United States Embassy, Vietnam (South), *Comments on South Vietnamese Government,* 1963, 2020, http://libproxy.temple.edu/login?url=https://search-proquest-com.libproxy.temple.edu/docview/1679127289?accountid=14270 (accessed, 30 April 2020); *Washington Post,* June 11, 1969; Westmoreland, *A Soldier Reports,* 85, 88-89, 97-98, 222-23, 320-21, 431, 434; Westmoreland II, 4; David Zebecki, e-mail to Author, May 11, 2020.

94 James D. Clayton, *Refighting the Last War: Crisis and Command in Korea* (New York: The Free Press, 1993), 2-3; Hay, *Tactical Innovations,* 5-9, 14, 20-22, 28, 33-34, 37-42, 52-54, 65, 67, Francis J. Kelly, *Vietnam Studies: U.S. Army Special Forces, 1961-1971* (Washington, D.C.: Center of Military History, 1973), 74; Krepinevich, *The Army and Vietnam,* 38-42; A. E. Lieberman, "Updating Impressions of the Military-Industry Complex," *California Management Review* 11 (Summer 1969): 51-62, http://libproxy.temple.edu/login?url=https://search-proquest-com.libproxy.temple.edu/docview/1301281610?accountid=14270 (accessed, 4 May 2020); Joseph A. McChristian, *Vietnam Studies: The Role of Military Intelligence, 1965-1967* (Washington, D.C.: Center of Military History, 1974), 105; Murray L. Weidenbaum, "The U.S. Aircraft Industry," *Financial Analysts Journal* 19 (1963): 49-53, www.jstor.org/stab le/4469470 (accessed, 4 May 2020).

95 John M. Carland, *The United States Army in Vietnam: Stemming the Tide, May 1965 to October 1966* (Washington, D.C.: Center of Military History, 2000), 4-5, 13-14, 17; Hay, *Tactical Innovations,* 103, 105; Grumman Aircraft Engineering Corporation, "The Mohawk Surveillance System," https://dk-media.s3.amazonaws.com/AA/AA/moomoo/downloads/288676/1963-06_Mohawk_Surveilance.pdf (accessed, 28 May 2020); Terry Love, *OV-1 Mohawk in Action* (Carrollton, TX: Signal Publications, 1989), 13; McChristian, *The Role of Military Intelligence,* 13, 97, 103, 152-53; Harry G. Summers, *On Strategy: A Critical Analysis of the Vietnam War* (Novato, CA: Presidio Press, 1982), 170; U.S. Government, Department of the Army, U.S. Army Intelligence Center and School, Fort Huachuca, "Sup R 45145 et al, OV-1 Mohawk Surveillance Systems, March 1974," 63-76; U.S. Government, Army Concept Team in Vietnam, "Mohawk Aircraft in the Target Acquisition Role," https://dk-media.s3.Amazonaws.com/AA/AA/moomoo/downloads/288678/1963-09_23RD_SWAD_Evaluation.pdf (accessed, 28 May 2020); Westmoreland, *A Soldier Reports,* 369.

96 Christopher Chant, *A Compendium of Armaments and Military Hardware* (New York: Routledge Revivals, 1987), 99; Department of the Army, *Technical Manual 11-5855-203-10: Operator's Manual for Night Vision Sight, Individual Served Weapon AN/PVS-2, AN/PVS-2A, and AN/PVS-2* (Washington, D.C.: Department of the Army, 1974), 2, 9; Department of the Army, *Technical Manual TM9-1345-200: Land Mines* (Washington, D.C.: Department of the Army, 1964), 29-34; Global Security.org, "M114 155mm Howitzer," https://www.globalsecurity.org/military/systems/ground/m114.htm (accessed, 6 May 2020); David Ott, *Vietnam Studies: Field Artillery, 1954*-1973 (Washington, D.C.: Center of Military History, 1975), 50.

97 Frederick E. Brazee, "The Mobile Riverine Force, Mekong Delta, Republic of Vietnam, 16 February 1967-10 January 1968: Personal Experiences of a Company Commander and Assistant Brigade S2," (class paper, Infantry Officer Advanced Course, U.S. Army Infantry

School, 1968), 11, https://www.mrfa.org/wp-content/uploads/2016/06/BrazeeFredericK0E.CPT_.pdf (accessed, 7 May 2020); Department of the Army, *Field Manual 31-75: Riverine* Warfare (Washington, D.C.: Department of the Army, 1971), 6-1, 6-8; https://www.youtube.com/watch?v=cIIhcA69Zt8 (accessed, 7 May 2020); William B. Fulton, "In Combat—Mobile Riverine Force," 20-22, 26-27, https://www.mrfa.org/wp-content/uploads/2016/06/General.Fulton.MRF_.pdf (accessed, 7 May 2020); William B. Fulton, *Vietnam Studies: Riverine Operations, 1966-1969* (Washington, D.C.: Department of the Army, 1973), 72-73, 137-38. Hereafter Fulton, *Vietnam Studies*; Ott, *Field Artillery*, 75-80; Stanton, *Order of Battle*, 77, 150; Westmoreland, *A Soldier Reports*, 271-74.

98 Hay, *Tactical Innovations*, 54.

99 Andrew J. Birtle, "PROVN, Westmoreland, and the Historians: A Reappraisal," *The Journal of Military History* 72 (October 2008): 1213-47, http://libproxy.temple.edu/login?url=https://search-proquest-com.libproxy.temple.edu/docview/195638949?accountid=14270 (accessed, 6 May 2020); Robert A. Doughty, *The Evolution of US Army Tactical Doctrine, 1946*-76 (Fort Leavenworth, Kansas: Combat Studies Institute, 1979), 37; *New York Times*, November 1, 1971; Ott, *Field Artillery*, 55; Palmer, *Summons of the Trumpet*, 184; Shelby L. Stanton, *The Rise and Fall of an American Army: U.S. Ground Forces in Vietnam, 1965-1973* (Novato, CA: Presidio Press, 1985), 343.

100 Vincent H. Demma, *American Military History*, ed. Stetson Conn (Washington, D.C.: Center of Military History, 1998), 633, 17-18; Fulton, *Vietnam* Studies, 26-41; Grau, *Riverine Mobile Force*, 61, 69; Kevin Keaveny, "Cavalry Afloat: The 39th Cavalry Platoon in the Mekong Delta," *Armor: The Professional Bulletin of Armor Branch*, 102 (July-August 1993): 12-19, https://www.benning.army.mil/armor/eARMOR/content/issues/1993/JUL_AUG/ArmorJulyAugust1993web.pdf (accessed, 9 May 2020); Mobile River Task Force Association, "Mobile River Task Force 117" https://www.mrfa.org/us-navy/us-navy-mo bile-riverine-force/task-force-117/ (accessed, 9 May 2020); U.S. Government, Department of the Army, Office of the Chief of Engineers, "Herbicides and Military Operations, Main Paper, April 11, 1972, 2-3, 13, https://apps.dtic.mil/dtic/tr/fulltext/u2/893214.pdf (accessed, 9 May 2020); Westmoreland, *A Soldier Reports*, 271-74, 367.

101 Cosmas, *MACV: The Joint Command*, 244, 482, 488; Lewis Sorley, *Thunderbolt—From the Battle of the Bulge to Vietnam and Beyond: General Creighton Abrams and the Army of His* Times (New York: Simon and Schuster, 1992), 185-86; U.S. Government, Telephone conversation # 12711, sound recording, LBJ and Robert McNamara, 2/12/1968, 8:29AM, Recordings and Transcripts of Telephone Conversations and Meetings, LBJ Presidential Library, https://www.discoverlbj.org/item/tel-12711 (accessed, 10 May 2020); U.S. Government, National Security Council, Folder, "[July 21, 1965 Meetings with Foreign Policy Advisors on Vietnam]," Meeting Notes Files, Box 1, 3-4, LBJ Presidential Library, https://www.discoverlbj.org/item/mtgnotes-b1-f19 (accessed, 10 May 2020); U.S. Government, National Security Council, Folder, "[NSC Meeting on] Deployment of Additional U. S. Troops in Vietnam, 7/27/1965, Volume 3, Tab 35," National Security Council Meetings Files, NSF, Box 1, 4, LBJ Presidential Library, https://www.discoverlbj.org/item/nsf-nscm-b1-f38 (accessed, 10 May 2020); Westmoreland II, 10; Westmoreland, *A Soldier Reports*, 163, 170, 181, 206-08, 235, 291, 298-99, 333, 351, 469-72, 542.

102 Carl von Clausewitz, *On War*, Michael Howard and Peter Paret, trans. (Princeton, NJ: Princeton University Press, 1984), 101-19; Daniel Ellsworth, "Counterinsurgency Tactics Led to Haditha, My Lai," *New Perspective Quarterly* 23 (2006): 49-53, https://onlinelibrary-wiley-com.libproxy.temple.edu/doi/abs/10.1111/j.1540-5842.2006.00828.x (accessed, 7 May 2020);

David Galula, *Counterinsurgency Warfare: Theory and Practice* (New York: Frederick A. Praeger, 1964), 50; Matthew Lippman, "War Crimes: The My Lai Massacre and the Vietnam War," *San Diego Justice Journal* 1 (Summer 1993), 295-364, https://advance-lexis-com.libproxy.temple.edu/api/document?collection=analytical-materials&id=urn:contentItem:3S3C-6050-00CV-V0TF-00000-00&context=1516831; (accessed, 7 May 2020); Prugh, 74-76, 100, 102, 113; Peter S. Kindsvatter, *American Soldiers: Ground Combat in the World Wars, Korea & Vietnam* (Lawrence, KS: University Press of Kansas, 2003), 205; Roger Trinquier, *Modern Warfare: A French View of Counterinsurgency* (Ft. Leavenworth, KS: Combat Studies Institute, 1985), 8, 19; Turse, 97, 104, 144, 157-72, 175-77, 185-91; U.S. Government, Department of Defense, Department of the Army, *Report of the Department of the Army Review into the Preliminary Investigation of the My Lai Incident, Volume I: Report of the Investigation, 14 March 1970,* Section 6, 1-20; Westmoreland, *A Soldier Reports,* 494-501.

103 Cosmas, *MACV: The Joint Command,* 45, 243, 398; William G. Bell, *Department of the Army History Summary, Fiscal Year* 1969 (Washington, D.C.: Department of the Army, 1973), 35-36; Demma, 642-43; Department of the Army, *Time Honored Professionals: The NCO Corps Since 1775* (Washington, D.C.: Center of Military History, 1989), 22; Anthony R. Herbert, Soldier (New York: Dell, 1973), 208, 224, 229, 259; David W. Hogan , Jr; Arnold G. Fisch, Jr; and Ronald K. Wright, Jr., *The Story of the Non-Commissioned Officer Corps: Backbone of the* Army (Washington, D.C.: Center of Military History, 2007), 47-48; John Ionoff, interview with Author May 10, 2020; Haynes Johnson and George C. Wilson, *Army in Anguish* (Washington, D.C.: Washington Post, 1972), 91-99; Kindsvatter, *American Soldiers,* 150, 154; Lt. Gen. William R. Peers, Report of the Department of the Army Review of the Preliminary Investigations into the My Lai Incident, 14 March 1970, Vol. I, Section 2, 9-13; Prugh, *Law at War,* 106-07, Appendix K; Pat Smith, interview with Author, May 12, 2020; Stanton, *The Rise and Fall of an American Army,* 26-27, 54, 70-71, 74-78, 95, 196; Richard W. Stewart, *American Military History, Volume II: The United States Army in a Global Era, 1917-2003* (Washington, D.C.: Center of Military History, 1989), 305, 350, 369; Westmoreland, *A Soldier Reports,* 386-91, 398, 489, 498; James W. Williams, *A History of Army Aviation from Its Beginnings to the War on Terror* (Lincoln, New York: i Universe, 2005), 96; United States Army War College, "Study on Military Professionalism, " (Carlisle, PA: United States Army War College, 1970); iii-v; 12-17; Jim Wright, e-mail to author, May 9, 2020; Zebecki e-mail.

104 Herbert, *Soldier,* 208, 224, 229, 259; Johnson and Wilson, *Army in Anguish,* 20, 25; United States Army War College, "Study on Military Professionalism," iii-v; 12-17.

105 William Boothby, *Weapons and the Law of Land* Warfare (Oxford, UK: Oxford University Press, 2016), 10, 62, 104-05, 325; Department of the Army, *Field Manual 27-10 with Change 1: The Law of Land* Warfare (Washington, D.C.: Department of the Army, 1976), 34; Headquarters, Department of the Army and Headquarters, United States Marine Corps, *Field Manual 6-27: The Commander's Handbook on the Law of Land Warfare* (Washington, D.C.: Department of the Army, United States Marine Corps, 2019), vii, 1-2, 1-7, 1-11, 7-1, 7-7, References 4; International Committee of the Red Cross, "Geneva Convention Relative to the Treatment of Prisoners of War, Article 4: Prisoners of War," https://www.icrc.org/en/doc/assets/files/publications/icrc-002-0173.pdf (accessed, 6 July 2020); Laura Perna, *The Formation of the Treaty Law of Non-International Conflicts* (Leiden, NL: Brill, 2006), 31-34.

Chapter 3

[1] Barbara W. Tuchman, *Stillwell and the American Experience in China, 1913-1945* (New York: McMillan, 1972), 132.

[2] "Casualty Status, as of 10:00 a.m. EST, January 18, 2021," https://www.defense.gov/casualty.pdf (accessed, 22 January 2021); Peter J. Schoomaker, interview with Author, April 15, 2019.

[3] Joseph J. Collins, *Understanding War in Afghanistan* (Washington, D.C.: National Defense University, 2011), 45; Yaroslav Trofimov, *Faith at War: A Journey to the Frontlines of Islam, From Baghdad to Timbuktu* (New York: Henry Holt, 2005), 204-5; U.S. Government, United States Congress, "*Iraqi Documents: A Glimpse into the Regime of Saddam Hussein: Hearing before the Subcommittee on Oversight and Investigation of the Committee on International Relations, House of Representatives,* 109th Cong, 203 (2006), https://heinonline-org.libproxy.temple.edu/HOL/Page?handle=hein.cbhear/fdsysafdy0001&id=1&c=ollection=congrec&index= (accessed, 17 July 2020); Charles H. Briscoe, Kenneth Finlayson, Robert W. Jones, Cherilyn A. Walley, et al, *All Roads Lead to Baghdad: Army Special Operations Forces in Iraq* (Fort Bragg, NC: US Army Special Operations Command History Office, 2006), 27-28, 407-8; Richard Kugler, Michael Baranick, and Hans Binnendijk, "Operation Anaconda: Lessons for Joint Operations" (Washington D.C., Center for Technology and National Security Policy, National Defense University, 2009), 48-50, https://permanent.fdlp.gov/gpo135804/DTP-060.pdf (accessed, 18 July 2020); U.S. Government, Library of Congress, Congressional Research Service by Kenneth Katzman, "Afghanistan: Current Issues and U.S. Policy: Updated March 31, 2003," 6-7, https://congressional-proquest-com.libproxy.temple.edu/congressional/result/congresultpage:pdfevent?rsId=172C7B57E1A&pdf=/app-bin/gis-congresearch/5/9/3/9/crs-2003-fdt-0521_from_1_to_39.pdf&uri=/app-gis/congersearch/crs-2003-fdt-0521 (accessed, 18 July 2020); U.S. Government, Library of Congress, Congressional Research Service by Richard P. Cronin, "Afghanistan: Challenges and Options for Reconstructing a Stable and Moderate State, May 10, 2002," https://congressional-proquest-com.libproxy.temple.edu/congressional/docview/t21.d22.crs-2002-fdt-0606?accountid=14270 (accessed, 17 July 2020); U.S. Government, Library of Congress, Congressional Research Service by Andrew Feickert, "U.S. Army's Modular Redesign: Issues for Congress Updated February 10, 2005," https://congressional-proquest-com.libproxy.temple.edu/congressionall/docview/t21.d22.crs-2005-fdt-0672?accountid=14270 (accessed, 17 July 2020); U.S. Government, Government Accountability Office, "Report to the Subcommittee on Personnel, Committee on Armed Services, U.S. Senate: Military Personnel--DoD Needs to Address Long-term Reserve Force Availability and Related Mobilization and Demobilization Issues," 2004, https://congressional-proquest-com.libproxy.temple.edu/congressional/result/congresultpage:pdfevent?rsId=172C36C150E&pdf=/app-bin/gis-congresearch/e/c/c/4/cmp-2004-gao-0481fro m_1_to_99.pdf&uri=/app-gis/congresearch/cmp-2004-gao-0481 (accessed, 17 July 2020).

[4] United Nations, "Security Council Resolution 1267 (1999) of 15 October 1999," https://undocs.org/S/RES/1267(1999) (accessed, 20 July 2020); U.S. House of Representatives, 105th Cong., 2nd Session, H. Con. Res 336, "Condemning the Taliban Regime and Supporting a Broad Based Government in Afghanistan," https://congressional-proquest-com.libproxy.temple.edu/congressional/result/congressional/congdocumentview?Accountid=14270&groupid=96626&parmId=172C7D71F5D&rsId=172C7D621AF (accessed, 19 July 202); U.S. House of Representatives, 106th Cong., 1st Session H. Res 187, "Expressing The Sense of the

House Of Representatives That the United States Should Seek to Prevent Any Taliban-Led Government In Afghanistan From Obtaining a Seat In the United Nations, And Should Refuse to Recognize Any Afghan Government, While Gross Violations of Human Rights Persist Against Women and Girls There," https://congressional-proquest-com.libproxy.temple.edu/congressional/docview/t01.d02.106_hres_187_ih?accountid=14270 (accessed, 18 July 2020); U.S. Government, U.S. House of Representatives, "House Document 106-90: A Response to the Actions and Policies of the Afghan Taliban," https://congressional-proquest-com.libproxy.temple.edu/congressional/result/congressional/congdocumentview?accountid=14270&groupid=96626&parmId=172C7EF19F6&rsId=172C7EBE4F6 (accessed, 18 July 2020); U.S. Government, Central Intelligence Agency, "The World Fact Book: Afghanistan," https://www.cia.gov/library/publications/the-world-factbook/geos/af.html (accessed, 18 July 2020); U.S. Government, U.S. Congress, 105th Cong., 2nd Session, "Country Reports on Human Rights Practices for 1997: Report Submitted to the Committee on International Relations U.S. House of Representatives and Time Committee on Foreign Relations U.S. Senate by the Department Of State," xvi, xviv, 1605-1606, 1608, 1612-1613, https://congressional-proquest-com.libproxy.temple.edu/congressional/result/pqpresultpage.gispdfhitspanel.pdflink/$2fapp-bin$2fgis-congresearch$2f5$2fc$2fb$2f8$2fcmp-1998-hir-0002_from_1_to_1756.pdf/entitlementkeys=1234%7Capp-gis%7Ccongresearch%7Ccmp-1998-hir-0002 (accessed, 18 July 2020); U.S. Government, Library of Congress, Congressional Research Service by Kenneth Katzman, "Afghanistan: Current Issues and U.S. Policy, Updated December 10, 1996," 13, https://congressional-proquest-com.libproxy.temple.edu/congressional/result/congresultpage:pdfevent?rsId=172C8954FCB&pdf=/app-bin/gis-congresearch/6/e/5/1/crs-1996-fnd-0240_from_1_to_20.pdf&uri=/app-gis/congresearch/crs-1996-fnd-0240 (accessed, 18 July 2020); U.S. Government, U.S. Senate, "106th Cong., 1st Session, S. RES. 68 Expressing the Sense of The Senate Regarding the Treatment of Women and Girls by the Taliban In Afghanistan," https://congressional-proquest-m.libproxy.temple.edu/congressional/docview/t01.d02.106_sres_68_ats?accountid=com14270 (accessed, 18 July 2020); House Document 106-90, "A Response to the Actions and Policies of the Afghan Taliban Communication from the President of the United States Transmitting His Declaration of a National Emergency with Respect to the Threat to the United States Posed by the Actions and Policies of the Afghan Taliban and an Executive Order to Deal with This Threat, Pursuant To 50 U.S.C. 1703(B)," https://congressional-proquest-com.libproxy.temple.edu/congressional/result/congressional/congdocumentview?accountid=14270&groupid=96626&parmId=172C8A203D6&rsId=172C8A0FF4E (accessed, 18 July 2020); Surekha Vijh, "Monuments of Destruction." *Insight on the News*, April 23, 2001, 24, https://link-gale-com.libproxy.temple.edu/apps/doc/A74011909/AONE?u=temple_main&sid=AONE&xid=1a7efe4b (accessed, 18 July 2020); *Los Angeles Times*, September 20, 2001; Nemathollah Nojumi, *The Rise of the Taliban in Afghanistan: Mass Mobilization, Civil War, and the Future of the Region* (New York: Palgrave, 2002), 127-33.

5 Department of the Army, *Field Manual 1-02/MCRP 5-12A: Operational Terms and Graphics* (Washington, D.C.: Department of the Army, 2004), 1-170.

6 Command Historian, "U.S. Army Field Support Command and Joint Munitions Command Support to Operation Iraqi Freedom, Phases I-III" (Rock Island Arsenal, IL: U.S. Army Field Support Command, 2005), 6-7; Gordon W. Rudd, *Humanitarian Intervention: Assisting the Kurds in Operation PROVIDE COMFORT* (Washington, D.C.: Center of Military History, 2004), 118-23, 215; William J. Allen, *Short of War: Major U.S. Air Force Contingency*

Operations, 1947-1997 (Montgomery, AL: Air Force Historical Agency, 2000), 189-190; United Nations, "Security Council Resolution 688 (1991), of 5 April 1991," https://undocs.org/en/S/RES/688(1991) (accessed, 21 July 2020); United Nations, "Security Council Resolution (1994) 949, of 15 October 1994, http://unscr.com/en/resolutions/doc/949 (accessed, 21 July 2020); Jeffrey A. Charlston, *Department of the Army Historical Summary, Fiscal Year 1999* (Washington, D.C.: Center of Military History, 2006), 53, 85, 140; W. Blair Hayworth, *Department of the Army Historical Summary, Fiscal Year 1998* (Washington, D.C.: Department of the Army, 2005), 51, 139; Paul K. White, "Airpower and a Decade of Containment," *Joint Force Quarterly* 27 (Winter 2000-2001): 35-40, https://media.defense.gov/2011/Aug/09/2001329928/-1/-1/ 0/a407897.pdf (accessed, 21 July 2020).

[7] Exec. Order No. 13289, 3 C.F.R., 2003 Comp. p.1; National Commission on Terrorist Attacks upon the United States, *The 9/11 Commission Report: Final Report of the National Commission on Terrorist Attacks Upon the United States* (New York: Knopf & Company, 2004), 63-67; Richard W. Stewart, *Operation Enduring Freedom: The United States Army in Afghanistan, October 2001-March 2002* (Washington, D.C.: Center of Military History, 2004), 7; Charles H. Briscoe, Richard L. Kiper, James A. Schroder, and Kalev I. Sepp, *Weapon of Choice: U.S. Army Special Operations Forces in Afghanistan* (Fort Leavenworth, KS: Combat Studies Institute, 2003), 34-36, 44, 77, 95; U.S. Central Command, "CENTCOM Coalition Nations," https://www.centcom.mil/AREA-OF-RESPONSIBILITY/CENTCOM-COALITION/ (accessed, 5 August 2020); Bruce R. Pirnie, John Gordon IV, and Richard R. Brennan, Jr., *Operation Iraqi Freedom: Decisive War, Elusive Peace* (Santa Monica, CA: The RAND Corporation, 2015), 64-65; Charlston, *Fiscal Year 1999*, 53, 85, 140; Steven E. Everett, *Department of the Army Historical Summary: Fiscal Year 1993* (Washington, D.C.: Center of Military History, 2002), 55-56; Richard W. Stewart, "War in the Persian Gulf: Operations Desert Shield and Storm, August 1990-March 1991" (Washington, D.C.: Center of Military History, 2005), 5-8, 17-23, 29-63; U.S. Government, Library of Congress, Congressional Research Service by Kenneth Katzman, "Iraq: Compliance, Sanctions and U.S. Policy, Updated May 21, 2001," https://congressional-proquest-com.libproxy.temple.edu/congressional/result/pqpresultpage.gispdfhitspanel.pdflink/$2fapp-bin$2fgis-congresearch$2f9$2fd$2ff$2f8$2fcrs-2001-fdt-0192_from_1_to_18.pdf/entitlementkeys=1234%7Capp-gis%7Ccongresearch%7Ccrs-2001-fdt-0192 (accessed, 21 July 2020); UNSCR 688; UNSCR 949; William J. Webb, Charles Anderson, Dale Andrade, et al, *Department of the Army Historical Summary: Fiscal Years 1990-1991* (Washington, D.C.: Center of Military History, 1997), 22-25, 42; Kathleen J. Nawyn, *Department of the Army Historical Summary, Fiscal Year 2003* (Washington, D.C.: Center of Military History, 2015), 58; William S. Wallace, Commanding General, V (US) Corps, e-mail to Author, July 21, 2020.

[8] Special Dictionary, "Kurdish Proverbs," http://www.special-dictionary.com/proverbs/source/k/kurdish_proverb/#:~:text=%20154%20Kurdish%20Proverbs%20%201%20A%20camel,good%20companion%20shortens%20the%20longest%20road.%20More%20 (accessed, 22 July 2020).

[9] Qamar Fatima, "Pakistan's Religious Elite and Afghanistan: Antecedents," *Journal of Political Studies* (2018): 39, https://link-gale-com.libproxy.temple.edu/apps/doc/A576524476/AONE?u=temple_main&sid=AONE&xid=0a097905 (accessed, 22 July 2020); Sajjad Nejatie, "Deobandi Sect: 1200 to 1900: South, Central, and West Asia." In *Cultural Sociology of the Middle East, Asia, & Africa: An Encyclopedia*, http://sk.sagepub.com.libproxy.temple.edu/reference/culturalsociologyemea/n681.xml (accessed, 21 July 2020); Nojumi, *The Rise of the Taliban in Afghanistan*, 119-21; U.S. Government, House of Representatives,

107 H. Con. Res. 145, https://congressional-proquest-com.libproxy.temple.edu/congressional/result/congressional/congdocumentview?accountid=14270&groupid=96626&parmId=172DD0764FD&rsId=172DD074566 (accessed, 21 July 2020); Gil Donovan, "Taliban to Require Hindus to Display Identity Labels," *National Catholic Reporter*, June 2001, http://search.ebscohost.com.libproxy.temple.edu/login.aspx?direct=true&db=rlh&AN=9502440&site=ehost-live&scope=site (accessed, 22 July 2020); Nojumi, *The Rise of the Taliban in Afghanistan*, 3, 5, 119-21, 224-45; Peter Mansfield, *A History of the Middle East* (New York: Viking Press, 1991), 157, 165-66, 182-83, 228, 289, 323, 325; Antonello Biagini and Giovanna Motta, eds., *Empires and Nations from the Eighteenth to the Twentieth Century: Volume 2* (Newcastle-upon-Tyne: Cambridge Scholars, 2014), 102-108; David Fromkin, *A Peace to End All Peace: The Fall of the Ottoman Empire and the Creation of the Modern Middle* East (New York: Avon, 1989), 415, 450-54; John Glubb, *A Short History of the Arab* Peoples (New York: Dorset Press, 1969), 279-280; Albert Hourani, *A History of the Arab* Peoples (New York: MFF Press, 1991) 329-31; Bernard Lewis, *Islam and the* West (New York: Oxford University Press, 1993), 152, 154; U.S. Government, Central Intelligence Agency, "The World Fact Book: Middle East: Iraq," https://www.cia.gov/library/publications/the-world-factbook/geos/iz.html (accessed, 23 July 2020); U.S. Government, 96th Congress, 2nd Session, "The Political Economy of the Middle East: 1973-1978, A Compendium of Papers Submitted to the Joint Economic Committee, Congress of the United States, April 21, 1980," 21-22, 64, 67, 79, 94, 215-16, 229, 283, 288-89, 335, 361, 429, 468-72, 532, https://congressional-proquest-com.libproxy.temple.edu/congressional/result/congresultpage:pdfevent?rsId=172E146B397&pdf=/app-bin/gis-congresearch/b/a/9/7/cmp-1980-ecj-0040_from_1_to_577.pdf&uri=/app-gis/congresearch/cmp-1980-ecj-0040 (accessed, 22 July 2020); Paula Kitching, "The Sykes-Picot Agreement and Lines in the Sand." *Historian* 128 (15, 2016): 18-22, http://libproxy.temple.edu/login?url=https://search-proquest-com.libproxy.temple.edu/docview/1787724312?accountid=14270 (accessed, 22 July 2020); Hourani, 318, 404-05; Guiditta Fontana, "Creating Nations, Establishing States: Ethno-Religious Heterogeneity and the British Creation of Iraq in 1919-23," *Middle Eastern Studies* 46 (2010): 1-16, www.jstor.org/stable/40647226 (accessed, 22 July 2020); Colbert C. Held, *Middle East Patterns: Places, Peoples, and* Politics (Boulder: Westview Press, 1994), 274-75; Michael Eppel, *Iraq From Monarchy to Tyranny: From the Hashemites to the Rise of Saddam* (Gainesville, FL: University Press of Florida, 2004), 147, 198, 201-8; Gabriel Ben-Dor, "Civilianization of Military Regimes in the Arab World." *Armed Forces & Society* 1 (April 1975): 317–27, https://journals-sagepub-com.libproxy.temple.edu/doi /abs/10.1177/0095327X7500100304 (accessed, 22 July 2020); John F. Devlin, "The Baath Party: Rise and Metamorphosis," *The American Historical Review* 96 (1991): 1396-407, https://www-jstor-org.libproxy.temple.edu/stable/2165277?seq=12#metadata_info_tab_contents (accessed, 22 July 2020); Bernard Lewis, *The Crisis of Islam: Holy War and Unholy* Terror (New York: Random House, 2004), 70-71; Roy E. Thoman, "Iraq Under Baathist Rule," *Current History* 62 (1972): 31, http://libproxy.temple.edu/login?url=https://search-proquest-com.libproxy.temple.edu/docview/1309775609?Accounted=14270 (accessed, 22 July 2020); Hanna Batatu, "Iraq's Underground Shi'a Movements: Characteristics, Causes and Prospects," *Middle East Journal* 35 (Fall 1981): 578, http://libproxy.temple.edu/login?url=https://search-proquest-com.libproxy.temple.edu/docview/1290822796?accountid=14270 (accessed, 22 July 2020); Lewis, *Islam and the* West, 164; Helen Metz, *Iraq: A Country Study* (Washington, D.C.: Library of Congress, 1990), 96-97; *The Observer* (London, UK), May 10, 1987; *Guardian* (Manchester, UK), June 28, 1983, U.S.

Government, 89th Cong., 1st Session, "Country Reports on Human Rights Practices, 1982: Report Submitted To The Committee On Foreign Relations, U.S. Senate And Committee On Foreign Affairs, U.S. House Of Representatives By The Department Of State In Accordance With Sections 116(D) And 502b(B) Of The Foreign Assistance Act Of 1901, As Amended," 1152, https://congressional-proquest-com.libproxy.temple.edu/conzgressional/result/congesultpage:pdfevent?rsId=172E146B397&pdf=/app-bin/gis-congresearch/7/e/7/9/cmp-1983-for-0008_from_1_to_1330.pdf&uri=/app-gis/congresearch/cmp-1983-for-0008 (accessed, 23 July 2020).

[10] Margaret K. Nydell, *Understanding Arabs: A Guide for Westerners* (Yarmouth, ME: Intercultural Press, 1987), 75-76; Raphael Patai, *The Arab Mind* (Long Island City, NY: Hatherleigh Press, 2002), 80-81; The American University, Richard Nyrop and Donald Seekins, *Afghanistan: A Country Study* (Washington, D.C.: Library of Congress Research Service, 1990), 3-4, 46, 92, 104, 108-13, 146; Andrea Chiovenda & Melissa Chiovenda (2018) "The Specter Of the 'Arrivant': Hauntology Of An Interethnic Conflict In Afghanistan," *Asian Anthropology* 17 (October 2018): 165-84, https://www-tandfonline-com.libproxy.temple.edu/doi/pdf/10.1080/1683478X.2018.1480917?needAccess=true (accessed, 24 July 2020); Robert Brown, *The Races of Mankind: Being a Popular Description of the Characteristics, Manners and Customs of the Principal Varieties of the Human Family, Vol. III.* (New York: Casell, Peter and Galpin, 1876), 254-55; David Gibbs, "The Peasant As Counter-Revolutionary: The Rural Origins Of The Afghan Insurgency," *Studies in Comparative International Development* 21 (1986): 36–59, https://link-springer-com.libproxy.temple.edu/content/pdf/10.1007/BF02717362.pdf (accessed, 24 July 2020);George S. Massanat, "Development and Diplomacy in Afghanistan," *Journal of Asian and African Studies* 4 (February 1969): 51-60, http://libproxy.temple.edu/login?url=https://search-proquest-com.libproxy.temple.edu/docview/1303206104?accountid=14270 (accessed, 25 July 2020); Richard S. Newell, *The Politics of Afghanistan* (Ithaca, NY: Cornell University Press, 1972), 80-84; Lester W. Grau, "Defeating Guerrilla Logistics: Soviet Operation 'Trap' in Western Afghanistan," *The Journal of Slavic Military Studies* 31 (February 2018): 65-79, https://www-tandfonline-com.libproxy.temple.edu/doi/citedby/10.1080/13518046.2018.1417241?scroll=top&needAccess=true (accessed, 1 October 2020); Theodore S. Gochenour "A New Try for Afghanistan," *Middle East Journal* 19 (Winter, 1965): 1-19, http://libproxy.temple.edu/login?url=https://search-proquest-com.libproxy.temple.edu/docview/1290681059?accountid=14270 (accessed, 25 July 2020); Vartan Gregorian, "Mahmud Tarzi and Saraj-Ol-Akhbar: Ideology of Nationalism and Modernization in Afghanistan," *Middle East Journal* 21 (Summer, 1967): 345, http://libproxy.temple.edu/login?url=https://search-proquest-com.libproxy.temple.edu/docview/1290704075?accountid=14270 (accessed, 25 July 2020); Ronen Zeidel, "Tribes in Iraq: A Negligible Factor in State Formation" in *Tribes and States in a Changing Middle East* (Oxford: Oxford Scholarship on Line, 2016), 171-86, https://www-oxfordscholarship-com.libproxy.temple.edu/view/10.1093/acprof:oso/9780190264925.001.0001/acprof-9780190264925-chapter-009?print=pdf (accessed, 29 July 2020); Robert B. Cunningham, Yasin K. Sarayrah, *Wasta: The Hidden Force in Middle Eastern Society* (Westport, CT: Praeger, 1993), 9, 12-13, 21; *Helen Lackner, The Political Economy of Wasta: Use and Abuse of Social Capital Networking*, Mohamed A. Ramady, ed. (Cham, Switzerland: Springer, 2016), 33-40.

[11] Briscoe, et al. *Weapon of Choice*, 40; Tommy Franks, *American Soldier* (New York: Harper-Collins, 2004), 126-27; Tommy Franks, opening remarks at the 2nd Infantry Division Pre-

Command Course, Camp Red Cloud, Korea, July 1996 (the author attended this course prior to assuming command of Company A, 102d Military Intelligence Battalion).

[12] Stewart, *Operation Enduring Freedom*, 8-10; Briscoe et al., *Weapon of Choice*, 42, 45, 74.

[13] Daniel Byman, Kenneth M. Pollack and Gideon G. Rose, "Beef Up the Taliban's Enemy," https://www.rand.org/blog/2001/09/beef-up-the-talibans-enemy.html (accessed, 10 September 2020); Apratim Mukarji, "Advent of the Northern Alliance in Afghanistan: US Policy Examined," *Himalayan and Central Asian Studies* 6 (January 2002): 35-40, http://libproxy.temple.edu/login?url=https://www-proquest-com.libproxy.temple.edu/docview/13 25025309?accountid=14270 (accessed, 10 September 2020).

[14] Franks, *American Soldier*, 280.

[15] Franks, *American Soldier*, 289; Doug Stanton, *Horse Soldiers: The Extraordinary Story of a Band of U.S. Soldiers Who Rode to Victory in Afghanistan* (New York: Scribner, 2009), 61.

[16] Jim Garamone, "America Launches Strikes Against Al Qaeda, Taliban," American Forces Press Service, October 7, 2001, https://archive.defense.gov/news/newsarticle.aspx?id=44680 (accessed, 13 September 2020); Jim Garamone, "Aircraft, Missiles Hammer Terrorists," American Forces Press Service, October 7, 2001, https://archive.defense.gov/news/newsarticle.aspx?id=44679 (accessed, 13 September 2020); Jim Garamone, "U.S. Crews Assault Al Qaeda, Taliban Support Structures," Armed Forces Press Service, October 9, 2020, https://archive.defense.gov/news/newsarticle.aspx?id=44666 (accessed, 13 September 2020); Benjamin S. Lambeth, *Airpower Against Terror: America's Conduct of Operation Enduring Freedom* (Santa Monica, CA: RAND Corporation, 2005), 78-81; Jim Garamone, "Washington Provides Food, Medicine to Afghans," Armed Forces Press Service, October 7, 2001, https://archive.defense.gov/news/newsarticle.aspx?id=44677 (accessed, 13 September 2020); Jim Garamone, "What Are Humanitarian Daily Rations?," Armed Forces Press Service, October 7, 2001, https://archive.defense.gov/news/newsarticle.aspx?id=44678 (accessed, 13 September 2020); Gerry J. Gilmore, "Air Force Fliers Continue Food Drop Operations," Armed Forces Press Service, October 12, 2001, https://archive.defense.gov/news/newsarticle.aspx?id=44649 (accessed, 13 September 2020); Randy Mitchell, "Afghan Food Drop Underscores Bush's Humanitarian Pledge," Armed Forces Press Service, October 9, 2001, https://archive.defense.gov/news/newsarticle.aspx?id=44669 (accessed, 13 September 2020); U.S. Government, Department of State, International Information Programs, "Humanitarian Aid to the Afghan People," https://web.archive.org/web/20070714002014/https://usinfo.state.gov/is/Archive_Index/Humanitarian_Aid_to_the_Afghan_People-4072c45bad525.html (accessed, 12 September 2020); Briscoe, et al., *Weapon of Choice*, 77-82.

[17] Stanton, *Horse Soldiers* 67-69; Briscoe, et al., *Weapon of Choice*, 77-82, 97-99; Stewart, *Operation Enduring Freedom*, 10, 14; Kathleen Rhem, "Two Soldiers Killed, Special Forces Assault Taliban Sites," Armed Forces Press Service, October 20, 2001, https://archive.defense.gov/news/newsarticle.aspx?id=44628 (accessed, 13 September 2020); Jim Garamone," U.S. Advisors Aid Northern Alliance, Build Cohesion," Armed Forces Press Service, November 6, 2001, https://archive.defense.gov/news/newsarticle.aspx?id=44481 (accessed, 13 September 2020).

[18] Lambeth, *Airpower*, xii, 57, 60; Briscoe, et al., *Weapon of Choice*, 99; Gerry L. Gilmore, "Northern Alliance Progressing With U.S. Help, Pace Says," Armed Forces Press Service, November 7, 2001, https://archive.defense.gov/news/newsarticle.aspx?id=44478 (accessed, 13 September 2020); Jim Garamone, "Wolfowitz Shares Special Forces' Afghanistan

Dispatches," Armed Forces Press Service, November 15, 2001, https://archive.defense.gov/news/newsarticle.aspx?id=44448 (accessed, 13 September 2020).

[19] Stewart, *Operation Enduring Freedom,* 13; Briscoe et al., *Weapon of Choice,* 102.

[20] Stewart, *Operation Enduring Freedom,* 11-14; Jim Garamone, "Northern Alliance, U.S. Special Forces at Gates of Kabul," Armed Forces Press Service, November 13, 2001, https://archive.defense.gov/news/newsarticle.aspx?id=44461 (accessed, 13 September 2020); Briscoe, et al., *Weapon of Choice,* 103, 159, 182; Jim Garamone, "Taliban Surrendering Kandahar, U.S. Forces in Firefights," Armed Forces Press Service, December 7, 2001, https://archive.defense.gov/news/newsarticle.aspx?id=44387 (accessed, 13 September 2020); Kathleen Rhem, "Third U.S. Servicemember Succumbs to 'Friendly Fire' Wounds," Armed Forces Press Service, December 5, 2001, https://archive.defense.gov/news/newsarticle.aspx?id=44393 (accessed, 13 September 2020); Rhem, "Two Soldiers Killed."

[21] Gregory Fontenot, E. J. Deegan, and David Tohn, *On Point: The United States Army in Operation Iraqi* Freedom (Fort Leavenworth, KS: Combat Studies Institute, 2004), 99-100; Memorandums by Donald H. Rumsfeld, September 19, 2001 and January 19, 2002, George W. Casey, Jr. Papers, Box 87, National Defense University Library Archives, Fort Lesley J. McNair, Washington, D.C.; Franks, 329; Richard H. Immerman, *Empire for Liberty: A History of American Imperialism from Benjamin Franklin to Paul Wolfowitz Book* (Princeton, NJ: Princeton University Press, 2010), 203, 223-27; Edward B. Atkeson, "War in Iraq: Shock and Awe?" *Army* 53, May 2003, 8-13, http://libproxy. temple.edu/login?url=https://www-proquest-com.libproxy.temple.edu/docview/237082729?accou n tid=14270 (accessed, 16 September 2020); J. Sean Curtin, Japanese Institute of Global Communications Social Trends #28, "Japanese Anti-War Sentiment on Iraq in Accord with Global Opinion," February 24, 2003, http://www.glocom.org/speci al_topics/social_trends/2 0030224_trends_s28/ (accessed, 14 September 2020); Ari Berman, "Taking on Taking on Iraq," *Editor & Publisher,* January 27, 2003, 31, https://link-gale-com.libproxy.temple.edu/apps/doc/A97895516/AONE?u=temple_main&sid=AONE&xid=31a7c2d8 (accessed, 14 September 2020); National Public Radio, "Interview: Greg Mitchell Discusses Newspapers' Editorial Opinions About the War in Iraq," https://infoweb-newsbank-com.libproxy.temple.edu/apps/news/document-view?p=AWNB&t=pubname%3AMEN2M%21Morning%2BEdition%2B%255BNPR%255D%2B%2528USA%2529/year%3A2003%212003/mody%3A0131%21January%2B31&action=browse&format=text&docref=news/1577CFBD36BA1DD8 (accessed, 14 September 2020); U.N. Security Council, Report of the Security Council 1 August 2002-31 July 2003, General Assembly Official Records, Fifty-eighth Session, Supplement No. 2 (A/58/2), 7, https://undocs.org/A/58/2(SUPP) (accessed, 14 September 2020); John T. Correll, "What Happened to Shock and Awe?," *Air Force* 96, November 2003, 52-57; Michael E. O'Hanlon, "Shinseki vs. Wolfowitz," March 4, 2003, https://www.brookings.edu/opinions/shinseki-vs-wolfowitz/ (accessed, 14 September 2020).

[22] *Dallas Morning News,* May 31, 2014.

[23] Harlan K. Ullman, *Shock and Awe: Achieving Rapid Dominance* (Washington, D.C.: National Defense University Press, 1996), 6; O'Hanlon, "Shinseki vs. Wolfowitz," Michael E. O'Hanlon, "History Will Credit Shinseki," June 19, 2003, https://www.brookings.edu/opinions/history-will-credit-shinseki/ (accessed, 15 September 2020); *New York Times,* February 28, 2003; Rumsfeld Foundation, "Donald Rumsfeld: Founder and President,' https://www.rumsfeldfoundation.org/about/staff/donald-rumsfeld (accessed, 16 September 2020); Jewish Virtual Library, "Paul Wolfowitz (1943-)," https://www.jewishvirtuallibrary.org/paul-wolfowitz (accessed, 16 September 2020); Nicolaus Mills, "The General Who

Understood Iraq from the Start," *Dissent,* April 8, 2008, https://www.dissentmagazine.org/online_articles/the-general-who-understood-iraq-from-the-start (accessed, 16 September 2020); *Washington Post,* May 31, 2003.

[24] Franks, 329.

[25] U.S. Government, Department of Defense, U.S. Central Command, Briefing: USCENTCOM Compartmented Planning Effort 15 August 2002, Papers of Gen. (Retired) George W. Casey, Jr., Box 109, National Defense University Archives, Fort Lesley J. McNair, Washington, D.C..

[26] John S. Brown, *Kevlar Legions: The Transformation of the U.S. Army, 1986-2005* (Washington, D.C.: Center of Military History, 2011), 231-32; USCENTCOM, 15 August 2002; Fontenot, et al., *On Point,* 55-56; Nawyn, *Fiscal Year 2003,* 42; U.S. Government, Department of the Army, General Officer Management Office, "General William S. Wallace," https://www.gomo.army.mil/ext/portal/Officer/Resumes.aspx?Ltr=L&Type=Retired (accessed, 8 December 2020); U.S. Government, Department of the Army, General Officer Management Office, "General David D. McKiernan," https://www.gomo.army.mil/ext/portal/Officer/Resumes.aspx?Ltr=L&Type=Retired (accessed, 8 December 2020).

[27] U.S. Government, U.S. Department of Defense, memorandum of December 20, 2002, Casey Papers, Box 109.

[28] Ibid.

[29] Jeff Ryals, Staff Intelligence Officer (S-2), 3rd Squadron, 7th Cavalry, 3rd Infantry Division (Mechanized), e-mail to Author, September 17, 2020; U.S. Government, Department of Defense, Department of the Army, 3rd Infantry Division (Mechanized) After Action Report, Operation Iraqi Freedom: Chapter 7, https://www.strategypage.com/articles/3IDAAR/chap7.asp (accessed, 17 September 2020); U.S. Government, U.S. Department of Defense, U.S. Army Europe, memorandum of LTG Colby Broadwater to LTG George W. Casey, Jr,: Subject: Turkish General Staff Role in Ongoing Events, March 3, 2003, Casey Papers, Box 109; Nawyn, *Fiscal Year 2003,* 42: Brown, *Kevlar Legions,* 231.

[30] Scott Wallace, e-mail to Author, July 21, 2020.

[31] Ibid.

[32] Brown, *Kevlar Legions,* 231; Everett, *Historical Summary, Fiscal Year 1993,* 44-45; Fontenot, Deegan and Tohn, *On Point,* 63.

[33] "Bush Ultimatum to Saddam: Text," March 17, 2003, http://news.bbc.co.uk/2/hi/americas/2859269.stm (accessed, 19 September 2020).

[34] Briscoe, et al., *All Roads,* 112; U.S. Air Force, "2003-Operation Iraqi Freedom," 2020, https://www.afhistory.af.mil/FAQs/Fact-Sheets/Article/458942/operation-iraqi-freedom/ (accessed, 29 September 2020); Department of the Army, *Field Manual 3-3: Chemical and Biological Contamination Avoidance* (Washington, D.C.: Department of the Army, 1992), 3-1 through 3-3; Fontenot, et al., *On Point,* 87; Wallace, e-mail to Author, July 21, 2020.

[35] Fontenot, et al., *On Point,* 80; Department of the Army, *Field Manual 3-3: Chemical and Biological Contamination Avoidance* (Washington, D.C.: Department of the Army, 1992), 3-1 through 3-3; Google Earth Pro distance measuring tool, September 20, 2020 (distance measured was from eastern edge of Lake Razazza to the west bank of the Euphrates River); Wallace, e-mail to Author, July 21, 2020.

[36] Everett, *Historical Summary Fiscal Year 1993,* 44; 173rd Airborne.com, "OPERATION NORTHERN DELAY," http://www.173rdairborne.com/manifest-iraq.htm (accessed, 22 September 2020); John Hall, "Combat Parachute into Bashur, Iraq—Fifteenth Anniversary," https://www.dvidshub.net/news/270671/combat-parachute-assault-into-bashur-iraq-15th-anniversary (accessed, 22 September 2020); Briscoe, et al., *All Roads,* 224.

[37] Jeff Ryals, e-mail to Author, September 17, 2020.
[38] Wallace, e-mail to Author, July 21, 2020.
[39] Fontenot, et al., *On Point,* 100, 269, 278; Wallace, e-mail to Author, July 21, 2020.
[40] Fontenot, et al., *On Point,* 258-60; Wallace, e-mail to Author, July 21, 2020.
[41] Wallace, e-mail to Author, July 21, 2020; Fontenot, et al., *On Point,* 336-37; Defense Casualty Analysis System, "U.S. Military Casualties - Operation Iraqi Freedom (OIF) Casualty Summary by Month and Service," https://dcas.dmdc.osd.mil/dcas/pages/report_oif_month.xhtml (accessed, 21 September 2020); ICasualties.org, "Iraq Fatalities Total: 4092," http://www.icasualties.org/App/Fatalities?page=1&rows=0 (accessed, 21 September 2020).
[42] Stewart, *War in the Persian Gulf,* 19, 27-28.
[43] West Point Center for Oral History, "Be Ready To Adjust Your Assumptions: Reflections On Adaptive Leadership From The 57th Superintendent, Franklin "Buster" Hagenbeck, https://www.westpointcoh.org/interviews/be-ready-to-adjust-your-assumptions-reflections-on-adaptive-leadership-from-the-57th-superintendent (accessed, 29 September 2020); Richard Kugler, *Operation Anaconda: A Case Study in Adaptation* (Washington, D.C.: Department of Defense, 2007), 11.
[44] Adam Geibel, "Operation Anaconda, Shah-i-Khot Valley, Afghanistan, 2-10 March 2002," *Military Review* 82 (May-June 2002), 72-77; Kugler, 6, 11; Gerry J. Gilmore, "U.S. Troops Describe All-Day Shahikhot Battle," https://archive.defense.gov/news/newsarticle.aspx?id=44277 (accessed, 29 September 2020).
[45] Stewart, *Operation Enduring Freedom,* 37-38, 41; Geibel; Austin Bay, "On Point: A Full Report on Operation Anaconda: America's First Battle of the 21st Century, A Complete After Action Interview with Colonel Weircinski, https://www.strategypage.com/on_point/20020627.aspx (accessed, 29 September 2020); James W. Williams, *A History of Army Aviation: From Its Beginnings to the War on Terror* (New York: IUniverse, 2005), 381-85; U.S. Government, Department of Defense, Department of the Air Force, *Operation Anaconda: An Airpower Perspective* (Washington, DC: Department of the Air Force, 2005): 80, 100, 111-12; Linda Kozaryn, "Operation Anaconda is Winding Down," https://archive.defense.gov/news/newsarticle.aspx?id=44258 (accessed, 29 September 2020); Kugler, 1, 15-17; Jim Garamone, "Anaconda Over, But Operations Continue in Afghanistan," March 18, 2002, https://archive.defense.gov/news/newsarticle.aspx?id=44239 (accessed, 29 September 2020); *New York Times,* March 6, 2002.
[46] Armed Forces Press Service, "Operation Mountain Thrust Continues Momentum in Afghanistan," June 19, 2006, https://archive.defense.gov/news/newsarticle.aspx?id=16009 (accessed, 29 September 2020; Armed Forces Press Service, "Patrol Kills 14 Extremists, Afghanistan Operations Continue," June 30, 2006, https://archive.defense.gov/news/newsarticle.aspx?id=26 (accessed, 29 September 2020); Bay; Trias-DeRyder, "Operation Mountain Thrust," *Army Logistician* 40 (January-February 2008): 12.
[47] Department of Defense Press Release 186-02, "DOD Identifies Soldiers Killed in Explosive Ordnance Accident," April 16, 2002, https://archive.defense.gov/Releases/Release.aspx?ReleaseID=3307 (accessed, 30 September 2020); iCasualty, "Afghanistan Fatalities," http://www.icasualties.org/App/AfghanFatalities (accessed, 30 September 2020).
[48] Briscoe, et al., *Weapon of Choice,* 265; Jim Garamone and Linda Kozaryn, "Still Plenty of Al Qaida in Afghanistan," December 17, 2001, https://archive.defense.gov/news/newsarticle.aspx?id=44348 (accessed, 30 September 2020); Deutsche Welt, "Another year of 'Enduring Freedom' for German Troops," June 11, 2002, https://www.dw.com/en/anoth

er-year-of-enduring-freedom-for-german-troops/a-671240 (accessed, 30 September 2020); Government of the Czech Republic, Ministry of Defence and Armed Forces, "Foreign Missions of the 601st Special Forces Group," http://www.601skss.cz/english/missions.html (accessed, 1 October 2020); Government of the Netherlands, Ministry of Defense, "Dutch Contribution to Operation Enduring Freedom in Afghanistan," https://english.defensie.nl/topics/historical-missions/mission-overview/2001/operation-enduring-fr eedom/dutch-contribution (accessed, 30 September 2020); Government of Romania, Ministry of National Defense "Historical Landmarks," https://english.mapn.ro/historical /ind ex.php (accessed, 30 September 2020); "Afghanistan: Bilan Mitigé pour les Forces Spéciales Françaises," http://secretdefense.blogs.liberation.fr/2007/12/22 /afghanistan-bil/ (accessed, 30 September 2020); Republic of Lithuania, Ministry of National Defense, "Lithuanian Special Force Troops Returned Home Upon Completing Mission in Afghanistan," http://kam.lt/en/news_1098/news_archives/news_archive_2004/news_a rchive_2004_-_12/lithuanian_special_force_troops_returned_home_upon_completing_ mission_in_afghanistan.html?pbck=0 (accessed, 1 October 2020); Exercito de Portugal, *Armas Combinas: Revista Militar de Escola mas Armas* (Mafra, Portugal: School of Arms, 2016), 34; SOFREP, "Polish SOF Reported in Iraq... "Again," https://sofrep.com/news /polish-sof-iraq/ (accessed, 30 September 2020); Nyrop and Seekins, *Country Study,* xiv; Stewart, *Operation Enduring Freedom,* 32-33.

49 Brian Neumann, Lisa Mundey, and Jon Mikolashek, *Operation Enduring Freedom: The United States Army in Afghanistan, March 2002- April 2005* (Washington, D.C.: Center of Military History, 2016), 17.

50 Briscoe, et al. *Weapon of Choice,* 350, 372; Harold Kennedy, "Pentagon Broadens duties for its Civil Affairs Teams," *National Defense,* February 2003, 36, https://link.gale.com/apps/doc/A99752760/AONE?u=temple_main&sid=AONE&xid=23b2c7a2 (accessed, 1 October 2020); Alexander Alexiev, *Inside the Soviet Army in Afghanistan* (Santa Monica, CA: Rand Corporation, 1988), vii, 20, 32.

51 Carter Malkasian and Gerald Meyerle, *Provincial Reconstruction Teams: How Do We Know They Work?* (Carlisle, PA: United States Army War College, 2009), vii.

52 Ibid., 6.

53 Ibid., 20, 25; Robert Perito, The U.S. Experience with Provincial Reconstruction Teams in Afghanistan: Lessons Identified," https://www.usip.org/publications/2005/10/us-ex perience-provincial-reconstruction-teams-afghanistan-lessons-identified (accessed, 2 October 2020).

54 Briscoe, et al., *Weapon of Choice,* 258; Herbert A. Friedman, "Psychological Operations in Afghanistan, Operation Enduring Freedom, 2001," https://www.psywar.org/content/afghanistan02 (accessed, 6 October 2020).

55 Timothy A. Carter and Daniel J. Veale, "Weather, Terrain and Warfare: Coalition Fatalities in Afghanistan," *Conflict Management and Peace Science* 30 (July 2013): 220-29, https://journals-sagepub-com.libproxy.temple.edu/doi/full/10.1177/0738894213484054 (accessed, 2 October 2020); "Cold Weather Said to Hinder Taleban Operations in Afghanistan," *BBC Monitoring South Asia,* February 20, 2005, http://libproxy.temple.edu /login?url=https://www-proquest-com.libproxy.temple.edu/docview/459924429?acco untid=1427 (accessed, 2 October 2020); U.S. Federal News Service, April 2, 2005; U.S. Department of Defense, "DoD Announce OEF/OIF Rotational Units," https://archive.de fense.gov/Releases/Release.aspx?ReleaseID=8061 (accessed, 2 October 2020); Kellard N. Townsend, Jonathan E. Marion, Joseph W. Boler, and Madison M. Carney, "Hoosier

Redlegs Train Afghan Kandaks: Semper Gumbi," *FA Journal* 73 (May 2005): 32-34, http://libproxy.temple.edu/login?url=https://www-proquest-com.libproxy.temple.edu/docview/218357257?accountid=14270 (accessed, 2 October 2020); Neumann, Mundey, and Mikolashek, 18, 40, 54, 63; Bob Haskell, "Guard Enhanced Brigade Prepares for Afghan Duty, " November 7, 2002, https://archive.defense.gov/news/newsarticle.aspx?id=27829 (accessed, 2 October 2020); R. W. White, "History of the 45th Infantry Brigade," https://signaleer.wordpress.com/2011/08/03/the-history-of-the-45th-infantry-brigade/ (accessed, 2 October 2020).

[56] Thomas H. Johnson, "Taliban Adaptations and Innovations," *Small Wars & Insurgencies* 24 (February 2013): 3-27, https://www-tandfonline-com.libproxy.temple.edu/doi/full/10.1080/09592318.2013.740228 (accessed, 6 October 2020); Anthony King, F. G. Hoffman, David J. Galbreath, and Theo Farrell, "Military Adaptation in Afghanistan," *International Politics Reviews* 1 (December 2013): 100-16, http://dx.doi.org.libproxy.temple.edu/10.1057/ipr.2013.19 (accessed, 6 October 2020).

[57] Wallace, e-mail to Author, July 21, 2020.

[58] Nora Bensahel, Olga Oliker, and Keith Crane, et al., *After Saddam Prewar Planning and the Occupation of Iraq* (Santa Monica, CA: RAND Corporation, 2008), 68.

[59] Ibid., 51, 66-69; Wallace, e-mail to Author, July 21, 2020.

[60] Kevin M. Woods, Michael Pease, Mark E. Stout, et al., *Iraqi Perspectives Project: A View of Operation Iraqi Freedom from Saddam's Senior Leadership* (Ft. Leavenworth, KS: Center for Army Lessons Learned, 2006), 3-4; Wallace, e-mail to Author, July 21, 2020; "Coalition Provisional Authority Order Number 2, May 23, 2003," https://permanent.fdlp.gov/lps45365/cpaord2.pdf (accessed, 29 October 2020); Trofimov, *Faith at War*, 177; GlobalSecurity.org, "Iraqi Insurgency Groups," accessed https://www.globalsecurity.org/military/ops/iraq_insurgency.htm (accessed, 29 October 2020); Timothy S. McWilliams and Kurtis P. Wheeler, eds., *Al-Anbar Awakening, U.S. Marines and Counterinsurgency in Iraq, Volume I: American Perspectives* (Quantico, VA: Marine Corps University, 2009), 2, 30, 33; Terrorism Research and Analysis Consortium, "Jaysh al Mahdi," https://www.trackingterrorism.org/group/jaysh-al-mahdi (accessed, 29 October 2020); Terrorism Research and Analysis Consortium, "Badr Corps," https://www.trackingterrorism.org/group/badr-corps (accessed, 29 October 2020); Bruce R. Pirnie, Edward O'Connell, *Counterinsurgency in Iraq, 2003-2006* (Santa Monica, CA: RAND Corporation, 2008), 4, 29-31; Nakissa Jahanbani, Center for Terrorism Studies, U.S. Military Academy, e-mail to Author, November 2, 2020; Walter Hamm, "Veterans Remember Historic 2nd BCT Deployment," https://go-gale-com.libproxy.t emple.edu/ps/i.do?p=AONE&u=temple_main&id=GALE|A368658350&v=2.1&it=r (accessed, 29 October 2020); Leon J. LaPorte, "Transforming an Alliance while Deterring Aggression" *Army* 55, October 2005, 89-92, http://libproxy.temple.edu/login?url=https://www-proquest-com.libproxy.temple.edu/docview/237074838?accountid=14270 (accessed, 29 October 2020); Nawyn, *Fiscal Year 2003*, 1; U.S. Government, Library of Congress, Congressional Research Service, Charles A. Henning, "Stop Loss: Key Questions and Answers, 2009" accessed October 29, 2020, https://heinonline-org.libpro xy.temple.edu/HOL/Page?collection=congrec&handle=hein.crs/crsabgl0001&id=1&men_tab=srchresults; William M. Donnelly, *Department of the Army Historical Summary Fiscal Year 2002* (Washington, D.C.: Department of the Army, 2011), 18.

[61] Metz, *A Country Study*, xviii; *Washington Post*, May 17, 2003; Department of the Army, 4th Infantry Division, "4th Infantry Division History," https://www.carson.army.mil/units/4ID.html (accessed, 30 October 2020); Walter L. Perry, Richard E. Darilek, Laurinda

L. Rohn, Jerry M. Sollinger, *Operation Iraqi Freedom: Decisive Victory, Elusive Peace* (Santa Monica, CA: RAND Corporation, 2015), 103; "Briefing from Mosul, Iraq, on the Involvement of the 101st Airborne Division in the Conflict and in Post-War Stabilization Efforts," *M2 Presswire*, May 14, 2003, https://link.gale.com/apps/doc/A101702566/ITOF?u=temple_main&sid=ITOF&xid=b1c09456 (accessed, 30 October 2020); Thomas A. Bruscino, Jr. in *Between the Rivers: Combat Action in Iraq, 2003-2005*, ed. John J. McGrath (Ft. Leavenworth, KS: Combat Studies Institute, 2012), 69; McWilliams & Wheeler, 4; "NATO: Poland Assumes Command of Multi-National Division in Iraq with NATO Support," *M2 Presswire*, Sep 03, 2003, http://libproxy.temple.edu/login?url=https://www-proquest-com.libproxy.temple.edu/docview/444657714?accountid=14270 (accessed, 30 October 2020); *New York Times*, May 9, 2003; Enacademic.com, "Multi-National Division South," https://enacademic.com/dic.nsf/enwiki/8773315 (accessed, 30 October 2020); Brown, *Kevlar Legions*, 251.

62 Stephen A. Carney, *Allied Participation in Operation Iraqi Freedom* (Washington, D.C.: Center of Military History, 2011), 12; "Biography of Lt. Gen. Ricardo Sanchez, United States Army (Retired), https://lbausa.com/wp-content/uploads/Biography-of-Lt-General-Ricardo-Sanchez-US-Army-_Retired_.pdf (accessed, 31 October 2020); Donnelly, 20; "Army Announces Unit Rotations." *Army* 53 (September 2003): 68-70,72; http://libproxy.temple.edu/login?url=https://www-proquest-com.libproxy.temple.edu/docview/237077321?accountid=14270 (accessed, 31 October 2020); "US DOD: Army Announces Mobilization of Two National Guard Brigades," *M2 Presswire*, September 29, 2003, http://libproxy.temple.edu/login?url=https://www-proquest-com.libproxy.temple.edu/docview/444647496?accountid=14270 (accessed, 31 October 2020); U.S. Government, Departments of the Army and Air Force, National Guard Bureau, *Annual Review of the Chief, National Guard Bureau, Fiscal Year 2003*, 32, 46. Hereafter *Annual Review, 2003;* Donald P. Wright, et al., *The United States Army in Operation IRAQI FREEDOM May 2003—January 2005, On Point II: Transition to the New Campaign* (Ft. Leavenworth, KS: Combat Studies Institute, 2008), 611, 613, 619-20.

63 Fred Kaplan, "Err War, The Army Buries Its Mistakes," https://slate.com/news-and-politics/2003/10/the-army-buries-its-mistakes.html (accessed, 30 October 2020).

64 Trofimov, *Faith at War*, 104, 115, 136, 159; W. Andrew Terrill, *Colloquium Report: U.S. Military Operations in Iraq: Planning, Combat, and Occupation* (Carlisle, PA: U.S. Army War College, 2006), 14, 15; Perry et al., *Decisive Victory, Elusive Peace*, 331.

65 Department of the Army, "Select Chronology of Army History for the Army's 240th Birthday," https://history.army.mil/html/faq/army_bday_chron.html (accessed, 31 October 2020); British Broadcasting Corporation, "How Saddam Hussein Was Captured," http://news.bbc.co.uk/2/hi/middle_east/3317881.stm (accessed, 31 October 2020); Perry et al., *Decisive Victory, Elusive Peace*, 331.

66 Frank Hall, e-mail to Author, September 29, 2020; David Clonts, e-mail to Author, October 30, 2020; Defense Visual Information Distribution Service, "Smoother Transitions Planned for OIF 04-06," November 12, 2004, https://www.dvidshub.net/news/593/smoother-transition s-planned-oif-04-06#:~:text=For%20OIF%202%2C%20the%20rotation%20of%20forces%20took,spread%20over%20six%20or%20seven%20months%2C%20Carra%20 said (accessed, 2 November 2020); Washington National Guard, "History of the 81st Brigade Combat Team," https://mil.wa.gov/about-the-81st (accessed, 2 November 2020); *M2 Presswire*, November 7, 2003, http://libproxy.temple.edu/login?url=https://www-proquest-com.libproxy.temple.edu/docview/443781947?accountid=14270 (accessed, 2 November 2020); First Cavalry Division Association, "256th Infantry Brigade History," https://1cda.org/history/256th-i

nfantry-brigade-history/http://libproxy.temple.edu/login?url=https://www-proquest-com.libproxy.temple.edu/docview/443781947?accountid=14270 (accessed, 2 November 2020); Marine Corps News, "155th BCT Hits the Ground Running," https://www.globalsecurity.org/military/library/news/2005/02/mil-050221-usmc01.htm (accessed, 2 November 2020); Department of Defense, "Army Announces the Alert of Two National Guard Enhanced Separate Brigades." *M2 Presswire,* July 28, 2003, http://libproxy.temple.edu/login?url=https://www-proquest-com.libproxy.temple.edu/docview/445780550?accountid=1427 (accessed, 2 November 2020); *Annual Review, 2003,* 46-47, 112, 114-19, 123-26, 129-39, 141-42, 144, 147-97.

[67] Clonts, e-mail to Author, October 30, 2020; Headquarters, U.S. Central Command News Release 04-02-08, February 5, 2004, https://www.globalsecurity.org/military/library/news/2004/02/mil-040205-centcom01.htm (accessed, 3 November 2020); Hall, e-mail to Author, September 29, 2020; McGrath, *Between the Rivers,* 19; Wright, et al., *Transition,* 173, 618, 605.

[68] Clonts, e-mail to Author, October 30, 2020.

[69] Remarks by President Bush announcing the end of major combat operations in Iraq, https://www.cbsnews.com/news/text-of-bush-speech-01-05-2003/ (accessed, 3 November 2020); John Sheehy, U.S. Army Recruiting Command Plans and Policy Office, interview with Author, November 3, 2020; Ann Roosevelt, "Aerostat to Guard Against Shoulder Fired Missiles, Mortars," *C4I News,* February 19, 2004, http://libproxy.temple.edu/login?url=https://www-proquest-com.libproxy.temple.edu/docview/232542058?accountId=14270 (accessed, 2 November 2020); Hall, e-mail to Author, September 29, 2020; Clonts, e-mail to Author, October 30, 2020; Charles Darwin, *The Origin of Species* (London, UK: John Murray, Albemarle Street, 1910), 3.

[70] Hall, e-mail to Author, September 29, 2020; Clonts, e-mail to Author, October 30, 2020; Wright, et al., *Transition,* 433-37.

[71] Wright, et al, *Transition,* 176, 627; Coalition Provisional Authority, Law of Administration for the State of Iraq for the Transitional Period, 8 March 2004, Article 59 (B), https://www.cesnur.org/2004/iraq_tal.htm (accessed, 5 November 2020); George W. Casey, Incoming Commander's Conference, "Acquired Insights" briefing slide, July 17, 2004, Casey Papers, Box 110; "Iraqi Tribes Seek Deal with Government to Delay Election." *BBC Monitoring Middle East,* January 5, 2005, http://libproxy.temple.edu/login?url=https://www-proquest-com.libproxy.temple.edu/d ocview/458695511?accountid=14270 (accessed, 4 November 2020).

[72] MNF-I, "Strategy for Attacking the Insurgency" briefing slide, ND, Casey Papers, Box 110.

[73] Wright, et al., *Transition,* 3, 299.

[74] Clonts, e-mail to Author, October 30, 2020; John R.S. Batiste and Paul R. Daniels, "The Fight for Samarra: Full-Spectrum Operations in Modern Warfare." *Military Review* (May-June 2005): 13-22, https://link.gale.com/apps/doc/A133371066/AONE?u=temple_main&sid=AONE&xid=19ee7881 (accessed, 4 November 2020); Wright and Reese, 41, 43-44, 625-27; Multi-National Force—Iraq, "Fact Sheet on the Incident at Forward Operating Base Marez, as of December 23, 2004, Casey Papers, Box 110.

[75] Hall, e-mail to Author, September 29, 2020; U.S. Government, Library of Congress, Congressional Research Service by Linwood B. Carter, "Iraq: Summary of U.S. Forces," 3-4, https://fas.org/sgp/crs/mideast/RL31763.pdf (accessed, 4 November 2020); Eric Durr, Public Affairs Office, 42nd Infantry Division, interview with Author, November 4, 2020; Departments of the Army and Air Force, *Annual Report of the Chief, National Guard Bureau, Fiscal Year 2005* (Washington, D.C: National Guard Bureau, 2005), 39, 44-48, 154-55.

[76] "Objectives for FY 2005," briefing slide dated January 4, 2020, Casey Papers, Box 110.
[77] U.S. Government, Library of Congress, Congressional Research Service by Kenneth Katzman, "Iraq: Elections, Government and Constitution, Updated May 22, 2006," 2-5, https://congressional-proquest-com.libproxy.temple.edu/congressional/result/congresultpage:pdfevent?rsId=174FE6CC407&pdf=/app-bin/gis-congresearch/1/7/b/6/crs-2006-fdt-0610_from_1_to_6.pdf&uri=/app-gis/congresearch/crs-2006-fdt-0610 (accessed, 5 November 2020); U.S. Department of Defense, "Measuring Stability and Security in Iraq February 2006 Report to Congress In accordance with the Department of Defense Appropriations Act 2006 (Section 9010)," 27, 38-40; "U.S. Government, Multi-National Force—Iraq by Eric Zimmerman, "Information for the CG (insurgent percentage of the population)," June 20, 2005, Casey Papers, Box 110; "State of the Insurgency," briefing slide dated January 4, 2020, Casey Papers, Box 110; U.S. Government, Library of Congress, Congressional Research Service by Kenneth Katzman, "Iran's Influence in Iraq, Updated June 13, 2006," 1-6, https://congressional-proquest-com.libproxy.temple.edu/congressional/result/pqpresultpage.gispdfhitspanel.pdflink/$2fapp-bin$2fgis-congresearch$2f0$2fe$2fe$2f1$2fcrs-2006-fdt-1063_from_1_to_6.pdf/entitlementkeys=1234%7Capp-gis%7Ccongresearch%7Ccrs-2006-fdt-1063 (accessed, 5 November 2020).
[78] DoD, "Measuring Stability and Security," 25-27.
[79] DoD, "Measuring Stability and Security," 27; McGrath, *Between the Rivers*, 38; Wright, et al., *Transition*, 340-43, 357, 419; Hall, e-mail to Author, September 29, 2020; Clonts, e-mail to Author, October 30, 2020.
[80] U.S. Department of Defense, Defense Casualty Analysis System, "U.S. Military Casualties - Operation Iraqi Freedom (OIF) Casualty Summary by Month and Service," https://dcas.dmdc.osd.mil/dcas/pages/report_oif_month.xhtml (accessed, 26 October 2020); U.S. Department of Defense, Defense Casualty Analysis System, "U.S. Military Casualties – Operation Enduring Freedom (OEF) Casualty Summary by Month and Service," https://dcas.dmdc.osd.mil/dcas/pages/report_oef_month.xhtml (accessed, 26 October 2020); U.S. Government, Library of Congress, Congressional Resarch Service by Hannah Fischer, "United States Military Casualty Statistics: Operation Iraqi Freedom and Operation Enduring Freedom, June 8, 2006," 4-5, accessed October 26, 2020, https://congressional-proquest-com.libproxy.temple.edu/congressional/result/congresultpage:pdfevent?rsId=174CAFA5595&pdf=/app-bin/gis-congresearch/9/0/6/f/crs-2006-ksg-0063_from_1_to_5.pdf&uri=/app-gis/congresearch/crs-2006-ksg-0063 (accessed, 26 October 2020).
[81] Hall, e-mail to Author, September 29, 2020; Clonts, e-mail to Author, October 30, 2020; U.S. Government, Multi-National Force—Iraq, by Dave Dreher, "Wheeled Vehicle Armoring in Iraq," June 18, 2005, Casey Papers, Box 110; William M. Donnelly, *Department of the Army Historical Summary, Fiscal Year 2004* (Washington, D.C.: Center of Military History, 2015), 12, 63-64; U.S. Government, Library of Congress, Congressional Research Service by Clay Wilson, "Improvised Explosive Devices (IEDs) in Iraq: Effects and Countermeasures, February 10, 2006," https://congressional-proquest-com.libproxy.temple.edu/congressional/result/congresultpage:pdfevent?rsId=174CBCC83CF&pdf=/app-bin/gis-congresearch/3/7/d/3/crs-2006-fdt-0388_from_1_to_6.pdf&uri=/app-gis/congresearch/crs-2006-fdt-0388 (accessed, 26 October 2020); U.S. Government, Library of Congress, Congressional Research Service by Andrew Feickert, "U.S. Army and Marine Corps Equipment Requirements: Background and Issues for Congress, December 20, 2006," 15-17, 19, https://congressional-proquest-com.libproxy.temple.edu/congressional/result/pqpresultpage.gispdfhitspanel.pdflink/$2fapp-bin$2fgis-congresearch$2fa$2f9$2f4$2f1

$2fcrs-2006-fdt-1729_from_1_to_39.pdf/entitlementkeys=1234%7Capp-gis%7Ccongresearch%7Ccrs-2006-fdt-1729 (accessed, 26 October 2020).

[82] Hall, e-mail to Author, September 29, 2020; ICasualties.org. (Afghanistan); ICasualties.org. (Iraq); U.S. Central Command, FOIA request 07-0154, "2004-2007 Significant Activity Reports," https://www3.centcom.mil/FOIALibrary/Search (accessed, 27 October 2020).

[83] U.S. Government, Library of Congress, Congressional Research Service by Clay Wilson, "Improvised Explosive Devices (IEDs) in Iraq and Afghanistan: Effects and Countermeasures, September 5, 2006," 2-3, https://www3.centcom.mil/FOIALibrary/cases/11-0292/FOIA%20Case%20(11-0292,%2021%20Nov%202011)%20%20Iraq%20Weekly%20Security%20incidents%20(Closed%20in%20FY12).pdf (accessed, 26 October 2020); *Washington Post*, January 3, 2020; "Major General Peter W. Chiarelli Holds A Defense Department News Briefing Via Teleconference From Iraq," *Political Transcript Wire*, Jan 05, 2005, http://libproxy.temple.edu/login?url=https://www-proquest-com.libproxy.temple.edu/docview/467188216?accountid=14270 (accessed, 27 October 2020); Hall, e-mail to Author, September 29, 2020; Clonts, e-mail to Author, October 30, 2020; U.S. Government, Library of Congress, Congressional Research Service by Audrey K. Cronin, "Terrorists and Suicide Attacks, August 28, 2003," 6-8, 17, https://congressional-proquest-com.libproxy.temple.edu/congressional/result/congresultpage:pdfevent?rsId=174CB8A36DA&pdf=/app-bin/gis-congresearch/7/8/7/b/crs-2003-fdt-0191_from_1_to_25.pdf&uri=/app-gis/congresearch/crs-2003-fdt-0191 (accessed, 26 October 2020); Madelyn Hsaio Rei-Hicks, Hamit Dardegan, and Peter M. Bagnall et al., "Casualties in Civilians and Coalition soldiers from Suicide Bombings in Iraq, 2003–10: A Descriptive Study," *The Lancet* 378 (3–9 September 2011): 906-14, https://www-sciencedirect-com.libproxy.temple.edu/science/article/pii/S0140673611610234 (accessed, 26 October 2020); Anne Speckhard, "Female Suicide Bombers in Iraq," *Democracy and Security* 5 (March 2009): 19-50, https://www-tandfonline-com.libproxy.temple.edu/doi/citedby/10.1080/17419160902723759?scroll=top&needAccess=true (accessed, 27 October 2020); U.S. Government, Department of Defense, "Defense Department Briefing, November 29, 2005,"https://search-proquest-com.libproxy.temple.edu/docview/190027264?accountid=14270&rfr_id=info%3Axri%2Fsid%3Aprimo (accessed, 27 October 2020).

[84] Edward T. Pound and Jennifer Jack, "The Iran Connection," *U.S. News & World Report*, November 22, 2004, 32–48, http://search.ebscohost.com.libproxy.temple.edu/login.aspx?direct=true&db=mth&AN=15060473&site=ehost-live&scope=site (accessed, 27 October 2020); Ivan Eland, Robert Newman, Jeffrey White, and W. Patrick Lang. "Occupied Iraq: One Country, Many Wars," *Middle East Policy* 12 (Fall 2005), 1-28, https://link.gale.com/apps/doc/A136895522/AONE?u=temple_main&sid=AONE&xid=e2345011 (accessed, 27 October 2020); "Defense Department Outlines Links between Iraq, Iran: U.S. Military Intelligence Official Discusses Iran's Strategy in Neighboring Iraq," Federal Information & News Dispatch, LLC, 2006, http://libproxy.temple.edu/login?url=https://www-proquest-com.libproxy.temple.edu/docview/189975438?accountid=14270 (accessed, 5 November 2020); U.S. Government, Library of Congress, Congressional Research Service by Kenneth Katzman, "Iran: U.S. Concerns and Policy Responses, Updated September 11, 2006, https://apps.dtic.mil/dtic/tr/fulltext/u2/a472427.pdf (accessed, 5 November 2020); Brian B. Shreve, *Operational and Medical Management of Explosive and Blast Incidents*, eds. David W. Callaway and Jonathan L. Burstein (Cham, SW: Springer, 2020), 101; "Senior American Military Officials Gave a Briefing in Baghdad on Iranian Weapons that have Killed 170 Americans and Wounded 620 Since 2004: EFPs, or Explosively Formed

Penetrators, that Release Armor-Cutting Molten Copper Balls," *National Review*, March 5, 2007, 8, https://link.gale.com/apps/doc/A165364444/AONE?u=temple_main&sid=AONE&xid=18dd701d (accessed, 27 October 2020); Pam Hess, "IEDs Fewer but more Deadly in South Iraq," *UPI Security & Terrorism*, November 4, 2005, https://link.gale.com/apps/doc/A138322666/ITOF?u=temple_main&sid=ITOF&xid=39ea0721 (accessed 27 October 2020).

[85] Paul Kamolnick, *Countering Radicalization and Recruitment to Al-Qaeda: Fighting the War of Deeds* (Carlisle, PA: United States Army War College Press, 2014), 27; U.S. Government, Department of Defense, "DoD Directive 2000.19E, Joint IED Defeat Organization, 14 February 2006, 2, https://www.hsdl.org/?search=&searchfield=title &all=Joint+IED+Defeat&collection=public&submitted=Search (accessed, 27 October 2020); Angel Rebassa, Peter Chalk, Kim Craigin, Heather S. Gregg, et al., *Beyond Al Qaeda: Part I, The Global Jihadist Movement* (Santa Monica: RAND Corporation, 2006), xviii, xxii, 13-15, 18-19, 164; Ahmed S. Hashim, *Insurgency and Counter-Insurgency in Iraq* (Ithaca, NY: Cornell University Press, 2011), 199.

[86] Domenico Tosini, "Al-Qaeda's Strategic Gamble: the Sociology of Suicide Bombings in Iraq," *Canadian Journal of Sociology* 35 (2010): 271+; https://link.gale.com/apps/doc/A230866901/AONE?u=temple_main&sid=AONE&xid=894319f8 (accessed, 13 April 2021); "U.S. Representative Peter Hoekstra (R-MI) Holds a Hearing on Al Qaida's use of the Internet for Strategic Communications," *Political Transcript Wire*, May 2004, http://libproxy.temple.edu/login?url=https://www-proquest-com.libproxy.temple.edu/wire-feeds/u-s-representative-peter-hoekstra-r-mi-holds/docview/467170325/se-2?accountid=14270 (accessed, 13 April 2021).

[87] William McCants, *Militant Ideology Atlas* (West Point, NY: Combatting Terrorism Center, U.S. Military Academy, 2006), 294; Brian M. Jenkins, "Is Al Qaida's Internet Strategy Working? Before the Committee on Homeland Security Subcommittee on Counterterrorism and Intelligence United States House of Representatives December 6, 2011," 1-2, file:///C:/Users/fogan/Downloads/RAND_CT371.pdf (accessed, 19 October 2020); Robert J. Bunker, *Networks, Terrorism and Global Insurgency* (London: Taylor and Francis Group, 2005), 139, 150, 166.

[88] Bensahel, et al., *After Saddam*, 67.

[89] Department of the Army, *Field Manual 100-20, Military Operations in Low Intensity Conflict* (Washington, D.C.: Departments of the Army and Air Force, 1990), 2-1-2-6; Department of the Army, *Field Manual 3-0, Operations* (Washington, D.C.: Headquarters, Department of the Army, 2001), 9-8 - 9-9; John A. Nagl, *Learning to Eat Soup with a Knife: Counterinsurgency Lessons from Malaya and Vietnam* (Chicago, IL: University of Chicago Press, 2002), 207.

[90] U.S. Government, Library of Congress, Congressional Research Service by Jeremy M. Sharp and Christopher M. Blanchard, "Post-War Iraq: Foreign Contributions to Training, Peacekeeping, and Reconstruction Updated November 1, 2005," 8-14, 17, 19-21, https://congressional-proquest-com.libproxy.temple.edu/congressional/result/congresultpage:pdfevent?rsId=174C03B3F26&pdf=/app-bin/gis-congresearch/3/f/0/9/crs-2005-fdt-1505_from_1_to_24.pdf&uri=/app-gis/congresearch/crs-2005-fdt-1505 (accessed, 24 October 2020); U.S. Government, Library of Congress, Congressional Research Service by Andrew Feickert, "U.S. and Coalition Military Operations in Afghanistan: Issues for Congress Updated December 11, 2006," 2, https://congressional-proquest-com.libproxy.temple.edu/congressional/result/congresultpage:pdfevent?rsId=174C0471D35&pdf=/app-bin/gis-congresearch/5/0/3/c/crs-2006-fdt-1696_from_1_to_18.pdf&uri=/app-gis/congresearch/crs-2006-fdt-1696 (accessed, 24 October 2020).

[91] Schoomaker interview, April 15, 2019.

Chapter 4

[1] Richard Cody, interview with Author, January 15, 2018.

[2] John W. Stewart, *The United States Army in Afghanistan: Operation Enduring Freedom, October 2001-March 2002* (Washington, D.C.: Center of Military History, 2004), 11, 27; Charles H. Briscoe, Kenneth Finlayson, et al., *All Roads Lead to Baghdad: Army Special Operations Forces in Iraq* (Ft. Bragg, NC: U.S. Army Special Operations Command History Office, 2006), 76.

[3] "President Bush Discusses Freedom in Iraq and the Middle East," https://georgewbush-whitehouse.archives.gov/news/releases/2003/11/20031106-2.html (accessed April 9, 2020); Richard Immerman, *Empire for Liberty: A History of American Imperialism from Benjamin Franklin to Paul Wolfowitz* (Princeton, NJ: Princeton University Press, 2010), 226.

[4] Kate Phillips, Shane Lauth, Erin Schenck, and W. Andrew Terrill, eds., "Colloquium Report: U.S. Military Operations in Iraq: Planning, Combat and Occupation" (Carlisle, PA: U.S. Army War College, 2006), 10-20; Federal Research Division, Library of Congress, *Afghanistan: A Country Study* (Washington, D.C.: Library of Congress, 2001), 118-20; *New York Times*, February 6, 2017, *New York Times*, June 20, 2004; Christopher Ulack, "Demography of Race and Ethnicity in Iraq," in *International Handbook of the Demography of Race and Ethnicity*, David G. Saenz, David G., David G. Embrick, and Nestor P. Rodrigues, eds. (Dordrecht, NL: Springer Netherlands, 2015), 367-69; Clinton Watts, "Beyond Iraq and Afghanistan: What Foreign Fighter Data Reveals About the Future of Terrorism," https://smallwarsjournal.com/blog/journal/docs-temp/49-watts.pdf (accessed, 16 March 2020); *San Francisco Chronicle*, August 26, 2003; Andrew J. Birtle, *U.S. Army Counterinsurgency and Contingency Operations doctrine, 1942–1976* (Washington, D.C.: Center of Military History, 2006), 100, 115; John P. McAfee, *A Slow Walk in a Sad Rain* (New York: Warner, 1993), 1.

[5] Department of the Army, Rapid Equipping Force, "About Us," https://www.ref.army.mil/AboutUs (accessed, 18 April 2019); Joseph Rozmeski, interview with Author, November 9, 2017.

[6] Christopher W. Hughes, interview with Author, February 13, 2020; Combatting Terrorism Technical Support Office, "About IWTSD," https://cttso.gov/AboutCTTS O.html (accessed, 21 February 2020); Department of Defense, The Joint Staff, Joint Staff Anti-Terrorism Directorate (J34), briefing: "Anti-Terrorism/Force Protection, Level III Training Objectives", slide 3, undated, collection of Christopher P. Hughes.

[7] William J. Clinton, Presidential Decision Directive-39, https://clinton.presidentiallibraries.us/items/show/12755 (accessed, 20 February 2020); William Donnelly, *Department of the Army Historical Summary, Fiscal Year 2004* (Washington, D.C.: Center of Military History, 2015), 32; Doug Sample, "DoD Joint Task Force Making Progress Against IED Threat," https://archive.defense.gov/news/newsarticle.aspx?id=31400 (accessed, 12 June 2019); Cody interview with Author, January 15, 2018.

[8] Cody interview with Author, January 15, 2018.

[9] Mike Bowman to Author, March 4, 2020; Peter J. Schoomaker, interview with Author, April 15, 2019.

[10] Robert Shaw, interview with Author, February 27, 2019.

[11] William G. Bell, *Commanding Generals and Chief of Staff* (Washington, D.C.: Center of Military History, 2013), 168; Schoomaker interview with Author, April 15, 2019; U.S. Congress, 10 U.S.C., The Army, §7033 (1956).

[12] Army Aviation Association of America, "GEN Richard A. Cody," https://www.quad-a.org/Public/Awards/Awardees/CodyRA.aspx (accessed, 15 April 2019); John J. Obrien to Author, April 16, 2019; Hughes interview with Author, February 13, 2020; Peter J. Schoomaker to Author, December 11, 2017.
[13] Schoomaker to Author, December 11, 2017.
[14] James Lovelace, interview with Author, January 15, 2018; Schoomaker to Author, December 11, 2017; Cody interview with Author, January 15, 2018.
[15] Cody interview with Author, January 15, 2018.
[16] Draft Operational and Organizational (O&O) Concept for the Asymmetric Warfare Group as of 2 Mar 05, 1, organizational files of the Asymmetric Warfare Group (AWG), Ft. George G. Meade, MD.
[17] Lovelace interview with Author, January 15, 2018; Cody interview with Author, January 15, 2018; Franklin Bohle, interview with Author, January 17, 2019; Carl von Clausewitz, *On War*, trans. and ed. Michael Howard and Peter Paret (Princeton: Princeton University Press, 1976), 119.
[18] Department of the Army, 20th CBRNE Command, 20th CBRNE Command History, https://www.cbrne.army.mil/history.html (accessed, 14 June 2019); U.S. Government, Department of Defense, Department of the Army, Richard D. Cody, "Memorandum for Director, Force Design Directorate, U.S. Army Training and Doctrine Command, Ft. Monroe, Virginia, 23651; Subject: Asymmetric Warfare Regiment Organizational Design," April 12, 2004, organizational files of the AWG.
[19] U.S. Government, Department of Defense, Department of the Army by Joseph L. Votel, "Memorandum for Director, Force Design Directorate, U.S. Army Training and Doctrine Command, Ft. Monroe, VA, 23651; Subject: Asymmetric Warfare Regiment Organizational Design," April 28, 2004, U.S. Army Training and Doctrine Command archives, Fort Eustis, VA.
[20] Department of Defense, Emergency Supplemental Appropriations to Address Hurricanes in the Gulf of Mexico, and Pandemic Influenza Act of 2006, Pub. L. No. 109-148, 119 Stat. 2708, Sec. 8043 (c) (2).
[21] U.S. Department of Defense, Department of the Army, Office of the Deputy Chief of Staff, G3, Army Campaign Plan, 12 April 2004, 17, https://digital.library.unt.edu/ark:/67531/metadc22340/m2/1/high_res_d/BRAC-2005_10368.pdf (accessed, 13 February 2020).
[22] Hughes interview with Author, February 13, 2020.
[23] Ibid.
[24] Ibid.
[25] Christopher P. Hughes, *War on Two Fronts: An Infantry Commander's War in Iraq and the Pentagon* (Havertown, PA: Casemate Publishing, 2007), 211; Michael Bowman to Author, May 16, 2019; Michael Bowman to Author, June 19, 2019; Tom Fields, interview with Author, May 24, 2019.
[26] Fields interview with Author, May 24, 2019.
[27] Fields interview with Author, May 24, 2019; National Special Forces Green Beret Memorial, "Major General (Ret.) James A. Guest," https://specialforcesgreenberetmemorial.org/major-general-ret-james-guest/ (accessed, 28 May 2019); Mark D. Sherry, The Army Command Post and Defense Reshaping 1987–1997 (Washington, D.C.: Center of Military History, 2008), 41; U.S. Army Pacific, "Major General Steven Silvasy, Jr.," https://www.usarpac.army.mil/history2/cg_silvasy.asp (accessed, 28 May 2019).
[28] Draft O&O, 2, organizational files of the AWG.

[29] U.S. Department of Defense, Department of the Army, Asymmetric Warfare Group, "151021 AWG Mission Statement Evolution Briefing," 1, organizational files of the AWG.
[30] Ibid., 2.
[31] Ibid., 2.
[32] Ibid., 3.
[33] Draft O&O, 2, organizational files of the AWG.
[34] Ibid., 15.
[35] Ibid., 2.
[36] Ibid., 8.
[37] U.S. Department of Defense, "Combatant Commands," www.defense.gov//Our-Story/Combatant-Commands/ (accessed, 13 April 2021).
[38] Draft O&O, March 2, 2005, 3-4, organizational files of the AWG; Fields interview with Author, May 24, 2019.
[39] Draft O&O, March 2, 2005, organizational files of the AWG, 11-12, 17.
[40] Department of the Army, *Field Manual 34-130: Intelligence Preparation of the Battlefield* (Washington, D.C.: Department of the Army, 1989), 1-1, 4-1-4
[41] Draft O&O, March 2, 2005, 5, organizational files of the AWG; Richard W. Stewart, ed. *American Military History, The United States Army in a Global Era, 1917-2003,* 2 vols. (Washington, D.C.: Center of Military History, 2005), 2: 239.
[42] Draft O&O, March 2, 2005, 5, organizational files of the AWG.
[43] Department of the Army, *FM 34-130,* 2-1-2, 6-8-9.
[44] Draft O&O, March 2, 2005, 6, organizational files of the AWG.
[45] Ibid.,6-7.
[46] Draft O&O, March 2, 2005, 3, organizational flies of the AWG; Fields interview with Author, May 24, 2019.
[47] Draft O&O, March 2, 2005, 5, organizational files of the AWG; 26-27, 37, 42.
[48] Jose Gordon, interview with Author, January 15, 2020.
[49] U.S. Department of Defense, Department of the Army, *Army Doctrinal Publication 3-90: Offense and Defense* (Washington, D.C.: Department of the Army, 2012), Glossary 1.
[50] Draft O&O, March 2, 2005, 30, organizational files of the AWG.
[51] Ibid., 31-36.
[52] Ibid.
[53] Shaw interview with Author, February 27, 2019.
[54] Draft O&O, March 2, 2005, 30, organizational files of the AWG; Fields interview with Author, May 24, 2019.
[55] Draft O&O, March 2, 2005, 10-14, organizational files of the AWG.
[56] Ibid., 16.
[57] Asymmetric Warfare Group, "151021 AWG Mission Statement Evolution Briefing," 1, organizational files of the AWG.
[58] Ibid., 16-17.
[59] Ibid., 17.
[60] Ibid., 16-17.
[61] Ibid., 19.
[62] Ibid., 2.
[63] Ibid., 19.

[64] U.S. Government, Department of Defense, Department of the Army, Asymmetric Warfare Group by Michael Bowman, ATAW-OPF, "Information Paper: Asymmetric Warfare Group History, June 16, 2017," organizational files of the Asymmetric Warfare Group; Pub. L. 109-148.
[65] U.S. Department of Defense, Department of the Army, Memorandum, dated 11 Aug 05, DAMO-FM (BG Formica) signature, "Subject: Concept Plan for Asymmetric Warfare Group (AWG)," organizational files of the Asymmetric Warfare Group.
[66] U.S. Department of Defense, Department of the Army, Department of the Army General Order Number 2, January 31, 2006, organizational files of the Asymmetric Warfare Group.
[67] Shaw interview, February 27, 2019.
[68] Ibid.
[69] William Donnelly, *Department of the Army Historical Summary—Fiscal Year 2002* (Washington, D.C.: Center of Military History, 2011), 3, 77-78.
[70] Robert Shaw to Author, July 3, 2019.

Chapter 5

[1] Robert Shaw, Asymmetric Warfare Group, interview with Author, February 27, 2019.
[2] Ibid.
[3] U.S. Department of Defense, Department of the Army, Asymmetric Warfare Group, briefing: "Honorable Francis J. Harvey, Secretary of the Army, AWG Final Stationing Briefing, 7 FEB 2007," 1-4, 6, Organizational Files of the Asymmetric Warfare Group (AWG), Fort Meade, MD.
[4] Department of the Army, "Fort Bragg, North Carolina," https://home.army.mil/bragg/index.php (accessed, 9 April 2020); "Fort Carson, Colorado," https://www.carson.army.mil/allunits.html (accessed, 9 April 2020); "Fort Hood, Texas," https://home.army.mil/hood/index.php/units-tenants (accessed, 9 April 2020); "Fort Knox, Kentucky," https://home.army.mil/knox/index.php/about/history (accessed, 9 April 2020); "Fort McCoy, Wisconsin," https://home.army.mil/mccoy/index.php/about/history (accessed, 9 April 2020); Harvey Briefing, 4, Organizational Files of the AWG, February 7, 2007.
[5] Harvey Briefing, 1-5, Organizational Files of the AWG, February 7, 2007.
[6] Dan Gaffney, interview with Author, July 2, 2019; Gustav Person, *BRAC at Fort Belvoir, 1988-2001* (Alexandria: Global Publishing, 2012), 20-21, 26; Brenda Stone, interview with Author, February 7, 2020.
[7] Draft Operational and Organizational (O&O) Concept for the Asymmetric Warfare Group (AWG) as of 2 Mar 05, 3, Organizational Files of the AWG; Department of the Army, "Fort George G. Meade History," https://home.army.mil/meade/index.php/abo ut/history (accessed, 10 April 2020).
[8] Department of the Army, "U.S. Army Garrison, Fort A. P. Hill History," https://home.army.mil/aphill/index.php/about/history (accessed, 16 January 2021).
[9] Shaw interview with Author, February 27, 2019; U.S. Department of Defense, Department of the Army, Asymmetric Warfare Group, "Information Paper: Asymmetric Warfare Group (AWG) History, October 15, 2018," 2, Organizational Files of the AWG, Fort Meade, MD.
[10] Eberhard Nordman to Author, July 25, 2019; Jim Mallon to Author, August 24, 2019; Robert Shaw, email to Author July 3, 2019; U.S. Department of Defense, Department of the Army, by Austin H. MacCormick, Chairman, *Report of the Special Civilian Committee for the Study of the United States Army Confinement System,* (Washington, D.C.: Department of the Army, 1970), 8.

[11] Shaw to Author, July 3, 2019.
[12] Ibid.
[13] Department of the Army, *Army Regulation 600-20: Army Command Policy* (Washington, D.C.: Department of the Army, 2006), 15-16.
[14] George E. "Eric" Zipperer, interview with Author, August 11, 2019.
[15] Ibid.
[16] Ibid.
[17] Ibid.
[18] Alfred G. Birch, interview with Author, August 5, 2019; U.S. Department of Defense, Headquarters, Department of the Army; United States Marine Corps, *FM 101-5-1, MCRP 5-2A: Operational Terms and Graphics* (Washington, D.C.: Headquarters, Department of the Army; United States Marine Corps, 1997), 1-17.
[19] U.S. Department of Defense, *Department of Defense Financial Management Regulation 7000.14-R, Volume 7, Chapter 8, March 2005,* https://comptroller.defense.gov/Portals/45/documents/fmr/archive/07aarch/07a_08b.pdf, 8-3 – 8-6 (accessed, 23 July 2019); Department of the Army, *Army Regulation 614-200, Enlisted Assignments and Utilization Management, 30 September 2004,* 27, https://www.p94008.eos-intl.net/eLibSQL10_P94008_Documents/Military%20Documents%2017/AR%20614-200,%2030%20September%202004.pdf, (accessed, 24 July 2019).
[20] *AR 614-200,* 22, 27; Bob Shaw to Author, June 2, 2020; U.S. Department of Defense, *Department of Defense Instruction (DODI) 1304.27,* 3-4, 3-20, https://biotech.law.lsu.edu/blaw/dodd/corres/pdf/i130427wch1_061496/i130427p.pdf, (accessed, 23 July 2019).
[21] Michael P. Bowman to Author, February 7, 2020; Shaw, July 3, 2019.
[22] Peter Arsenault to Author, February 5, 2020; Peter Arsenault, interview with Author, February 6, 2020.
[23] Michael P. Bowman to Author, June 25, 2019.
[24] Franklin Bohle, interview with Author, Fort Meade, MD, January 17, 2019.
[25] U.S. Department of Defense, Department of the Army, Army Asymmetric Warfare Office, *Army Asymmetric Warfare Office (AAWO) 2007 Annual Review* (Washington, D.C.: Department of the Army, 2007), iv. Here after cited as AAWO 2007 Review.
[26] U.S. Department of Defense, Department of the Army, Asymmetric Warfare Group, briefing, "Azimuth Check for COA Development to BG (P) Halverson and BG Wendell, December 7, 2007," 3, Organizational Files of the AWG, U.S. Department of Defense, Department of the Army, HQDA EXORD 158-06 ISO Army Asymmetric Warfare Program Implementation, 13 April 2006, Organizational Files of the AWG.
[27] U.S. Department of Defense, Department of the Army, Office of the Vice Chief of Staff, by Richard Cody, "Memorandum for See Distribution, Subject: Approved Asymmetric Warfare Group (AWG) Organizational Design, June 23, 2004," Organizational Files of the AWG.
[28] Acquisitionnotes.com, "PPBE Process: Program Budget Decision," http://acqnotes.com/acqnote/acquisitions/program-budget-decision-pbd (accessed, 29 August 2019).
[29] Cody, June 23, 2004 memorandum; Department of the Army, *Army Regulation 1-1: Planning, Programming, Budgeting and Execution* (Washington, D.C.: Headquarters, Department of the Army, 2016), 17.
[30] Department of Defense, Emergency Supplemental Appropriations to Address Hurricanes in the Gulf of Mexico, and Pandemic Influenza Act of 2006, Pub. L. No. 109-148, 119 Stat. 2708, Sec. 8043 (c) (2).

[31] Department of the Army, General Officer Management Office. Lt. Gen. James J. Lovelace. https://www.gomo.army.mil/ext/portal/Officer/Resumes.aspx?Ltr=L&Type=Retired (accessed, 21 December 2019).
[32] Bohle interview, January 17, 2019; U.S. Department of Defense, Department of the Army, U.S. Army War College, *How the Army Runs: A Senior Leader Reference Handbook* (Carlisle, PA: United States Army War College, 2001), 14, 246-47; Pub. L. 109-148.
[33] Shaw interview with Author, February 27, 2019.
[34] Arsenault interview with Author, February 6, 2020; Birch interview with Author, August 5, 2019.
[35] U.S. Department of Defense, Department of the Army, Asymmetric Warfare Group, "The Assessment and Selection Course," https://www.awg.army.mil/Recruiting/AWG-Recruiting/ (accessed, 30 July 2019); U.S. Department of Defense, Department of the Army, "How to Become an Army Ranger," https://www.goarmy.com/ranger/join-the-rangers/requirements-to-join.html (accessed, 30 July 2019). Department of the Army, "Special Forces: Training—The Road to Becoming a Green Beret," https://www.goarmy.com/special-forces/training.html (accessed, 30 July 2019).
[36] Birch interview with Author, August 5, 2019; Zipperer interview with Author, August 11, 2019.
[37] Birch interview with Author, August 5, 2019; Zipperer interview with Author, August 11, 2019.
[38] Bruce W. Tuckman, ""Developmental Sequence in Small Groups," *Psychological Bulletin* 63 (June 1965): 384–99.
[39] Bob Shaw to Author, August 14, 2019.
[40] Ibid.
[41] *AR 600-20*, 5, 49, 52; Shaw, August 14, 2019; Bob Shaw, email to Author, October 22, 2019; Shaw interview with Author, February 27, 2019.
[42] Birch interview with Author, August 5, 2019.
[43] Birch interview with Author, August 5, 2019; Zipperer interview with Author, August 11, 2019.
[44] Mike Bowman, interview with Author, April 18, 2018.
[45] Bowman interview with Author, April 18, 2018; *AAWO 2007*, 25; Arsenault interview with Author, February 6, 2020.
[46] U.S. Department of Defense, Department of the Army, *Army Doctrinal Publication 3-0: Operations* (Washington, D.C.: Department of the Army, 2016), 10.
[47] Arsenault interview with Author, February 6, 2020; Michael Oates, interview with Author, February 4, 2020.
[48] *AAWO 2007*, 25; U.S. Department of the Army, Department of the Army, *Army Doctrinal Publication 7-0: Training* (Washington, D.C.: Department of the Army, 2019), 1-2.
[49] *AAWO 2007*, 25-26.
[50] Jim Schwitters, interview with Author, August 19, 2019; U.S. Department of Defense, Department of the Army, Fort Jackson, "Victory Starts Here," https://home.army.mil/jackson/ (accessed, 6 September 2019).
[51] Scwhitters interview with Author, August 19 2019.
[52] Ibid.
[53] Ibid.
[54] Department of the Army, U.S. Army Center for Lessons Learned, *Handbook 03-35: Operation Enduring Freedom II: Tactics, Techniques, and Procedures* (Ft. Leavenworth, KS: U.S. Army Combined Arms Center, 2003), 2.
[55] Theresa Reece-Austin to Author, August 19, 2019; Luis Enriquez, interview with Author, August 16, 2019; Zipperer interview with Author, August 11, 2019.

[56] Zipperer interview with Author, August 11, 2019.
[57] Ibid.
[58] Birch interview with Author, August 5, 2019.
[59] Department of the Army, *Field Manual 34-81-1: Battlefield Weather Effects* (Washington, D.C.: Department of the Army, 1992), K-1; Department of the Army, *Technical Manual TM 43-0001-29, Army Ammunition Data Sheets for Grenades* (Washington, D.C.: Department of the Army, 1994), 2-23-27.
[60] *AAWO 2007*, 26.
[61] Zipperer interview with Author, August 11, 2019.
[62] Robert Shaw, "Memorandum for Director, The Institute of Heraldry, Subject: SSI Request for the Asymmetric Warfare Group, "June 12, 2006, The Institute of Heraldry, organizational history files for the AWG.
[63] Department of the Army, *Field Manual 3-90: Smoke* Operations (Washington, D.C.: Department of the Army, 1990), 89-90.
[64] Airborne Ranger in The Sky, "MSG William "Chief" Carlson from Southern Pines, NC," http://arits.org/biographies/256-msg-william-chief-carlson (accessed, 16 July 2019); Shaw interview with Author, February 27, 2019; U.S. Department of Defense, Department of the Army, by Charles V. Mugno, Director, The Institute of Heraldry, Memorandum for Department of the Army G1, Uniform Policy Section, Subject: SSI Request for AWG, August 23, 2006, The Institute of Heraldry, organizational history files for the AWG; U.S. Department of Defense, Department of the Army, The Institute of Heraldry (TIOH), undated memorandum "Suggested Design, Shoulder Sleeve Insignia, Asymmetric Warfare Group," 74, TIOH organizational history files of the AWG, Ft. Belvoir, VA.
[65] Mugno memorandum, August 23, 2006, Organizational Files, TIOH.
[66] U.S. Department of Defense, Department of The Army, Jane Helton, *Memorandum for Commander, Asymmetric Warfare Group, Subject: Shoulder Sleeve Insignia (SSI) for the Asymmetric Warfare Group, August 29, 2006*, 77, TIOH organizational history files.

Chapter 6

[1] Center of Military History, Historical Resources Branch, "United States Army Chiefs of Staff," https://history.army.mil/faq/FAQ-CSA.htm (accessed, 18 December 2019); U.S. Department of Defense, Department of the Army, by William G. Bell, Commanding Generals and Chiefs of Staff of the United States Army, 1775-2013: Portraits and Biographical Sketches of the United States Army's Senior Officer (Washington, D.C.: Center of Military History, 2013), 170.
[2] George W. Casey to Author, January 25, 2020; U.S. Department of Defense, Department of the Army, by George W. Casey and Peter Geren," A Statement on the Posture of the United States Army to the Committees and Subcommittees of the United States Senate and the House of Representatives, 2d session, 110th congress, 26 February 2008" (Washington, D.C.: Headquarters, Department of the Army, 2008), 6.
[3] George W. Casey, Jr. to Author, January 25, 2020.
[4] Richard J. Cody, interview with Author, January 15, 2018.
[5] Cody interview with Author, January 15, 2018; Department of the Army, General Officer Management Office, "General Peter J. Chiarelli," https://www.gomo.army.mil/ext/portal/Officer/Resumes.aspx?Ltr=L&Type=Retired (accessed, 21 December 2019); Bob Shaw to Author, December 18, 2019.

[6] GOMO Resume, Chiarelli; Shaw, December 18, 2019; Jenny Stripling, "Chiarelli: Mental Wounds No 'Phantom Issue'," https://www.army.mil/article/28431/chiarelli_mental_wounds_no_phantom_issue (accessed, 13 April 2020).
[7] Hank Kinnison, interview with Author, February 3, 2020; James Mis, interview with Author, December 5, 2019; Shaw, December 18, 2019.
[8] Department of the Army, General Officer Management Office, "General James D. Thurman," https://www.gomo.army.mil/ext/portal/Officer/Resumes.aspx?Ltr=L&Type=Retired (accessed, 21 December 2019); Rodney Johnson, "U.S. Army Central Takes a Big Step Towards Being Full-Spectrum Operations Capable," Defense Visual Information Service, March 13, 2008, https://www.dvidshub.net/news/printable/17337 (accessed, 19 December 2019).
[9] James D. Thurman to Author, February 4, 2020.
[10] U.S. Department of Defense, Department of the Army, Asymmetric Warfare Group, "Program for the Change of Command Ceremony: Asymmetric Warfare Group, 14 July 2009," Organizational Files of the Asymmetric Warfare Group (AWG), Fort George G. Meade, MD.
[11] Program for the Change of Command, July 14, 2009; Mis interview with Author, December 5, 2019.
[12] Program for the Change of Command, July 14, 2009; James Mis to Author, December 11, 2019.
[13] Melanie Casey, northwestmilitary.com, "Following in a Father's Footsteps: Keeping Military Service in the Family," February 16, 2011, http://www.northwestmilitary.com/news/focus/2011/02/northwest-military-ranger-airlifter-newspaper-JBLM-following-in-fathers-footsteps-military-service-i/ (accessed, 13 April 2020); Ray Lombardi, "Do you ever wonder where your role models end up? I do. Here are a few I tracked down," https://karlsendlessstories.blogspot.com/2008/11/do-you-ever-wonder-where-your-role.html (accessed, 13 April 2020).
[14] Mis interview with Author, December 5, 2019.
[15] Blakeslee Connors to Author, December 16, 2019; Michael Bowman Author, December 17, 2019; Texas A&M University Corps of Cadets, "Blake Connors, Colonel, United States Army, Professor of Military Science," https://corps.tamu.edu/staff/pms/ (accessed, 2 January 2020); Mark L Bradley, Department of the Army Historical Summary, Fiscal Year 2009 (Washington, D.C.: Center of Military History, 2015), 12.
[16] Bowman, December 17, 2019; Department of the Army, Asymmetric Warfare Office Annual Report, 2007 (Washington, D.C.: Department of the Army, 2007), 24, 26; Franklin Bohle, interview with Author, January 17, 2019; U.S. Department of Defense, Department of the Army, Center for Army Lessons Learned, Handbook No. 10-41 Assessment and Measures of Effectiveness in Stability Ops (Fort Leavenworth, KS: Combined Arms Center, 2010), 69.
[17] Mis interview with Author, December 5, 2019.
[18] Ibid.
[19] Ibid.
[20] Ibid.
[21] Jose Gordon, interview with Author, January 15, 2020.
[22] Mis interview with Author, December 5, 2019.
[23] Mis interview with Author, December 5, 2019; Bob Shaw to Author, April 15, 2020; U.S. Department of Defense, Department of the Army, ALARACT 078/2005 DTG R 202200Z APR 05, SUB: ARMY COMBAT UNIFORM (ACU) ENSEMBLE, Section 3, https://www.ar15.com/forums/general/Wear-and-Appearance-of-the-Army-ACU/77-1006213/&page=1 (accessed, 13 April 2020).

[24] Jim Mis to Author, April 12, 2020.
[25] Jim Mis, to Author, April 19, 2020.
[26] Mis interview with Author, December 5, 2019.
[27] Gordon interview with Author, January 15, 2020.
[28] S. L. A. Marshall, The Soldier's Load and the Mobility of the Nation, (Arlington, VA: Association of the United States Army, 1950), 5.
[29] Charlie Company Vietnam 1966-1972, Charlie Company, 1st Battalion, 22nd Infantry, 4th Infantry Division, "What Did Vietnam Soldiers Carry?", https://charliecompany.org/2013/01/19/what-did-vietnam-soldiers-carry/ (accessed, 21 January 2020); U.S. Department of Defense, Department of the Army, Eighth United States Army, Korean Service Corps Battalion, "History," https://ksc.korea.army.mil/ (accessed, 21 January 2020).
[30] Christopher E. Howard, "This Vest May Save Your Life: U.S. Army Body Armor From World War II to the Present," https://arsof-history.org/articles/19oct_body_armor_page_ 1.html (accessed, 14 April 2020); Department of the Army, Program Executive Office Soldier, briefing slides Interceptor Body Armor and Interceptor Body Armor (Capability Enhancements), Papers of General (Retired) George W. Casey, Jr., Chief of Staff, United States Army, Box 35, National Defense University Library, Fort Lesley J. McNair, Washington, D.C; Department of the Army, Office of the Quartermaster General, "Armored Vest Fact Sheet," https://qmmuseum.lee.army.mil/korea/armored_vest.htm (accessed, 14 April 2020).
[31] Gordon interview with Author, January 15, 2020.
[32] Ibid.
[33] Casey Howard to Author, January 24, 2020; Mike Bowman to Author, April 15, 2020; Gordon interview with Author, January 15, 2020; National Defense Authorization Act for Fiscal Year 2010, Pub. L. No. 111-84 (2009); William J. Webb, Charles Anderson, William Andrade, et al., Department of the Army Historical Summary, Fiscal Years 1990-1991 (Washington, D.C.: Center of Military History, 1997), 58; William M. Donnelly, Department of the Army Historical Summary, Fiscal Year 2004 (Washington, D.C.: Center of Military History, 2015), 68.
[34] U.S. Department of the Defense, Department of the Army, Asymmetric Warfare Group by Tomy Wright, briefing: "AWG PAO Requested, Solution Development 2006 – 2016, S32 Search Results (Post 2/26/2016), Pre-FRAGO Results, (as of 3/14/2016)," AWG Organizational Files; Department of the Army, Field Manual 100-2-3, (Washington, D.C: Headquarters, Department of the Army, 1990), 10-21.
[35] Department of the Army, Field Manual 1-100: Doctrinal Principles for Army Aviation in Combat Operations (Washington, D.C.: Headquarters, Department of the Army, 1989), 1-17.
[36] Thomas Beyerl to Author, January 22, 2020.
[37] Jim Mis to Author, February 2, 2020; Mike Akridge to Author, January 19, 2019.
[38] Colombia Reports, "FARC," https://colombiareports.com/farc/ (accessed, 20 February 2020).
[39] Colombia Reports, Gordon interview with Author, January 15, 2020.
[40] British Broadcasting Corporation, "Top FARC Rebel Leader Alfonso Cano Killed in Colombia," November 5, 2011, https://www.bbc.com/news/world-15604456 (accessed, 2 February 2020).
[41] Daniel P. Boger to Author, September 9, 2019; Mis interview with Author, December 5, 2019.
[42] Daniel P. Bolger Author, February 1, 2020; U.S. Department of Defense, Department of the Army G-3/5/7, briefing: "Asymmetric Warfare Group Update, as of 20 October 2010," 1, AWG Organizational Files.

[43] AWG Update-20 OCT 10, 1; Department of the Army, Field Manual 7-1 (25-101): Battle Focused Training (Washington, D.C.: Department of the Army, 2003), 3-6.
[44] Daniel P. Bolger Author, January 29, 2020; Michael Oates, interview with Author, February 3, 2020; U.S. Army Materiel Command, "The Army's Materiel Integrator," https: //www.amc.army.mil/ (accessed, 13 April 2020); U.S. Army Forces Command, "About FORSCOM," https://www.forscom.army.mil/Pages/About (accessed, 13 April 2020); U.S. Army Training and Doctrine Command, "About Us," https://www.tradoc.army.mil/ About-Us/ (accessed, 13 April 2020).
[45] AWG Update-20 OCT 10, 4.
[46] Daniel P. Bolger to Author, February 1, 2020.
[47] Middletown (NY) Times Herald-Record, September 20, 2019.
[48] Bell, Commanding Generals and Chiefs of Staff of the United States Army, 172; Bolger, September 9, 2019; Peter Chiarelli, "The 100 Most Influential People: Martin Dempsey," Time, April 16, 2015, https://time.com/collection-post/3823200/martin-dempsey-2015-time-100/; (accessed, 14 April 2020).
[49] Department of the Army, General Officer Management Office, "Lieutenant General Daniel P. Bolger" https://www.gomo.army.mil/ext/portal/Officer/Resumes.aspx?Ltr=L& Type=Retired; (accessed, 21 December 2019); Bolger to Author, September 9, 2019.
[50] U.S. Government, Department of Defense, Department of the Army, Asymmetric Warfare Group by Mike Bowman, "Information Paper, Subject: Asymmetric Warfare Group History, June 16, 2017," AWG Organizational Files; Bolger, February 1, 2020; Michael Oates, interview with Author, February 3, 2020; Wright, AWG PAO, 16-19; U.S. Government, Department of Defense, Office of the Secretary of Defense Comptroller/CFO, United States Department of Defense Fiscal Year 2012 Budget Request: Program Acquisition Costs by Weapon System (Washington, D.C.: Department of Defense, 2011), 1-10, 3-5.
[51] Bolger, January 20, 2020; Mis interview with Author, December 5, 2019; William G. Bell, Secretaries of War and Secretaries of the Army: Portraits and Biographical Sketches (Washington, D.C.: Center of Military History, 2011), 172.
[52] Bolger, September 9, 2020; U.S. Department of Defense, Department of the Army, Asymmetric Warfare Group, briefing: "Honorable Francis J. Harvey, Secretary of the Army, AWG Final Stationing Briefing, 7 FEB 2007", 9, AWG Organizational Files; Department of the Army, 20th Support Command, "Mission," https://www.army.mil/20thcbrne#org-about (accessed, 14 April 2020).
[53] Department of the Army, Field Manual 1-02: Operational Terms and Graphics (Washington, D.C.: Department of the Army, 2004), 2-6; Department of the Army, 20th Support Command, "History," accessed January 28, 2020, https://www.cbrne.army.mil/ history.html (accessed, 28 January 2020).
[54] Department of the Army, briefing slide: "FY 09 Command Organization, 20th Support Command," Casey Papers, Box 35.
[55] U.S. Government, Department of Defense, Department of the Army, Asymmetric Warfare Group, "AWG Task Organization C2 COA", 6-8, AWG Organizational Files.
[56] Bell, Commanding Generals and Chiefs of Staff, 172; Bolger to Author, February 1, 2020; U.S. Department of Defense, Department of the Army, Asymmetric Warfare Group, "Information Paper: Asymmetric Warfare Group History, 16 June 2017," AWG Organizational Files.
[57] U.S. Department of Defense, Department of the Army, Asymmetric Warfare Group, briefing: "SECARMY Decision Brief: Future of the Asymmetric Warfare Group, XX MAY 2013," 8, AWG Organizational Files.

[58] Wright, AWG PAO, 3-6, 13-15.
[59] Mis interview with Author, December 5, 2019.
[60] Mis interview with Author; Michael Oates, interview with Author, February 3, 2020.
[61] Pat Mahaney, interview with Author, January 7, 2020; U.S. Department of Defense, Department of the Army, Asymmetric Warfare Group, "Program for the Change of Command Ceremony for the Asymmetric Warfare Group, 14 July 2011," AWG Organizational Files.
[62] Mahaney interview with Author, January 7, 2020; U.S. Government, Department of Defense, Department of the Army G-3/5/7, briefing slide: "G-3/5/7, Organization and Functions as of 22 JUN 10," Collection of Daniel P. Bolger.
[63] Bruce W. Bennet, "Managing Catastrophic North Korean Risks," *Korea Herald,* January 21, 2010; Toufic Haddad, "Insurgent Infrastructure: Tunnels of the Gaza Strip," Middle East - Topics & Arguments 10 (June 2018): 71-85, https://doi.org/10.17192/meta.2018.10.7594 (accessed, 15 April 2020); Mahaney interview with Author, January 7, 2020; Department of the Army, Field Manual 3-0, Operations (Washington, D.C.: Headquarters, Department of the Army, 2011), 4-3-12; Judith Warner, U.S. Border Security: A Reference Handbook (Santa Barbara: ABC-CLIO, LLC, 2010), 75.
[64] Mahaney interview with Author, January 7, 2020.
[65] U.S. Government, Department of Defense, Department of the Army, "HQDA EXORD 304-11, 211813SEP11Z," AWG Organizational Files.
[66] U.S. Department of Defense, Department of the Army, "General Order 2011-06, dated September 21, 2011", AWG Organizational Files.
[67] Kinnison interview with Author, February 3, 2020.
[68] EXORD 304-11, AWG Organizational Files.
[69] Ibid.
[70] Ibid.
[71] U.S. Government, Department of Defense, Department of the Army, by John M. McHugh, Secretary of the Army, "Memorandum for Commanding General, U.S. Army Training and Doctrine Command, Assistant Chief of Staff G-3/5/7, Subject: Asymmetric Warfare Group, dated September 23, 2011," AWG Organizational Files.
[72] TRADOC TASKORD IN 1279-212, 061410OCT11: "Asymmetric Warfare Group Assignment, 06 OCT 11," AWG Organizational Files.
[73] Ibid.
[74] "TASKORD IN 1279-212 dated 061401OCT11", AWG Organizational Files; Department of the Army, U.S. Army Training and Doctrine Command, "FRAGO 1 to TASKORD IN 1279-212 dated 14 OCT 11," AWG Organizational Files.
[75] Pat Mahaney to Author, February 8, 2020.
[76] Casey, January 25, 2020.
[77] Bolger, September 9, 2020; Casey, January 25, 2020.
[78] Bolger to Author, September 9, 2020; Bolger to Author, November 27, 2017; Oates interview with Author, February 3, 2020; Wright, AWG PAO, 8-18.

Conclusions

[1] U.S. Government, Department of Defense, Department of the Army, "Army Asymmetric Warfare Office (AAWO), 2007 Annual Review," cover.
[2] U.S. Government, Department of Defense, "Department of Defense Background Briefing on Enhancing Counter-ISIL Operations, October 30, 2015," https://www.defen

se.gov/Newsroom/Transcripts/Transcript/Article/626814/department-of-defense-back ground-briefing-on-enhancing-counter-isil-operations/ (accessed, 19 December 2020); U.S. Central Command, "Combined Joint Task Force-Operation Inherent Resolve (CJTF-OIR)," www.centcom.mil/OPERATIONS-AND-EXERCISES/OPERATION-INHE RENT-R ESOLVE/ (accessed, 11 April 2021); Barack Obama, "Address to the Nation by the President, December 6, 2015," https://www.obamawhitehouse.archives.gov/the-press-office/2015 /12/06/address-nation-president; (accessed, 16 December 2020); Combined Joint Task Force—Operation Inherent Resolve Media Desk to Author, December 10, 2020.

3 Department of the Army, "Army Futures Command," www.army.mil/standto/archive/ 2018/03/28 (accessed, 10 December 2020).

4 Department of the Army, "Army Modernization," January 16, 2018, https://www.army. mil/standto/archive/2018/01/16/?dmd (accessed, 10 December 2020).

5 Army Modernization;" U.S. Army Europe and Africa, "Mission and History," https:// www.europeafrica.army.mil/Mission-History/ (accessed, 10 December 2020); U.S. Army Europe and Africa, "ATLANTIC RESOLVE," https://www.europeafrica.army.mil/Atlantic Resolve/ (accessed, 10 December 2020); U.S. Army Europe and Africa, "Press Release: Ceremony to Establish V Corps Headquarters (Forward)," November 19, 2020, https://w ww.europeafricat.army.mil/ArticleViewPressRelease/Article/2420593/press-release-cer emony-to-establish-v-corps-headquarters-forward/ (accessed, 10 December 2020).

6 Bob Shaw to Author, February 5, 2021; Scott Ackerly to Author, December 9, 2020.

7 Ibid.

8 Department of the Army, "1st SFAB Hosts Activation Ceremony; Heraldry Announced, 8 February 2018, https://www.army.mil/article/200244/1st_sfab_hosts_activation_cerem ony_heraldry_announced (accessed, 16 December 2020).

9 Department of the Army, "Multi-Doman Operations," October 10, 2018, accessed December 16, 2020, https://www.tradoc.army.mil/Publications-and-Resources/Article-Display/Articl e/1655556/multi-domain-operations (accessed, 16 December 2020); Department of the Army, "Security Force Assistance Command, 2nd Security Force Assistance Brigade activate at Fort Bragg," December 7, 2018, https://www.army.mil/article/214605/security_force_assista nce_command_2nd_security_force_assistance_brigade_activate_at_fort_bragg (accessed, 16 December 2020); Gerhard Spinney, "SFABs Are Crucial New Tools," Army 71, February 2021, 13-14.

10 Tom Beyerl to author, October 29, 2019.

11 Bob Shaw to Author, February 5, 2021.

12 Jim Mis to Author, February 5, 2021.

13 Pat Mahaney to Author, February 9, 2021.

14 Department of the Army, "Permanent Order 092-02, dated 2 April 2014," https://histo ry.army.mil/html/forcestruc/HRC/2014/092-02_20140402_HRCMD.pdf (accessed, 1 November 2019).

15 "Permanent Order 092-02;" Department of the Army, *Army Regulation 600-8-22, Military Awards* (Washington, D.C.: Department of the Army, 2019), 84.

16 Carl von Clausewitz, Michael Howard, Peter Paret, eds./trans. *On War* Michael Howard and Peter Paret ed. and trans. (Princeton, New Jersey: Princeton University Press, 1984), 89.

17 Ibid., 89.

18 Center of Military History, "Index of Operational Code Names," https://www.history.ar my.mil/CODE.htm (accessed, 16 December 2020); "Multi-National Force and Observers," https://www.mfo.org/about-us (accessed, 16 December 2020); Gordon W. Rudd,

Humanitarian Intervention: Assisting the Iraqi Kurds in Operation PROVIDE COMFORT, 1991 (Washington, D.C.: Center of Military History, 2004), vii; GlobalSecurity.org, "Operation Support Hope," https://www.globalsecurity.org/military/ops/support_hope.htm; (accessed, 16 December 2020); R. Cody Phillips, *Operation Joint Guardian: The U.S. Army in Kosovo* (Washington, D.C.: Center of Military History, 2007) 5-9; R. Cody Phillips, *Bosnia-Hercegovina: The U.S. Army's Role in Peace Enforcement Operations, 1995-2004* (Washington, D.C.: Center of Military History, 2015), 5, 19-21.

[19] National Security Act, Pub. L. No. 235, 61 Stat. 496 (1947); National Security Act Amendments of 1949, Pub. L. 81-216, 61 Stat. 496.

[20] Richard Stewart, *American Military History, Volume 1: The United States Army and the Forging of a Nation, 1775-1917* (Washington, D.C.: Department of the Army, 2009), 134, 148; Stephen J. Lofgren, *The Highest Level of Skill and Knowledge: A Brief History of U.S. Army Civilians, 1775-2015* (Washington, D.C.: Department of the Army, 2016), 12; Benjamin Jensen, *Forging the Sword: Doctrinal Change in the U.S. Army* (Stanford, CA: Stanford University Press, 2016), 16-17.

[21] GlobalSecurity.org, "US Army 1990 Divisions," https://www.globalsecurity.org/military/agency/army/division-90.htm; (accessed, 26 September 2019); Stephen Y. L. Gammons and William Donnelly, *Department of the Army Historical Summary, Fiscal Year 1995* (Washington, D.C.: Department of the Army, 2004), 3, 55.

[22] Department of the Army, 177th Armored Brigade and the Operations Group of the National Training Center, *The Iraqi Army: Organization and Tactics* (Fort Irwin, CA: National Training Center, 1991), 146.

[23] Phillips, *Bosnia-Hercegovina,* 19.

[24] Phillips, *Bosnia-Hercegovina,* 18; John F. Gordon IV, Bruce Narduli, and Walter L. Perry, "The Operational Challenges of Task Force Hawk," *Joint Forces Quarterly* 29 (Autumn-Winter 2001-2002): 52-57, https://www.ndupress.ndu.edu/portals/68/Documents/jfq/jfq-29.pdf. (accessed, 14 January 2021).

[25] Department of the Army, by W. Blair Hayworth, *Department of the Army Historical Summary, Fiscal Year 2000* (Washington, D.C.: Center of Military History, 2011), 75; GlobalSecurity.org, "Task Force Falcon/IFOR/SFOR Deployments," https://www.globalsecurity.org/military/ops/sfor-deploy.htm (accessed, 1 October 2019); Department of the Army, by Jeffrey Charlston, *Department of the Army Historical Summary, Fiscal Year 1999* (Washington, D.C.: Center of Military History, 2006), 52.

[26] "Army Issues Notice of Crusader Contract Cancellation; United Defense Awarded Contract to Develop New Cannon Artillery System for the U.S. Army." *PR Newswire,* August 09, 2002, http://libproxy.temple.edu/login?url=https://www-proquest-com.libproxy.temple.edu/wire-feeds/army-issues-notice-crusader-contract-cancellation/docview/450197373/se-2?accountid=14270 (accessed, 17 March 2021); "Northrop Grumman Press Statement on Cancellation Announcement of RAH-66 Comanche Helicopter Program," *PR Newswire,* February 23, 2004, https://link.gale.com/apps/doc/A131536326/AONE?u =temple_main&sid=AONE&xid=b2146bca (accessed, 22 January 2021); Marjorie Censer, "DOD Formally Cancels FCS, ADM Details Four Follow-on Efforts," *Inside the Pentagon* 25, June 25, 2009, 13-14, https://signin-lexisnexis-com.libproxy.temple.edu/lnaccess/app/signin?back=https%3A%2F%2Fadvance.lexis.com%3A443%2Furl-api%2Flaapi%2Fsearch%3Fsource%3DMTA1N TMxMQ&aci=la (accessed, 22 January 2021).

[27] Sean M. DiGiovanna, Ann Markusen, and Michael C. Leary, *From Defense to Development?: International Perspectives on Realizing the Peace Dividend* (Abingdon, UK: Taylor & Francis Group, 2003), 15-39; Mike Moore, "So Where's the Peace Dividend?" *Bulletin of the Atomic Scientists* 51 (1995): 30-51, https://link.gale.com/apps/doc/A17290207/AONE?u=temple_main&sid=AONE&xid=efb8f4c9. (accessed, 22 January 2021).
[28] Department of the Army, *Army Doctrine Publication Number 5-0, The Operations Process* (Washington, D.C.: Department of the Army, 2019), v.
[29] Mike Oates, interview with Author, February 3, 2020; Frank Bohle, interview with Author, January 17, 2019; Jim Mis, interview with Author, January 7, 2020; Peter Arsenault, interview with Author, February 6, 2020; Daniel P. Bolger to Author, November 27, 2017.
[30] 75th Rangers.com, "Discipline, Honor and Respect Must be Earned—They Cannot be Issued," https://www.75thrangers.com/#:~:text=The%20Black%20Beret%20has%20been%20worn%20exclusively%20by,Army%20were%20to%20remain%20silent%20on%20this%20issue (accessed, 16 December 2020); Department of the Army, "History of the U.S. Army Beret," https://www.webarchive.org/web/20180704012922/ (accessed, 4 February 2021); "U.S. Army Rangers," https://www.army.mil/rangers (accessed, 4 February 2021).
[31] Karl E. Cocke, William G. Bell, Young G. Chang, et al, *Department of the Army Historical Summary: Fiscal Year 1982* (Washington, D.C.: Center of Military History, 1988), 62; U.S. Government, Department of Defense, Department of Defense Instruction 6496.01, accessed December 23, 2020, DoD Directive 6495.01, January 23, 2012; Incorporating Change 1, April 30, 2013 (sapr.mil); U.S. Government, Department of Defense, Department of the Army, "Report of the Fort Hood Independent Review Committee, November 6, 2020," ii-x, 11-81, https://www.army.mil/e2/downloads/rv7/forthoodreview/2020-12-03_FHIRC_report_redacted.pdf (accessed, 16 December 2020).
[32] U.S. Special Operations Command, "Introduction to U.S. Special Operations Command, 2006" https://www.socom.mil/FactBook/2006%20Fact%20Sheet.pdf (accessed, 16 December 2020); U.S. Special Operations Command, "Special Operations Command Factbook, 2010," https://www.socom.mil/FactBook/2010%20Fact%20Book.pdf (accessed, 16 December 2020).
[33] Hank Kinnison, interview with Author, February 6, 2020; Arsenault, interview with Author, February 6, 2020; David Flynn, letter to LTG Mulholland, 5th SFG (A), AWG, Melissa Pittman, "Subject: The Battle for Bakersfield, 30 July 2010," collection of Peter Arsenault.
[34] Mike Hastleton to Author, February 18, 2020.
[35] Patrick Hamilton, remarks at the Texas A & M University Army Reserve Officer Training Corps commissioning ceremony, December 19, 2019.
[36] Pat Mahaney to Author, February 9, 2021.

Selected Bibliography

177th Armored Brigade and the Operations Group of the National Training Center. *The Iraqi Army: Organization and Tactics.* Fort Irwin, CA: National Training Center, 1991.

Addington, Larry H. *The Patterns of War Since the Eighteenth* Century. Bloomington, IN: Indiana University Press, 1994.

Akande, Dapo and Lawrence Hill-Cawthorne. "The Lieber Code and the Regulation of Civil War in International Law." *Columbia Journal of Transnational Law* 53: 638–51, June 2015.

Allen, William H. "The Election of 1900." *The Annals of the American Academy of Political and Social Science* 17 (January 1901): 53–73.

Allen, William J. *Short of War: Major U.S. Air Force Contingency Operations, 1947-1997.* Montgomery, AL: Air Force Historical Agency, 2000.

Alexiev, Alexander. *Inside the Soviet Army in Afghanistan.* Santa Monica, CA: RAND Corporation, 1988.

Arnold, James R., ed. *Americans at War: Eyewitness Accounts from the American Revolution to the 21st Century, 2.* Santa Barbara, CA: ABC-CLIO, 2018.

Association of the United States Army. "Army Announces Unit Rotations." A*rmy* 53 (September 2003): 68-70,72.

Association of Graduates, United States Military Academy. *Sixty-First Annual Report of the Association of Graduates of the United States Military Academy at West* Point. Newburgh, NY: Moore Printing, 1930.

————. *Sixty-Second Annual Report of the Association of Graduates of the United States Military Academy at West* Point. Crawfordsville, IN: Lakeside Press, 1931.

————. *Sixty-Third Annual Report of the Association of Graduates of the United States Military Academy at West* Point. Newburgh, NY: Moore Printing, 1932.

————. *Sixty-Fourth Annual Report of the Association of Graduates of the United States Military Academy at West* Point. Newburgh, NY: Moore Printing, 1933.

————. *Sixty-Fifth Annual Report of the Association of Graduates of the United States Military Academy at West Point. Newburgh,* NY: Moore Printing, 1934.

————. *Sixty-Sixth Annual Report of the Association of Graduates of the United States Military Academy at West* Point. Newburgh, NY: Moore Printing, 1935.

———. *Sixty-Seventh Annual Report of the Association of Graduates of the United States Military Academy at West* Point. Newburgh, NY: Moore Printing, 1936.

———. *Sixty-Eighth Annual Report of the Association of Graduates of the United States Military Academy at West* Point. Newburgh, NY: Moore Printing, 1937.

———. *Sixty-Ninth Annual Report of the Association of Graduates of the United States Military Academy at West* Point. Newburgh, NY: Moore Printing, 1938.

———. *Seventieth Annual Report of the Association of Graduates of the United States Military Academy at West* Point. Newburgh, NY: Moore Printing, 1939.

———. *Seventy-First Annual Report of the Association of Graduates of the United States Military Academy at West* Point. Newburgh, NY: Moore Printing, 1940.

———. *Seventy-Second Annual Report of the Association of Graduates of the United States Military Academy at West* Point. Newburgh, NY: Moore Printing, 1941.

Atkeson, Edward B. "War in Iraq: Shock and Awe?" A*rmy* 53: 8-13. May 2003.

Bacevich, Andrew J. *The Pentomic Era: The U.S. Army Between Korea and Vietnam.* Washington, DC: National Defense University Press, 1986.

Ballard, Ted. *Staff Ride Guide: Battle of Antietam.* Washington, DC: Center of Military History, 2008.

Bacon, George W. *The Life and Administration of Abraham Lincoln: Presenting His Early History, Political Career, Speeches, Messages, Proclamations, Letters, etc., with a General View of His Policy as President of the United States... Also, the European Press on his Death.* London, UK: S. Low, Son, and Marston, 1865.

Barno, David and Nora Benashel. *Adaptation Under Fire: How Militaries Adapt in Wartime.* New York: Oxford University Press, 2020.

Batatu, Hanna. "Iraq's Underground Shi'a Movements: Characteristics, Causes and Prospects." *Middle East Journal* 35 (Fall, 1981): 578-94.

Batiste, John R. S. and Paul R. Daniels. “The Fight for Samarra: Full-Spectrum Operations in Modern Warfare.” *Military Review* 85 (May-June 2005): 13-21.

Bell, William G. *Commanding Generals and Chiefs of Staff, 1775-2012.* Washington, DC: Center of Military History, 2013.

———. *Secretaries of War and Secretaries of the Army: Portraits and Biographical Sketches.* Washington, DC: Center of Military History, 1992.

———. *Department of the Army Historical Summary, Fiscal Year 1969.* Washington, DC: Center of Military History, 1973.

———. *Department of the Army Historical Summary, Fiscal Year 1970.* Washington, DC: Center of Military History, 1973.

Bell, William G. and Cocke, Carl E., eds. *Department of the Army Historical Summary, Fiscal Year 1972.* Washington, DC: Center of Military History, 1977.

Ben-Dor, Gabriel. "Civilianization of Military Regimes in the Arab World." *Armed Forces & Society* 1 (April 1975): 317–27.

Bensahel, Nora, Olga Oliker, and Keith Crane. *After Saddam Prewar Planning and the Occupation of Iraq.* Santa Monica, CA: RAND Corporation, 2008.

Berman, "Taking on Taking on Iraq," *Editor & Publisher,* January 27, 2003: 31.

Biagini, Antonello and Giovanna Motta, eds. *Empires and Nations from the Eighteenth to the Twentieth Century: 2.* Newcastle-upon-Tyne: Cambridge Scholars, 2014.

Birtle, Andrew J. "PROVN, Westmoreland, and the Historians: A Reappraisal." *The Journal of Military History* 72 (October 2008): 1213-47.

———. *U.S. Army Counterinsurgency and Contingency Operations Doctrine, 1860-1941.* Washington, DC: Center of Military History, 2003.

Boldan, Edith M., ed. *Department of the Army Historical Summary, Fiscal Year 1979.* Washington, DC: Center of Military History, 1982.

Bolzak, Jerry R., ed. *The Collected Works of the Thirty-Second Chief of Staff of the United States Army, Gordon R. Sullivan, General, United States Army, June 1991-June 1995.* Washington, DC: Center of Military History, undated.

Boothby, William. *Weapons and the Law of Land* Warfare. Oxford: Oxford University Press, 2016.

Bradley, Mark L. *Department of the Army Historical Summary, Fiscal Year 2009.* Washington, DC: Center of Military History, 2015.

Brackett, Albert G. "The United States Cavalry, Past and Present." U*nited Service; A Quarterly Review of Military and Naval Affairs* 4 (August 1890): 8, 114-19.

Bradley, Mark L. *The Army and Reconstruction,* 1865-1877. Washington, DC: Center of Military History, 2015.

Brill, Charles J. *Conquest of the Southern Plains: Uncensored Narrative of the Battle of the Washita and Custer's Southern Campaign.* Oklahoma City, OK: Golden Saga Publishers, 1938.

Brazee, Frederick. "The Mobile Riverine Force, Mekong Delta, Republic of Vietnam, 16 February 1967-10 January 1968: Personal Experiences of a Company Commander and Assistant Brigade S2." Class paper, Infantry Officer Advanced Course, U.S. Army Infantry School, 1968.

Briscoe, Charles H., Kenneth Finlayson, Robert W. Jones, Jr., Cherilyn A. Walley, A. Dwayne Aaron, Michael R. Mullins and James A. Schroeder. *All Roads Lead to Baghdad: Army Special Operations Forces in Iraq.* Fort Bragg, NC: US Army Special Operations Command History Office, 2006.

Brown, Dee. *Bury My Heart at Wounded Knee: An Indian History of the American West.* New York: Holt, Reinhart, and Winston, 1970.

Brown, John S. *Kevlar Legions: The Transformation of the U.S. Army, 1989-2005.* Washington, DC: Center of Military History, 2005.

Brown, Robert. *The Races of Mankind: Being a Popular Description of the Characteristics, Manners and Customs of the Principal Varieties of the Human Family, 3.* New York: Casell, Peter and Galpin, 1876.

Bruscino, Thomas A. In *Between the Rivers: Combat Action in Iraq 2003-2005,* edited by John McGrath. Ft. Leavenworth, KS: Combat Studies Institute Press, 2012.

Bunker, Robert J. *Networks, Terrorism and Global Insurgency.* London, UK: Taylor and Francis Group, 2005.

Byler, Charles. "Pacifying the Moros: American Military Government in the Southern Philippines, 1899-1913." *Military Review* 85 (May 2005): 41-45.

California National Guard. *Military Support of Law Enforcement During Civil Disturbances: A Report Concerning the California National Guard's Part in Suppressing the Los Angeles Riot,* August 1965.

Carland, John M. *The United States Army in Vietnam: Stemming the Tide, May 1965 to October 1966.* Washington, DC: Center of Military History, 2000.

Carney, Stephen A. *Allied Participation in Operation Iraqi Freedom.* Washington, DC: Center of Military History, 2011.

Carter, Donald A. *The U.S. Army Before Vietnam, 1953-1965.* Washington, DC: Center of Military History, 2015.

Carter Timothy and Daniel J. Veale, "Weather, Terrain and Warfare: Coalition Fatalities in Afghanistan," *Conflict Management and Peace Science* 30 (July 2013): 220-29.

Carter, Linwood. "Iraq: Summary of U.S. Forces." Washington, DC: Congressional Research Service, 2005.

Casey, George W., Jr., and Peter Geren. "A Statement on the Posture of the United States Army to the Committees and Subcommittees of the United States Senate and the House of Representatives, 2d session, 110th congress, 26 February 2008." Washington, DC: Headquarters, Department of the Army, 2008.

Cathcart, K. Patrick. "TACFIRE Deployment and Training." *Field Artillery Journal* 49 (January-February 1981): 8-14.

Center for Army Lessons Learned. Center for Army Lessons Learned, *Handbook No. 10-41 Assessment and Measures of Effectiveness in Stability Ops.* Fort Leavenworth, KS: Combined Arms Center, 2010), 69.

———. *Handbook 03-35: Operation Enduring Freedom II: Tactics, Techniques, and Procedures.* Fort Leavenworth, KS: U.S. Army Combined Arms Center, 2003.

Chant, Christopher. *A Compendium of Armaments and Military Hardware.* New York: Routledge Revivals, 1987.

Chiovenda, Andrea and Melissa Chiovenda. "The Specter Of the 'Arrivant': Hauntology Of An Interethnic Conflict In Afghanistan. *Asian Anthropology* 17 (2018): 165-184.

Chapman, Anne W. *The National Training Center Matures, 1985-1993.* Ft. Monroe, VA: United States Army Training and Doctrine Command, 1997.

Chapman, Anne W., Lilly J. Carol, John L. Romjue and Susan Canedy. *Prepare the Army for War: A Historical Overview of the Training and Doctrine Command, 1973-1998.* Ft. Monroe, VA: U.S. Army Training and Doctrine Command, 1998.

Charlston, Jeffrey. *Department of the Army Historical Summary, Fiscal Year 1999.* Washington, DC: Center of Military History, 2006.

Cherry, Timothy D. "Future Command and Control Systems: IVIS and B^2C^2 Only Scratch the Surface." *Armor* 103 (November-December 1994): 16-19.

Chief, National Guard Bureau. *Annual Review of the Chief, National Guard Bureau, 1993.* Washington, DC: National Guard Bureau, 1993.

———. *Annual Review of the Chief, National Guard Bureau, 1995.* Washington, DC: National Guard Bureau, 1995.

———. *Annual Report of the Chief, National Guard Bureau, 2003.* Washington, DC: National Guard Bureau, 2003.

———. *Annual Report of the Chief, National Guard Bureau, 2005.* Washington, DC: National Guard Bureau, 2005.

Chiarelli, Peter. "The 100 Most Influential People: Martin Dempsey." *Time,* April 16, 2015.

Chivington, John M. "To the People of Colorado. Synopsis of the Sand Creek Investigation, Denver, Colorado, June 1865."

Cima, Robert J., ed. *Vietnam: A Country Study.* Washington, DC: Department of the Army, 1989.

Clark, Forest J, "TACFIRE and the Maneuver Commander." *Field Artillery Journal* 49 (January-February 1981): 14-20.

Clarke, Jeffrey J. *The United States Army in Vietnam, Advice and Support: The Final Years, 1965-1973.* Washington, DC, Center of Military History, 1988.

Clayton, James D. *Refighting the Last War: Crisis and Command in Korea.* New York: Free Press, 1993.

Clendenen, Clarence C. *Blood on the Border: The United States Army and the Mexican Irregulars.* New York: McMillan, 1969.

Clinton, William J. "Exec. Order No. 13289." 3 C.F.R., 2003 Comp.

————. Presidential Decision Directive 39, 1995.

Coakley, Robert W. "The Army as a Flexible Instrument of National Policy." In *National Security in the Nuclear Age: Basic Facts and Theories,* edited by Gordon B. Turner and Richard Challener. New York: Praeger, 1960.

————. *Department of the Army Historical Summary, Fiscal Year 1982.* Washington, DC: Center of Military History, 1988.

Cody, Phillip R. *Operation Joint Guardian: The U.S. Army in Kosovo.* Washington, DC: Center of Military History, 2007.

Coffman, Edward M. *The Embattled Past: Reflections on Military History.* Lexington, KY: University Press of Kentucky, 2014.

Collins, James L., Jr. *The Development and Training of the South Vietnamese Army.* Washington, DC: Center of Military History, 1975.

Collins, Joseph J. *Understanding War in Afghanistan.* Washington, DC: National Defense University, 2011.

Conn, Stetson, ed. *American Military History.* Washington, DC: Center of Military History, 1988.

Continental Army Command. General Orders No. 21, 23 February 1973.

————. General Orders No. 59, 19 March 1973.

Correll, "What Happened to Shock and Awe?" *Air Force* 96: 52-57. November 2003.

Cosmas, Graham A. *MACV The Joint Command in the Years of Escalation, 1962–1967.* Washington, DC: Center of Military History, 2006.

————. *The United States Army in Vietnam: MACV, The Joint Command in the Years of Withdrawal, 1968-1973.* Washington, DC: Center of Military History, 2006.

Cosmas, Graham, ed. *Correspondence Relating to the War with Spain, Including the Insurrection in the Philippine Islands and the China Relief Expedition, April 15, 1898 to July 30, 1902.* Washington, DC: Center of Military History, 1993.

Cothren, Sarah and Alexander Barnes. "Long Distance Logistics: The Mexican Expedition." *Army Sustainment,* 47: 45-49. May/June 2015.

Crandall, Warren D. *History of the Ram Fleet and the Mississippi Marine Brigade in the War for the Union on the Mississippi and its Tributaries: The Story of the Ellets and their Men.* St. Louis, MO: Press of Buschart Brothers, 1907.

Craven, David. "Lineages of the Mexican Revolution (1910–1940)." *Third Text* 28: 223-34. July 2014.

Crockatt, Richard. *The Fifty Years War: The United States and The Soviet Union in World Politics, 1941-1991*. New York: Routledge, 1995.

Cronin, Audrey K. Audrey K. Cronin, "Terrorists and Suicide Attacks, August 28, 2003." Washington, DC: Congressional Research Service, 2003.

Cronin, Richard P. "Afghanistan: Challenges and Options for Reconstructing a Stable and Moderate State, May 10, 2002." Washington, DC: Congressional Reseaech Service, 2002.

Cunningham, Robert B. and Yasin K. Sarayrah. *Wasta: The Hidden Force in Middle Eastern Society*. Westport, CT: Praeger, 1993.

Curtin, J. Sean. "Japanese Institute of Global Communications Social Trends #28, Japanese Anti-War Sentiment on Iraq in Accord with Global Opinion," 24 February 2003.

Darwin, Charles. *The Origin of Species*. London, UK: John Murray, Albemarle Street, 1910.

Daugherty, Paul J. and Herbert C. Eidt. "Wound Ballistics: Minie Ball vs. Full-Metal Jacketed Bullets: A Comparison of Civil War and Spanish-American War Firearms." *Military Medicine* 174: 403-09. April 2009.

Department of the Army. "ALARACT 078/2005 DTG R 202200Z APR 05, SUB: ARMY COMBAT UNIFORM (ACU). Washington, DC: Department of the Army, 2005.

———. "HQDA EXORD 304-11, 211813SEP11Z."

———. Army Concept Team in Vietnam. *Mohawk Aircraft in the Target Acquisition Role*. San Francisco, CA: Army Concept Team.

———. "Horseblanket on Transformation." Undated.

———. Message DTG 241907MAR72Z, Subject: Installation Organization Charts.

———. *Technical Manual 11-666, Antennas and Radio Wave Propagation*. Washington, DC: Headquarters, Department of the Army, 1953.

———. *Field Manual 100-5, Field Service Regulations, Operations*. Washington, DC: Headquarters, Department of the Army, 1954 with Change 3, 1958.

———. *Field Manual 100-5, Operations*. Washington, DC: Headquarters, Department of the Army, 1962.

———. *Technical Manual TM9-1345-200: Land Mines*. Washington, DC: Department of the Army, 1964.

———. *Artillery Trends: Reference Data Issue*. Ft. Sill, OK: U.S. Army Artillery and Missile School, 1966.

———. *Report of the Department of the Army Review into the Preliminary Investigation of the My Lai Incident, Volume I: Report of the Investigation, 14 March 1970.* Washington, DC: Department of the Army, 1970.

———. *Report of the Department of the Army Review into the Preliminary Investigation of the My Lai Incident, Volume I: Report of the Investigation, 14 March 1970.* Washington, DC: Department of the Army, 1970.

———. *Field Manual 31-75, Riverine Warfare.* Washington, DC: Department of the Army, 1971.

———. Memorandum from Brigadier General Jones to Chief of Staff, CONARC, March 15, 1972.

———. Operation Steadfast Detailed Plan, Volume B, 20 July 1972.

———. Reorg one detailed plan briefing for cdc commanders' conference, 25 Julio 1972.

———. Talking Paper, Subject: TRADOC and CDC, 28 November 1972.

———. Memorandum, Subject: Guidance for Implementation, CONUS Reorganization 1973.

———. *Technical Manual 11-5855-203-10: Operator's Manual for Night Vision Sight, Individual Served Weapon AN/PVS-2, AN/PVS-2A, and AN/PVS-2.* Washington, DC: Department of the Army, 1974.

———. *Field Manual 100-5, Operations.* Washington, DC: Headquarters, Department of the Army, 1976.

———. *Field Manual 27-10 with Change 1, The Law of Land* Warfare. Washington, DC: Department of the Army, 1976.

———. *Field Manual 100-5, Operations.* Washington, DC: Headquarters, Department of the Army, 1982.

———. *Field Manual 17-15 (Test), Tank Platoon: Division 86.* Washington, DC: Headquarters, Department of the Army, 1984.

———. *Field Manual 100-5, Operations.* Washington, DC: Headquarters, Department of the Army, 1986.

———. *Field Manual 25-100, Training the Force.* Washington, DC: Headquarters, Department of the Army, 1988.

———. *Field Manual 100-15, Corps Operations.* Washington, DC: Department of the Army, 1989.

———. *Time Honored Professionals: The NCO Corps Since 1775.* Washington, DC: Center of Military History, 1989.

———. *Field Manual 100-20, Military Operations in Low Intensity Conflict.* Washington, D.C.: Departments of the Army and Air Force, 1990.

——————. *Field Manual 1-100, Doctrinal Principles for Army Aviation in Combat Operations.* Washington, DC: Headquarters, Department of the Army, 1989.

——————. *Field Manual 34-130, Intelligence Preparation of the Battlefield.* Washington, DC: Headquarters, Department of the Army, 1989.

——————. *Time Honored Professionals: The NCO Corps Since 1775.* Washington, DC: Center of Military History, 1989.

——————. *Field Manual 31-20, Doctrine for Special Forces Operations.* Washington, DC: Department of the Army, 1990.

——————. *Field Manual 3-90, Smoke Operations.* Washington, DC: Department of the Army, 1990.

——————. *Field Manual 71-100, Division Operations.* Washington, DC: Headquarters, Department of the Army, 1990.

——————. *Field Manual 100-2-3, The Soviet Army: Troops, Organization and Equipment.* Washington, DC: Department of the Army, 1991.

——————. *Field Manual 3-3, Chemical and Biological Contamination Avoidance.* Washington, DC: Department of the Army, 1992.

——————. *Field Manual 34-81-1, Battlefield Weather Effects.* Washington, DC: Department of the Army, 1992.

——————. *Reader's Guide: The United States Army in World War II.* Washington, DC: Center of Military History, 1992.

——————. *Field Manual 100-5, Operations.* Washington, DC: Headquarters, Department of the Army, 1993.

——————. *Technical Manual TM 43-0001-29, Army Ammunition Data Sheets for Grenades.* Washington, DC: Department of the Army, 1994.

——————. *Field Manual 101-5-1, MCRP 5-2A: Operational Terms and Graphics.* Washington, DC: Headquarters, Department of the Army; United States Marine Corps, 1997.

——————. *Army Regulation 600-20, Army Command Policy.* Washington, DC: Department of the Army, 1999.

——————. *Operational and Organizational Concept for Future Combat Systems, Chapter 1: The Army's Transformation Strategy, 2 November 1999 Draft* (Washington, D.C.: Department of the Army, 1999.

——————. Briefing: "Army Transformation, 17 October 2000." 2000.

——————. Briefing: "The Army Vision." Washington, D.C.: Office of the Secretary of the Army and the Chief of Staff, 2000.

——————. *Field Manual 3-0, Operations.* Washington, DC: Headquarters, Department of the Army, 2001.

———. *Field Manual 7-1 (25-101): Battle Focused Training.* Washington, DC: Department of the Army, 2003.

———. *Army Regulation 614-200, Enlisted Assignments and Utilization Management.* Washington, DC: Department of the Army 2004.

———. *Field Manual 1-02: Operational Terms and Graphics.* Washington, DC: Department of the Army, 2004.

———. General Order Number 2, January 31, 2006.

———. HQDA EXORD 158-06 ISO Army Asymmetric Warfare Program Implementation, 13 April 2006.

———. *Army Asymmetric Warfare Office (AAWO) 2007 Annual Review.* Washington, D.C.: Department of the Army, 2007.

———. *Field Manual 3-0, Operations.* Washington, DC: Headquarters, Department of the Army, 2011.

———. General Order 2011-06, dated September 21, 2011.

———. Memorandum for Commanding General, U.S. Army Training and Doctrine Command, Assistant Chief of Staff G-3/5/7, Subject: Asymmetric Warfare Group, dated September 23, 2011.

———. *Army Doctrinal Reference Publication 3-90, Offense and Defense.* Washington, DC: Headquarters, Department of the Army, 2012.

———. *Army Regulation 350-1: Army Training and Leader Development.* Washington, DC: Department of the Army, 2014.

———. *Army Regulation 1-1: Planning, Programming, Budgeting and Execution.* Washington, DC: Headquarters, Department of the Army, 2016

———. *Army Doctrine Publication Number 5-0, The Operations Process.* Washington, DC: Department of the Army, 2019.

———. *Army Doctrinal Publication 7-0: Training.* Washington, DC: Department of the Army, 2019.

———. *Army Regulation 600-8-22, Military Awards.* Washington, DC: Department of the Army, 2019.

———. *Field Manual 6-27, The Commander's Handbook on the Law of Land Warfare.* Washington, DC: Department of the Army, United States Marine Corps, 2019.

———. *"Report of the Fort Hood Independent Review Committee, November 6, 2020. Washington, DC: Department of the Army, 2020.*

Department of Defense. *Department of Defense Reorganization Order 10,* dated 10 January 1962. Arlington, VA: Department of Defense, 1962.

———. *Department of Defense Instruction (DODI) 1304.27.* Washington, DC: Department of Defense, 1996.

———. *Department of Defense Financial Management Regulation 7000.14-R, Volume 7, Chapter 8, March 2005.* Washington, DC: Department of Defense, 2005.

———. "Measuring Stability and Security in Iraq February 2006 Report to Congress In accordance with the Department of Defense Appropriations Act 2006 (Section 9010)." Washington, DC: Department of Defense, 2006.

———. Department of Defense, "DoD Directive 2000.19E, Joint IED Defeat Organization, 14 February 2006." Washington, DC: Department of Defense, 2006.

———. DoD Directive 6495.01, Sexual Assault Prevention and Response Program. Washington, DC: Department of Defense, 2013.

Department of the Interior. *Letter of the Secretary of the Interior, Communicating, In Compliance with the Resolution of the Senate of the 14th Ultimo, Information Relating to the Late Battle on the Washita River, January 6, 1869.*

Department of State. *Papers Relating to the Foreign Relations of the United States with the Address of the President to Congress December 5, 1916.* Washington, DC: U.S. Government Printing Office, 1925.

———. *Papers Relating to the Foreign Relations of the United States, The Lansing Papers, 1914-1920.* Washington, DC: U.S. Government Printing Office, 1939.

———. *Foreign Relations of the United States, 1952–1954, East Asia and the Pacific, Volume XII, Part 1.* Washington, DC: Department of State, 1984.

Devlin, John F. "The Baath Party: Rise and Metamorphosis." *The American Historical Review* 96: 1396-407. 1991.

DiGiovanna, Sean M., Ann Markusen, and Michael C. Leary. *From Defense to Development? International Perspectives on Realizing the Peace Dividend.* Abingdon, Oxon: Taylor & Francis Group, 2003.

Dippie, Brian W. *The Vanishing American: White Attitudes and U.S. Indian Policy.* Middletown, CT: Wesleyan University Press, 1982.

Dobak, Wiliam A. *Freedom by the Sword: The U.S. Colored Troops, 1862-1867.* Washington, DC: Center of Military History, 2011.

Dolan, R. *Philippines: A Country Study.* Washington, DC: Federal Research Division, Library of Congress, 1993.

Donaldson, William H., ed. *Biographical Register of the Officers and Graduates Of The U. S. Military Academy At West Point, New York Since Its Establishment In 1802: Supplement, Volume VII 1920-1930.* Chicago, IL: R. R. Donnelley & Sons Company, Lakeside Press, 1930.

Donnelly, William M. *Department of the Army Historical Summary, Fiscal Year 2002.* Washington, DC: Center of Military History, 2013.

———. *Department of the Army Historical Summary, Fiscal Year 2004.* Washington, DC: Center of Military History, 2015.

Donovan, Gil. "Taliban to Require Hindus to Display Identity Labels," *National Catholic Registerer,* 3 June 2001.

Doubler, Michael D. *I Am the Guard: A History of the Army National Guard, 1636*-2000. Washington, DC: National Guard Bureau, 2001.

Doughty, Robert A. *The Evolution of U.S. Army Tactical Doctrine, 1946-1947.* Ft. Leavenworth, KS: Combat Studies Institute, 1979.

Downs, Gregory D. *After Appomattox: Military Occupation and the Ends of War.* Cambridge: Harvard University Press, 2015.

Egerton, Douglas R. *The Wars of Reconstruction: The Brief, Violent History of America's Most Progressive Era.* New York: Bloomsbury, 2014.

Eland, Ivan, Robert Newman, and Jeffrey White. Occupied Iraq: One Country, Many Wars." *Middle East Policy* 12 (Fall 2005), 1-28.

Ellsworth, Daniel. "Counterinsurgency Tactics Led to Haditha, My Lai." *New Perspective Quarterly* 23 (2006): 49-53.

Eppel, Michael. *Iraq From Monarchy to Tyranny: From the Hashemites to the Rise of Saddam.* Gainesville, FL: University Press of Florida, 2004.

Exercito de Portugal, *Armas Combinas: Revista Militar de Escola mas Armas.* Mafra, PG: Escola mas Armas, 2016.

Fatima, Qamar. "Pakistan's Religious Elite and Afghanistan: Antecedents." *Journal of Political Studies* (2018): 39.

Federal Research Division, Library of Congress. *Afghanistan: A Country Study.* Washington, DC: Library of Congress, 2001.

Fehrenbach, T. R. *This Kind of War: A Study in Unpreparedness.* Washington, DC: Brassey's, 1986.

Feickert, Andrew. *CRS Report for Congress, U.S. Army's Modular Redesign: Issues for Congress (Updated February 10, 2005).* Washington, DC: Congressional Research Service, 2005.

———. *CRS Report for Congress, U.S. Army and Marine Corps Equipment Requirements: Background and Issues for Congress.* Washington, DC: Congressional Research Service, 2006.

———. "U.S. and Coalition Military Operations in Afghanistan: Issues for Congress Updated December 11, 2006." Washington, DC: Congressional Research Service, 2006.

Ferrell, Robert. *Reminiscences of Conrad S. Babcock: The Old U.S. Army and the New, 1898-1918.* Columbia, MO: University of Missouri Press, 2012.

Feuer, A. B. *America at War: The Philippines, 1898-1913.* Westport, CT: Greenwood Publishing Group, 2002.

Fischer, Hannah. "United States Military Casualty Statistics: Operation Iraqi Freedom and Operation Enduring Freedom, June 8, 2006." Washington, DC: Congressional Research Service, 2006.

Fontana, Guiditta. "Creating Nations, Establishing States: Ethno-Religious Heterogeneity and the British Creation of Iraq in 1919-23." *Middle Eastern Studies* 46, (2010): 1-16.

Fontenot, Gregory, J. C. Egan, and David Tohn. *On Point: The U.S. Army in Operation Iraqi Freedom.* Ft. Leavenworth, KS: Combat Studies Institute, 2004.

Foulois, Benjamin with C. V. Glines. *From the Wright Brothers to the Astronauts: The Memoirs of Major General Benjamin D. Foulois.* New York: McGraw-Hill Book Company, 1968.

Fox, J. Ronald. *Defense Acquisition Reform, 1960-2009: An Elusive Goal.* Washington, DC: Center of Military History, 2011.

Franks, Tommy. *American Soldier.* New York: Harper Collins, 2004.

Frederiksen, John C. *The U.S. Air Force: A Chronology.* Santa Barbara, CA: Greenwood Publishing, 2011.

Fritz, David L. "Before the "Howling Wilderness": The Military Career of Jacob Hurd Smith, 1862-1902." *Military Affairs* 43 (1979): 186-90.

Fromkin, David. *A Peace to End All Peace: The Fall of the Ottoman Empire and the Creation of the Modern Middle* East. New York: Avon, 1989.

Fulton, William B. *Vietnam Studies: Riverine Operations, 1966-1969.* Washington, DC: Department of the Army, 1973.

Galula, David. *Counterinsurgency Warfare: Theory and Practice.* New York: Praeger, 2005.

Gammons, Stephen Y. L. and William Donnelly. *Department of the Army Historical Summary, Fiscal Year 1995.* Washington, DC: Department of the Army, 2004.

Gates, John M. *Schoolbooks and Krags: The United States Army in the Philippines, 1898*-1902. Westport, CT: Greenwood Press, 1973.

———. "War-Related Deaths in the Philippines, 1898-1902." *Pacific Historical Review* 53: 367-78. 1984.

Gavin, Francis J. "The Myth of Flexible Response: United States Strategy in Europe during the 1960's." *The International History Review* 23: 847-75. December 2001.

Gawrych, George W. *Leavenworth Papers Number 21, The 1973 Arab-Israeli War: The Albatross of Decisive Victory.* Ft. Leavenworth, KS: Combat Studies Institute, 1996.

Geibel, Adam. "Operation Anaconda, Shah-i-Khot Valley, Afghanistan, 2-10 March 2002," *Military Review* 82: 72-77. May-June 2002.

Gerhard, Spinney. “SFABs Are Crucial New Tools.” *Army* 71: 13-14. February 2021.

Government Accountability Office. *Military Transformation: Army's Evaluation of Stryker and M113-A3 Infantry Carrier Vehicles Provided Sufficient Data for Statutorily Mandated Comparison.* Washington, DC: General Accountability Office, 2003.

———. “Report to the Subcommittee on Personnel, Committee on Armed Services, U.S. Senate: Military Personnel-DoD Needs to Address Long-term Reserve Force Availability and Related Mobilization and Demobilization Issues.” Washington, DC: General Accountability Office, 2004.

Gibbs, David. “The Peasant as Counter-Revolutionary: The Rural Origins of The Afghan Insurgency.” *Studies in Comparative International Development* 21: 26-59. 1986.

Gilhadi, Rotem. *International Review of the Red Cross* 94 (885) (Occupation): 81-116. 1994.

Gillett, Mary C. *The Army Medical Department, 1865*-1917. Washington, DC: Center of Military History, 1995.

Glubb, John. *A Short History of the Arab* Peoples. New York: Dorset Press, 1969.

Gochenour, Theodore S. "A New Try for Afghanistan." *Middle East Journal* 19: 1-19. 1965.

Gole, Harry G. “Don't Kill the Messenger: Vietnam War Reporting in Context.” *Parameters* 26 (Winter 1996-1997): 148-53.

Gordon, John F., Bruce Narduli, and Walter L. Perry. “The Operational Challenges of Task Force Hawk.” *Joint Forces Quarterly* 29 (Autumn-Winter): 52-57. 2001-2002.

Grau, Lester. “Defeating Guerrilla Logistics: Soviet Operation ‘Trap’ in Western Afghanistan.” *The Journal of Slavic Military Studies* 31: 65-79. February 2018.

Greene, Richard A., Jr., “Assessing the Army National Guard's Enhanced Brigade Concept: Searching For Readiness And Relevance.” Ft. Leavenworth, KS: U.S. Army Command and General Staff College, 2003.

Gregorian, Vartan. "Mahmud Tarzi and Saraj-Ol-Akhbar: Ideology of Nationalism and Modernization in Afghanistan.” *Middle East Journal* 21: 345-68. 1967.

Gueda, Virginia. "The Process of Mexican Independence.” *American Historical Review* 105: 116-30. 2000.

Haddad, Toufic. “Insurgent Infrastructure: Tunnels of the Gaza Strip.” *Middle East - Topics & Arguments* 10: 71-85. 2018.

Hammond, William N. *Public Affairs: The Military and the Media, 1962-1968.* Washington, DC: Center of Military History, 1990.

Harper, James W. "The El Paso-Juárez Conference of 1916." *Arizona and the West* 20: 231-44. 1978.

Hart, Gideon M. "Military Commissions and the Lieber Code: Toward a New Understanding of the Jurisdictional Foundations of Military Commissions." *Military Law Review* 203: 1–77. 2010.

Hashim, Ahmed. *Insurgency and Counter-Insurgency in Iraq.* Ithaca, NY: Cornell University Press, 2011.

Hatch, David. "The Punitive Expedition: Military Reform and Communications Intelligence." *Cryptolgia* 31: 38-45. 2007.

Hay, John M. *Vietnam Studies: Tactical and Material Innovations.* Washington, DC: Center of Military History, 1974.

Haynes, Mary L. *Department of the Army Historical Summary, Fiscal Year 1987.* Washington, DC: Center of Military History, 1995.

Hayworth, R. Blair. *Department of the Army Historical Summary, Fiscal Year 1998.* Washington, DC: Department of the Army, 2005.

———. *Department of the Army Historical Summary, Fiscal Year 2000.* Washington, DC: Center of Military History, 2011.

Held, Colbert C. *Middle East Patterns: Places, Peoples, and* Politics. Boulder, CO: Westview Press, 1994.

Henning, Charles A. "Stop Loss: Key Questions and Answers." Washington, DC: Congressional Research Service, 2009.

Hewes, James E. *Special Studies: From Root to McNamara, Army Organization and Administration.* Washington, D.C: Center of Military History, 1975.

Hendrix, John W. "Commandant's Note: Infantry Force XXI—Facing the Challenge." *Infantry* 85: 1-2. 1995.

Herbert, Anthony. *Soldier.* New York: Dell, 1973.

Herbert, Paul H., *Deciding What Has to Be Done: General William E. DePuy and the 1976 Edition of FM 100-5, Operations.* Fort Leavenworth, KS: Combat Studies Institute, 1988.

Hogan, David W. Jr., Arnold G. Fisch, and Ronald K. Wright. *The Story of the Non-Commissioned Officer Corps: Backbone of the* Army. Washington, DC: Center of Military History, 2007.

Holden, W. N. "The Role of Geography in Counterinsurgency Warfare: The Philippine American War, 1899–1902." *GeoJournal* 85: 423-37. 2020.

Holman, Victor. "Negro Indians, Macabebes, and Civilian Irregulars: Models for the Future Employment of Indigenous Forces." Master's thesis, USACGSC, 1995.

Hollister, O. James and Richard Harwell, ed. *Colorado Volunteers in New Mexico: 1862.* Chicago, IL: R.R. Donnelly, 1962.

Hourani, Albert. *A History of the Arab Peoples.* New York: MFF Press, 1991.

Hsaio Rei-Hicks, Madelyn, Hamit Dardegan, Peter M. Bagnall, and John A. Sloboda. "Casualties in Civilians and Coalition soldiers from Suicide Bombings in Iraq, 2003–10: A Descriptive Study." *The Lancet* 378: 906-14. 2011.

Hughes, Christopher P. *War on Two Fronts: An Infantry Commander's War in Iraq and the Pentagon.* Havertown, PA: Casemate Publishing, 2007.

Hunt, Geoffrey. *Colorado's Volunteer Infantry in the Philippine Wars, 1898*-1899. Albuquerque, NM: University of New Mexico Press, 2006.

Hutchinson, W. H. *With General Crook in the Indian* Wars. Palo Alto, CA: Lewis Osbourne, 1968.

Hurst, James W. *Pancho Villa and Blackjack* Pershing. Westport, CT: Praeger, 2008.

Immerman, Richard H. *Empire for Liberty: A History of American Imperialism from Benjamin Franklin to Paul Wolfowitz.* Princeton, NJ: Princeton University Press, 2010.

Ireland, Alleyne. "The United States in the Philippines." T*he Atlantic Monthly*, November 1904.

Jensen, Benjamin M. *Forging the Sword: Doctrinal Change in the U.S. Army.* Stanford: Stanford University Press, 2018.

Johnson, Haynes and George C. Wilson George C. *Army in Anguish.* New York: Simon and Schuster, 1971.

Johnson, Thomas. "Taliban Adaptations and Innovations." *Small Wars & Insurgencies* 24: 3-27. 2013.

Joint Chiefs of Staff. Joint Chiefs of Staff, JCS Permanent Order J-1SO-0280-88, 18 August 1988, "Award of the Joint Meritorious Unit Award," to the 160th Special Operations Aviation Group (Airborne). 1988.

————. *Joint Publication 1-02, Department of Defense Dictionary of Military and Associated Terms, 8 November* 2010. Arlington, VA: United States Department of Defense, 2010, As Amended Through 15 February 2016.

————. *Joint Publication 1-02, Department of Defense Dictionary of Military and Associated Terms, January 2019.*

Kamolnick, Paul. *Countering Radicalization and Recruitment to Al-Qaeda: Fighting the War of Deeds.* Carlisle, PA: United States Army War College Press, 2014.

Karnow, Stanley. *Southeast Asia.* New York: Time, 1962.

————. *In Our Image: America's Empire in the Philippines.* New York: Random House, 1989.

Katzman, Kenneth. "Iraq: Compliance, Sanctions and U.S. Policy, Updated May 21, 2001." Washington, DC: Congressional Research Service, 2001.

———. "Afghanistan: Current Issues and U.S. Policy: Updated March 31, 2003," Washington, DC: Congressional Research Service, 2003.

———. "Afghanistan: Post-Taliban Governance, Security and U.S. Policy." Washington, DC: Congressional Research Service, 2003.

———. "Iran: U.S. Concerns and Policy Responses, Updated September 11, 2006. Washington, DC: Congressional Research Service, 2006.

———. "Iran's Influence in Iraq, Updated June 13, 2006. Washington, DC: Congressional Research Service, 2006.

———. "Iraq: Elections, Government and Constitution, Updated May 22, 2006." Washington, DC: Congressional Research Service, 2006.

Keaveny, Kevin. "Cavalry Afloat: The 39th Cavalry Platoon in the Mekong Delta." *Armor: The Professional Bulletin of the Armor Branch* 102 (July-August 1993): 12-19. 1993.

Kelly, Francis J. *Vietnam Studies: U.S. Army Special Forces: 1961-1971.* Washington, DC: Center of Military History, 1973.

Kennedy, Harold. "Pentagon Broadens Duties for its Civil Affairs Teams." *National Defense* 87, February 2003.

Kersey, Harry A. and Michael Petersen. "Was I a Member of Congress...": Zachary Taylor's Letter to John J. Crittenden, January 12, 1838, concerning the Second Seminole War." *Florida Historical Quarterly* 75: 447-61. 1997.

Kim, Byung-Kook and Eric Vogel, eds. *The Park Chung Hee Era: The Transformation of South Korea.* Cambridge: Harvard University Press, 2011.

Kindsvatter, Peter S. *American Soldiers: Ground Combat in the World Wars, Korea* & Vietnam. Lawrence: University Press of Kansas, 2003.

King, Benjamin, Richard Riggs, and Eric R. Criner. *Spearhead of Logistics: A History of the United States Army Transportation Corps.* Washington, DC: Center of Military History, 2001.

King, Anthony, F. G. Hoffman, David J. Galbreath, and Farrel Theo. "Military Adaptation in Afghanistan." *International Politics Reviews* 1: 100-16. December 2013.

Kirby, Reid. "Operation Snoopy: The Chemical Corps' People Sniffer. "*CML: Army Chemical Revie,* (January-June 2007): 20-22. Ft. Leonard Wood, MO: U.S. Army Maneuver Support Center, 2007.

Kitching, Paula. "The Sykes-Picot Agreement and Lines in the Sand." *Historian* 128: 18-22. 2018-2016.

Kolko, Gabriel. *Anatomy of a War: Vietnam, the United States, and the Modern Historical Experience.* New York: The New Press, 1985.

Koontz, Christopher P. *Department of the Army Historical Summary, Fiscal Year 2001.* Washington, DC: Department of the Army, 2011.

Krepenevich, Andrew F., Jr. *The Army and Vietnam.* Baltimore, MD: The Johns Hopkins University Press, 1986.

Kretchik, Walter and John T. Fishel. *Invasion, Intervention, "Intervasion:" A Concise History of the U.S. Army in Operation Uphold Democracy.* Fort Leavenworth, KS: Combat Studies Institute, 1997.

Kugler, Richard. *Operation Anaconda: A Case Study in Adaptation,* Washington, DC: Department of Defense, 2007.

Kugler, Richard, Michael Baranick, and Hans Binnendijk, "Operation Anaconda: Lessons for Joint Operations" Washington D.C., Center for Technology and National Security Policy, National Defense University, 2009.

Lackner, Helen. In *The Political Economy of Wasta: Use and Abuse of Social Capital Networking,* edited by Mohamed A. Ramady. Cham, SW: Springer, 2016.

Lambeth, Benjamin S. *Airpower Against Terror: America's Conduct of Operation Enduring Freedom.* Santa Monica, CA: RAND Corporation, 2005.

Lang, Andrew F. *In the Wake of the War: Military Occupation, Emancipation and Civil War* America. Baton Rouge, LA: Louisiana State University Press, 2017.

LaPorte, Leon J. "Transforming an Alliance while Deterring Aggression" A*rmy* 55: 89-92. October 2005.

Latham, Michael. *The Right Kind of Revolution: Modernization, Development and U.S. Foreign Policy from the Cold War to the* Present. Cornell, NY: Cornell University Press, 2011.

Lathi, Janne. *Wars of Empire: Apaches, The United States, and the Southwest Borderlands.* Norman: University of Oklahoma Press, 2017.

Larson, Stanley R. and James L. Collins, Jr. *Vietnam Studies: Allied Participation in Vietnam.* Washington, DC: Center of Military History, 1975.

Lawrence, Mark A. *The Vietnam War: A Concise International History.* Oxford: Oxford University Press, 2008.

Lechie, Robert. *The Wars of* America. New York: Harper and Row, 1981.

Lewis, Bernard. *Islam and the* West. New York: Oxford University Press, 1993.

———. *The Crisis of Islam: Holy War and Unholy Terror.* New York: Random House, 2004.

Lieberman, A. E. "Updating Impressions of the Military-Industry Complex." *California Management Review* 11 (Summer, 1969): 51.

Library of Congress, Federal Research Division. *Historical Attempts to Reorganize the Reserve Components, A Report prepared by the Federal Research Division, Library of Congress Under an Interagency Agreement with the Commission on the National Guard and Reserves, October 2007.* Washington, DC: Library of Congress, 2007.

Lieber, Francis. *Guerilla Parties, Considered with Reference to the Laws and Usages of War, Written at the Request of Major-General Henry W. Halleck, General-in-Chief of the Army of the United States.* New York: Baker and Godwin Printers, 1862.

Linn, Brian M. *Guardians of Empire: The U.S. Army and the Pacific, 1902-1940.* Chapel Hill: University of North Carolina Press, 1997.

———. *The Philippine War, 1899-1902.* Lawrence, KS: The University Press of Kansas, 2000.

———. *The U.S. Army and Counterinsurgency in the Philippine War.* Lawrence, KS: University Press of Kansas, 2000.

———. *The Echo of Battle: The Army's Way of War.* Cambridge: Harvard University Press, 2007.

Lieber, Francis. *Guerilla Parties, Considered with Reference to the Laws and Usages of War, Written at the Request of Major-General Henry W. Halleck, General-in-Chief of the Army of the United States.* New York: Baker and Godwin Printers, 1862.

Lippman, Matthew. "War Crimes: The My Lai Massacre and the Vietnam War." *San Diego Justice Journal* 1 (Summer, 1993): 295-364.

Lofgren, Stephen J. *The Highest Level of Skill and Knowledge: A Brief History of U.S. Army Civilians, 1775-2015.* Washington, DC: Department of the Army, 2016.

Logevall, Fredrik. *Choosing War: The Lost Chance for Peace and Escalation in Vietnam.* Berkeley: University of California Press, 1999.

Love, Terry. *OV-1 Mohawk in Action.* Carrollton, TX: Signal Publications, 1989.

MacCormick, Austin H., Chairman. *Report of the Special Civilian Committee for the Study of the United States Army Confinement System.* Washington, DC: Department of the Army, 1970.

Mackey, Robert R. *The Uncivil War: Irregular Warfare in the Upper South, 1861-1865.* Norman: University of Oklahoma Press, 2004.

Maggart, Lon E., "Your Mind is Your Primary Weapon." *Armor: The Professional Bulletin of the Armor Branch* 105: 5. 1996.

Magid, Paul. *The Gray Fox: George Crook and the Indian* Wars. Norman: University of Oklahoma Press, 2015.

Mahon, John. *History of the Second Seminole War, 1835-1842.* Gainesville: University Press of Florida, 1967.

Malkasian, Carter and Gerald Meyerle. *Provincial Reconstruction Teams: How Do We Know They Work?* Carlisle, PA: United States Army War College, 2009.

Mangold, Tom and John Penycate. *The Tunnels of Cu Chi: The Untold Story of Vietnam.* New York: Random House, 1985.

Mansfield, Peter. *A History of the Middle East.* New York: Viking Press, 1991.

Marshall, S. L. A. *The Soldier's Load and the Mobility of the Nation.* Arlington, VA: Association of the United States Army, 1950.

Mason, Gregory. Mason, "The Man Hunt in Mexico: The First of a Series of Articles on the Mexican Expedition," *Outlook,* April 19, 1916.

Massanat, George S. "Development and Diplomacy in Afghanistan." *Journal of Asian and African Studies* 4: 51-60. 1969.

Mason, Gregory. "The Man Hunt in Mexico: The First of a Series of Articles on the Mexican Expedition." *Outlook,* April 19, 1916.

May, Glen A. "Why the United States Won the Philippine-American War, 1899-1902." *Pacific Historical Review* 52: 353-78. 1983.

Mayo, Lida. *United States Army in World War II—The Technical Services: The Ordnance Department: On Beachhead and Battlefront.* Washington, DC: Center of Military History, 1968.

McAfee, John P. *A Slow Walk in a Sad Rain.* New York: Warner, 1993.

McCallus, Joseph P. *Forgotten Under A Tropical Sun: War Stories by American Veterans in the Philippines, 1898-1913.* Kent, OH: The Kent State University Press, 2017.

McCants, William, ed. *Militant Ideology Atlas.* West Point, NY: Combatting Terrorism Center, U.S. Military Academy, 2006.

McChristian, Joseph A. *Vietnam Studies: The Role of Military Intelligence, 1965-1967.* Washington, DC: Center of Military History, 1974.

McCoy, Alfred. "Policing the Imperial Periphery: The Philippine-American War and the Origins of U.S. Global Surveillance." *Surveillance & Society* 1: 4-26. 2015.

McKenney, Janice E. *The Organizational History of Field Artillery, 1775-2003.* Washington, DC: Center of Military History, 2007.

McLynn, Frank. *Villa and Zapata: A History of the Mexican Revolution.* New York: Basic Books, 2000.

McPherson, James. *Battle Cry of Freedom: The Civil War Era.* New York: Oxford University Press, 1988.

Menoher, Paul E. "Force XXI: Redesigning the Army Through Warfighting Experiments." *Military Intelligence Professional Bulletin* 22: 6-8. 1996.

McWilliams, Timothy and Kurtis P. Wheeler, eds., *Al-Anbar Awakening, U.S. Marines and Counterinsurgency in Iraq, Volume I: American Perspectives* (Quantico, VA: Marine Corps University, 2009.

Metz, Helen, ed. *Iraq: A Country* Study. Washington, DC: Library of Congress, 1990.

Michaels, Jeffrey H. “Helpless or Deliberate Bystander: American Policy towards South Vietnam's Military Coups, 1954–1975.” *Small Wars & Insurgencies* 25: 560-83. 2014.

Millet, F. D. "The Military Position in the Philippines." *The Speaker: The Liberal Review* 1 (October 7, 1899).

Millett, Allan R. “The General Staff and Cuban Intervention of 1906.” Military Affairs 31: 113-19. 1967.

Millett, Allan R., Peter Maslowski, and William B. Feis. *For the Common Defense: A Military History of the United States from 1607 to 2012.* New York: Free Press, 2012.

Miller, Roger. *A Preliminary to War: The First Aero Squadron and the Mexican Punitive Expedition, 1916.* Washington, DC: U.S. Air Force History and Museum Programs, 2003.

Milligan, John D. “Charles Ellet and His Steam Ram.” *Civil War History* 9: 121-32. 1963.

Moenk, Jean R. and Brooks R. Kleber. *Operation Steadfast Historical Summary: A History of the Reorganization of the U.S. Continental Army Command, 1972-1973.* Fort McPherson, GA and Fort Monroe, VA: Office of the Deputy Chief of Staff for Operations, U.S. Army Forces Command and Historical Office, Office of the Chief of Staff, U.S. Army Training and Doctrine Command, 1974.

Moore, Mike "So Where's the Peace Dividend?" *Bulletin of the Atomic Scientists* 51: 30-51. 1995.

Morai-Young, Cheryl, ed. *Department of the Army Historical Summary, Fiscal Year 1983.* Washington, DC: Department of the Army, 1990.

Morris, Edmund. *Theodore Rex.* New York: Random House, 2001.

Morrison, Samuel E. *The Oxford History of the American People.* New York: Oxford University Press, 1965.

Morrison, Samuel E., Frederick Merk, and Frank Freidel. *Dissent in Three* Wars. Cambridge: Harvard University Press, 2013.

Mossman, B. C. and Morris J. McGregor, ed. *American Military History.* Washington, DC: Center of Military History, 1989.

Moyar, Mark. “Villager Attitudes During the Final Decade of the Vietnam War.” Texas Tech University 1996 Vietnam Symposium, “After the Cold War, Reassessing Vietnam.” 1996.

———. “Political Monks: The Militant Buddhist Movement During the Vietnam War.” *Modern Asian Studies* 38: 749–84. 2004.

Mukarji, Apartim. "Advent of the Northern Alliance in Afghanistan: US Policy Examined.” *Himalayan and Central Asian Studies* 6: 35-40. 2002.

Murphey, Dwight D. "Relevant Today: Lessons from the Spanish-American War." *Journal of Social, Political, and Economic Studies* 38: b 80-98. 2013.

Nagl, John A. *Learning to Eat Soup with a Knife: Counterinsurgency Lessons from Malaya and Vietnam.* Chicago: University of Chicago Press, 2002.

National Commission on Terrorist Attacks upon the United States. *The 9/11 Commission Report: Final Report of the National Commission on Terrorist Attacks Upon the United States.* New York: Knopf & Company, 2004.

National Guard Association of the United States. "National Guard Association of the United States Report to Secretary of the Army Resor, July 28, 1967." Washington, DC: National Guard Association of the United States, 1967.

Nawyn, Kathleen J. *Department of the Army Historical Summary, Fiscal Year 2003.* Washington, DC: Center of Military History.

Nejatie, Sajjad. "Deobandi Sect: 1200 to 1900: South, Central, and West Asia." In *Cultural Sociology of the Middle East, Asia, & Africa: An Encyclopedia,* edited by Andrea L. Stanton.

Neumann, Brian F., Lisa Mundey, and Jon Mikoalshek. *Operation Enduring Freedom: The United States Army in Afghanistan, March 2002-April 2005.* Washington, DC: Center of Military History, 2012.

Newell, Richard S. *The Politics of Afghanistan.* Ithaca, NY: Cornell University Press, 1972.

Newsweek. "NATO's War Party." *Newsweek,* May 2, 1999.

Newton, Thorpe W., ed. *Federal and State Constitutions Colonial Charters, and Other Organic Laws of the States, Territories, and Colonies Now or Heretofore Forming the United States of America.* Washington, DC: Government Printing Office, 1909.

Niblack, A. P. "Operations of the Navy and Marine Corps in the Philippine Archipelago." *Proceedings* 30 (October 1904): 745-53.

Nojumi, Nematollah. *The Rise of the Taliban in Afghanistan: Mass Mobilization, Civil War, and the Future of the Region.* New York: Palgrave, 2002.

North, Don. "Holding the Line at Credibility Gap." *Vietnam,* 24 June 2011.

Nydell, Margaret. *Understanding Arabs: A Guide for Westerners.* Yarmouth, ME: Intercultural Press, 1987.

Nyrop, Richard and Donald Seekins, ed. *Afghanistan: A Country Study.* Washington, DC: Library of Congress Research Service, 1990.

Office of the Chief of Engineers, *Herbicides and Military Operations, Vol. I, Main Paper, April 11, 1972.*

Office of the Secretary of Defense, Comptroller/CFO. *United States Department of Defense Fiscal Year 2012 Budget Request: Program Acquisition Costs by Weapon System.* Washington, DC: Department of Defense, 2011.

Oland, Dwight and David W. Hogan. *Department of the Army Historical Summary, Fiscal Year 1992.* Washington, DC: 2001.

O'Rourke, Lindsey A. *Covert Regime Change: America's Secret Cold War.* Ithaca, NY: Cornell University Press, 2018.

Otsuko, Umio. "Coalition Coordination During the Boxer Rebellion: How Twenty-Seven "Councils of Senior Naval Commanders" Contributed to the Conduct of Operations." *Naval War College Review* 71: 111-30. 2018.

Ott, David. *Vietnam Studies: Field Artillery, 1954*-1973. Washington, DC: Center of Military History, 1975.

Palmer, David R. *Summons of the Trumpet: A History of the Vietnam War from a Military Man's* Viewpoint. New York: Ballantine Books, 1978.

Palmer, Frederick. *John J. Pershing, General of the Armies: A Biography.* Westport, CT: Greenwood Publishers, 1970.

Patai, Raphael. *The Arab Mind.* Long Island City, NY: Hatherleigh Press, 2002.

Perna, Laura. *The Formation of the Treaty Law of Non-International Conflicts.* Leiden, NL: Brill, 2006.

Pernin, Christopher, G., and Elliot Axelband. *Lessons from the Army's Future Combat Systems Program.* Santa Monica, CA: RAND Corporation, 2012.

Lt. Gen. Peers, William R. Report of the Department of the Army Review of the Preliminary Investigations into the My Lai Incident, Vol. I. 14 March 1970.

Perry, Walter L.; Richard E. Darilek, and Laurinda L. Rohn. *Iraqi Freedom: Decisive Victory, Elusive* Peace. Santa Monica, CA: RAND Corporation, 2015.

Pershing, John J., and John T. Greenwood, ed. *My Life Before the World War, 1860-1917.* Lexington, Kentucky: University Press of Kentucky, 2013.

Person, Gustav. *BRAC at Fort Belvoir, 1988-2001.* Alexandria: Global Publishing, 2012.

Phillips, Kate, Shane Lauth, Erin Schenk and Terrill W. Andrew, ed. "Colloquium Report: U.S. Military Operations in Iraq: Planning, Combat and Occupation. Carlisle, PA: U.S. Army War College, 2006.

Phillips, R. Cody. *Operation Just Cause: The Incursion into Panama.* Washington, DC: Center of Military History, 2004.

————. *Operation Joint Guardian: The U.S. Army in Kosovo.* Washington, DC: Center of Military History, 2007.

————. *Bosnia-Herzegovina: The U.S. Army Role in Peace Enforcement Operations, 1995-2004.* Washington, DC: Center of Military History, 2015.

Pirnie, Bruce; John Gordon IV and Richard R. Brennan, Jr. *Operation Iraqi Freedom: Decisive War, Elusive Peace.* Santa Monica, CA: The RAND Corporation, 2015.

Pirnie, Bruce and Edward O'Connell. *Counterinsurgency in Iraq, 2003-2006.* Santa Monica, CA: RAND Corporation, 2008.

Pound, Edward T. and Jennifer Jack. "The Iran Connection." *U.S. News & World Report* 137, 32–48. 22 November 2004.

Program Manager, Future Combat System. "White Paper: To Provide an Overview of the Future Combat Systems (FCS) Program, 30 September 2004." Warren, MI: Program Manager, Future Combat Systems, 2004.

Prieto, Julie I. *The Mexican Expedition: 1916-1917.* Washington, DC: Center of Military History, 2016.

Price, George F. *Across the Continent with the Fifth Cavalry.* New York: Antiquarian Press, 1959.

Pritchett, Charles B. "End-of-Tour Interview with Joseph W. Kinzer, Advisory Team 163, Liaison Officer, 3d Airborne Brigade, Airborne Division, 11 June 1968." 1968.

Prugh, George S. *Vietnam Studies: Law at War, 1964*-1973. Washington, DC: Center of Military History, 1975.

Raines, Rebecca R. *Getting the Message Through: A Branch History of the U.S. Army Signal Corps.* Washington, DC: Department of the Army, 1996.

Reardon, Mark J., and Jeffrey A. Charlston. *From Transformation to Combat: The First Stryker Brigade at War.* Washington, DC: The Center of Military History, 2007.

Rebassa, Angel, Peter Chalk, Kim Craigin, Heather S. Gregg, et al. *Beyond Al Qaeda: Part I, The Global Jihadist Movement.* Santa Monica: RAND Corporation, 2006.

Reimer, Dennis J., and James J. Carfano, ed. *Soldiers are our Credentials: The Collected Works and Selected Writings of the Thirty-Third Chief of Staff, United States Army.* Washington, DC: Center of Military History, 2000.

Romjue, John L. *From Active Defense to Airland Battle: The Development of Army Doctrine, 1973-1982.* Fort Monroe, VA: Historical Office, United States Army Training and Doctrine Command and Center of Military History, 1984.

———. *The Army of Excellence: The Development of the 1980's Army.* Fort Monroe, VA and Washington, D.C: Office of the Command Historian, United States Army Training and Doctrine Command and Center of Military History, 1997.

Rudd, Gordon. *Humanitarian Intervention: Assisting the Iraqi Kurds in Operation Provide Comfort, 1991.* Washington, DC: Center of Military History, 2004.

Sawicki, James P. *Infantry Regiments of the United States Army.* Dumfries, VA: Wyvern Publications, 1981.

———. *Cavalry Regiments of the United States Army.* Dumfries, VA: Wyvern Publications, 1985.

Sawyer, Robert K. *Military Advisors in Korea: KMAG in Peace and* War. Washington, DC: Office of the Chief of Military History, 1961.

Scarlata, Paul. "United States .38 Cal. Military Revolvers Part I: Long Denigrated Rut Still Serving Proudly. Like Other Nations, The USA Moved Its Sidearms Down in Caliber at the end of the 19th Century. Results Were Mixed, To Say The Least." *Firearms News*, April 10, 2017.

———. "United States .38 Cal. Military Revolvers Part I: Long Denigrated Rut Still Serving Proudly. Like Other Nations, The USA Moved Its Sidearms Down in Caliber at the end of the 19th Century. Results Were Mixed, To Say The Least." *Firearms News*, April 10, 2017.

Scheips, Paul J. *The Role of Federal Forces in Domestic Disorders, 1945-1992.* Washington, DC: Center of Military History, 2012.

Schoultz, Lars. *Beneath the United* States. Cambridge: Harvard University Press, 1998.

———. "Blessings of Liberty: The United States and the Promotion of Democracy in Cuba." *Journal of Latin American Studies* 34: 397–425. 2002.

Schubert, Frank N., and Teresa L. Krause, eds. *The Whirlwind War: The United States Army in Operations Desert Shield and Desert Storm.* Washington, DC: Center of Military History, 1995.

Scott, Joseph C. "The Infernal Balloon: Union Aeronautics During the American Civil War." *Army History: The Professional Bulletin of Army History* PB 20-14-4: 6-24. 2014.

Segovia, Lazaro and Frank de Thoma, trans. *The Full Story of Aguinaldo's Capture.* Manila: MCS Enterprises, 1969.

Sherry, Mark D. The Army Command Post and Defense Reshaping 1987–1997. Washington, DC: Center of Military History, 2008.

Sharp, Jeremy M. and Christopher M. Blanchard, "Post-War Iraq: Foreign Contributions to Training, Peacekeeping, and Reconstruction Updated November 1, 2005." Washington, DC: Congressional Research Service, 2005.

Shreve, Brian. In *Operational and Medical Management of Explosive and Blast Incidents*, edited by David W. Callaway and Jonathan L. Burstein. Cham, SW: Springer, 2020.

Singletary, Otis A. *Negro Militia and Reconstruction.* Westport, CT: Greenwood Press, 1984.

Skidmore, Thomas E. and Peter H. Smith. *Modern Latin America.* Oxford: Oxford University Press, 1984.

Skrabec, Quentin R., Jr. "Highland Park Ford Assembly Line (1910)." *The 100 Most Significant Events in American Business: An Encyclopedia.* Santa Barbara, CA: Greenwood, 2012.

Smiley, Will. "Lawless Wars of Empire? the International Law of War in the Philippines, 1898–1903." *Law and History Review* 36: 511-50. 2018.

Smyth, Daniel. "Avoiding Bloodshed? U.S. Journalists and Censorship in Wartime." *War and Society* 32: 64-94. 2013.

Smythe, Donald. *Guerrilla Warrior: The Early Life of John J. Pershing.* New York: Charles Scribner's Sons, 1973.

Sorley, Lewis. *Thunderbolt, From the Battle of the Bulge to Vietnam and Beyond: General Creighton Abrams and the Army of his Times.* New York: Simon and Schuster, 1994.

Speckhard, Anne. "Female Suicide Bombers in Iraq." *Democracy and Security* 5: 19-50. 2009.

Spiller, Roger J. *"Not War, But Like War": The American Intervention in Lebanon.* Ft. Leavenworth, KS: Combat Studies Institute, 1981.

Stampp, Kenneth M. *The Era of Reconstruction: 1865-1877.* New York: Alfred A. Knopf, 1966.

Stanton, Doug. *Horse Soldiers: The Extraordinary Story of a Band of U.S. Soldiers Who Rode to Victory in Afghanistan.* New York: Scribner, 2009.

Stanton, Shelby. *Vietnam Order of Battle.* New York: New Books, 1981.

———. *World War II Order of Battle.* New York: Galahad Press, 1984.

———. *The Rise and Fall of an American Army: U.S. Ground Forces in Vietnam, 1965-1973.* Novato, CA: Presidio Press, 1985.

Starita, Joe. "The Case of Standing Bear: Establishing Personhood under the Law." *Court Review* 45: 4–11. 2008.

Stewart, Richard W., ed. *The United States Army in Somalia, 1992-1994.* Washington, DC: The Center of Military History, 2002.

———. *Operation Enduring Freedom: The United States Army in Afghanistan, October 2001-March 2002.* Washington, DC: The Center of Military History, 2004.

———. *American Military History, Volume I: The United States Army and the Forging of a Nation, 1775-1917.* Washington, DC: The Center of Military History, 2005.

———. *War in the Persian Gulf: Operations Desert Shield and Storm, August 1990-March 1991.* Washington, DC: Center of Military History, 2005.

———. *American Military History, Volume II: The United States Army in a Global Era, 1917-2008.* Washington, DC: The Center of Military History, 2010.

Stewart, Richard W. and Edgar Raines. *Operation Urgent Fury: The Invasion of Grenada.* Washington, DC: Center of Military History, 2008.

Storey, Margaret M. "The Military and Reconstruction, 1862–77." In *A Companion to American Military History,* edited by J.C. Bradford. Hoboken, NJ: Wiley, 2009.

Summers, Harry. *On Strategy: A Critical Analysis of the Vietnam War.* Novato, CA: Presidio Press, 1982.

Summers, Mark W. *The Ordeal of the Reunion: A New History of Reconstruction.* Chapel Hill: NC University of North Carolina Press, 2014.

Sutherland, Daniel E. *A Savage Conflict: The Decisive Role of Guerrillas in the American Civil War.* Chapel Hill: University of North Carolina Press, 2009.

———. *American Civil War Guerrillas: Changing the Rules of* Warfare. Santa Barbara, CA: Praeger, 2013.

Taylor, John R. M. *The Philippine Insurrection Against the United States: A Compilation of Notes and Introduction with an Introduction by Renato Constanatino.* Pasay City, Philippines: Eugenio Lopez Foundation, 1971.

Taylor, Maxell D. *Swords and Plowshares.* New York: W.W. Norton and Company, 1972.

The Institute of Heraldry. "Memorandum for Department of the Army G1, Uniform Policy Section, Subject: SSI Request for AWG, August 23, 2006."

———. Undated memorandum, Suggested Design, Shoulder Sleeve Insignia, Asymmetric Warfare Group.

———. Memorandum for Commander, Asymmetric Warfare Group, Subject: Shoulder Sleeve Insignia (SSI) for the Asymmetric Warfare Group, August 29, 2006.

Theis, Wallace J. "Twenty-first Century Army." *Parameters* 21: 62-76. 1991.

Times Magazine. "The Press: Farewell to the Follies." February 12, 1973.

Toczek, David M. "They Did Everything But Learn From It: The Battle of Ap Bac, 1963." Master's thesis, Texas Tech, 1999.

Thoman, Roy E. "Iraq Under Baathist Rule," *Current History* 62: 31. 1972.

Townsend, Kellard N., Jonathan E. Marion, Joseph W. Boeler, and Madison Carney. "Hoosier Redlegs Train Afghan Kandaks: Semper Gumbi." *FA Journal* 73: 32-34. 2005.

Trauschweizer, Ingo. *The Cold War U.S. Army: Building Deterrence for Limited War* (Lawrence: University Press of Kansas), 180-83. 2008.

Trofimov, Yaroslav. *Faith at War: A Journey to the Frontlines of Islam, From Baghdad to Timbuktu.* New York: Henry Holt, 2005.

Tuckman, Bruce W. "Developmental Sequence in Small Groups." *Psychological Bulletin* 63: 384–399. 1965.

Tosini, "Al-Qaeda's Strategic Gamble: The Sociology of Suicide Bombings in Iraq," *Canadian Journal of Sociology* 35: 271+. 2010.

Turse, Nicholas. *Kill Anything That Moves.* New York: Metropolitan, 2013.

Ulack, Christopher. "Demography of Race and Ethnicity in Iraq." In *The International Handbook of the Demography of Race and Ethnicity,* edited by

David Saenz, David G. Embrick and Nestor P. Rodriguez. Dordrecht, NL: Springer Netherlands, 2015.

Ullman, Harlan K. *Shock and Awe: Achieving Rapid Dominance.* Washington, DC: National Defense University Press, 1996.

U.N. Security Council, Report of the Security Council 1 August 2002-31 July 2003, General Assembly Official Records, Fifty-eighth Session, Supplement No. 2 (A/58/2). New York: United Nations, 2003.

Urwin, Gregory J. W. *The United States Infantry: An Illustrated History, 1775-1918.* London: Blandford Press, 1988.

U.S. Army Asymmetric Warfare Group. "Draft Operational and Organizational (O&O) Concept for the Asymmetric Warfare Group as of 2 Mar 05." Ft. Meade, MD: USAAWG, 2005.

————. Briefing: "Honorable Francis J. Harvey, Secretary of the Army, AWG Final Stationing Briefing, 7 February 2007." 2007.

————. Briefing, "Azimuth Check for COA Development to BG (P) Halverson and BG Wendell, December 7, 2007." 2007.

————. Briefing: "AWG PAO Requested, Solution Development 2006 – 2016, S32 Search Results (Post 2/26/2016), Pre-FRAGO Results, (as of 3/14/2016)." 2016.

————. "Information Paper, Subject: Asymmetric Warfare Group History, June 16, 2017." 2017.

————. "Program for the Change of Command Ceremony: Asymmetric Warfare Group, 14 July 2009." 2009.

————. AWG Update-20 OCT 10. 2010.

————. "SECARMY Decision Brief: Future of the Asymmetric Warfare Group, XX MAY 2013." 2013.

U.S. Air Force. *Operation Anaconda: An Airpower Perspective.* Washington, DC: Department of the Air Force, 2005.

U.S. Army Field Support Command, Command Historian. "U.S. Army Field Support Command and Joint Munitions Command Support to Operation Iraqi Freedom, Phases I-III." Rock Island Arsenal, IL: U.S. Army Field Support Command, 2005.

U.S. Army War College. *Study on Military Professionalism.* Carlisle: United States Army War College, 1970.

————. *How the Army Runs: A Senior Leader Reference Handbook, 2011-2012.* Carlisle, PA: U.S. Army War College, 2011.

U.S. Army Training and Doctrine Command. Briefing: Force XXI Heavy Division "F" Series TOEs—As of 3 MAR 97. Ft. Monroe: U.S. Army Training and Doctrine Command, 1997.

———. Briefing: *The Operational Environment—Where We've Come From,* September 8, 2007. Ft. Monroe, VA: U.S. Army Training and Doctrine Command, 2007.

———. TRADOC TASKORD IN 1279-212, 061410OCT11. 2011.

———. FRAGO 1 to TASKORD IN 1279-212 dated 14 OCT 11. 2011.

U.S. Congress. 42nd Cong. 1st Sess. *S. Rep 1-41, Select Committee of the Senate to Investigate Alleged Outrages in the Southern States Ordered Printed March 10, 1871.*

———. 42nd Cong. 2nd Sess. *Testimony Taken by the Joint Select Committee to Enquire into the Condition of Affairs in Late Insurrectionary States: Alabama, 3 (1872).*

———. 42nd Cong. 2nd Sess. *Testimony Taken by the Joint Select Committee to Enquire into the Condition of Affairs in Late Insurrectionary States: Mississippi, (1872).*

———. 42nd Cong. 2nd Sess. *Testimony Taken by the Joint Select Committee to Enquire into the Condition of Affairs in Late Insurrectionary States: North Carolina, (1872)*

———. 42nd Cong. 2nd Sess. *Testimony Taken by the Joint Select Committee to Enquire into the Condition of Late Insurrectionary States: South Carolina, (1871).*

———. 55th Cong. 3rd Sess. *Ch. 352. An Act for Increasing the Efficiency of the Army of the United States, and for Other Purposes.*

———. 56th Cong. 2nd Sess. *H. Rep. Doc. No. 2, Annual Reports of the War Department for the Fiscal Year Ended June 30, 1900.*

———. 56th Cong. 2nd Sess. *CHS, 191-92 (1901).*

———. 56th Cong. 2nd Sess. *H. Rep. 2010: Efficiency of the Military Establishment.*

———. 57th Cong. 1st Sess. *H. Rep. Doc. No. 2, Annual Reports of the War Department for the Fiscal Year Ended June 30, 1901: Report of the Philippine Commission.*

———. 58th Cong. 2nd Sess. *H. Rep. Doc. No. 2, Annual Reports of the War Department for the Fiscal Year Ended June 30, 1903: Volume VI, Report of the Philippine Commission.*

———. 89th Cong. 1st Sess. *Country Reports on Human Rights Practices, 1982: Report Submitted To The Committee On Foreign Relations, U.S. Senate And Committee On Foreign Affairs, U.S. House Of Representatives By The Department Of State In Accordance With Sections 116(D) And 502b(B) Of The Foreign Assistance Act Of 1901, As Amended.*

———. 96th Cong. 2nd Sess. *The Political Economy of the Middle East: 1973-1978, A Compendium of Papers Submitted to the Joint Economic Committee, Congress of the United States, April 21, 1980.*

———. 105th Cong. 2nd Sess. *Country Reports on Human Rights Practices For 1997: Report Submitted To The Committee On International Relations U.S. House Of Representatives And Time Committee On Foreign Relations U.S. Senate By The Department Of State.*

———. 106th Cong. 1st Sess., S. RES. 68 (1999) *Expressing the Sense of the Senate Regarding the Treatment of Women and Girls by the Taliban in Afghanistan.*

———. Congressional Budget Office. "An Analysis of the U.S. Military's Ability to Sustain an Occupation of Iraq, September 3, 2003."

———. House of Representatives. 105th Cong. 2nd Session, *H. Con. Res 336, Condemning the Taliban Regime and Supporting a Broad Based Government in Afghanistan.*

———. House of Representatives, 106th Cong., 1st Sess. *H. Res 187, Expressing the Sense of the House Of Representatives that the United States Should Seek to Prevent any Taliban-Led Government in Afghanistan from Obtaining A Seat In The United Nations, and Should Refuse to Recognize any Afghan Government, While Gross Violations of Human Rights Persist Against Women and Girls There.*

———. House of Representatives, House Document 106-90: *A Response to the Actions and Policies of the Afghan Taliban.*

———. House of Representatives, *107 H. Con. Res. 145.*

———. *Iraqi Documents: A Glimpse into the Regime of Saddam Hussein: Hearing before the Subcommittee on Oversight and Investigation of the Committee on International Relations, House of Representatives, 109th Cong., 203 (2006).*

U.S. Government. "Text of the First Reconstruction Act, March 2, 1867." 1867.

———. Government Accountability Office. *Pub. L. No. 235, 61 Stat. 496 (1947): National Security Act.* 1947.

———. *Pub. L. No. 81-216, 61 Stat 496 (1949): National Security Act Amendments of 1949.* 1949.

———. U.S. Congress, 10 U.S.C., The Army, §7033. 1956.

———. *Report to the Chairman, Committee on Armed Services, House of Representatives: Kosovo Air Operations—Army Resolving Lessons Learned Regarding the Apache Helicopter.* Washington, DC: Government Accountability Office, 2001.

———. *Pub. L. No. 109-148 (2006): Department of Defense, Emergency Supplemental Appropriations to Address Hurricanes in the Gulf of Mexico, and Pandemic Influenza Act, 2006.* 2006.

———. 119 Stat. 2708, Sec. 8043 (c) (2).

———. *Pub. L. No. 111-84 (2010): National Defense Authorization Act for Fiscal Year 2010.* 2010.

United States Senate. *Hearing Before the Committee on the Philippines, United States Senate, May 5, 1902.*

Van Deman, Ralph H. and Ralph E. Weber, ed. *The Final Memoranda: Major General Ralph H. Van Deman, USA (Ret), 1865-1952, The Father of U.S. Military Intelligence.* Wilmington, DE: SR Books, 1988.

Vick, Aland, David Orlsesky. *The Stryker Brigade Combat Team: Rethinking Strategic Responsiveness and Assessing Deployment Options.* Santa Monica, CA: RAND Corporation, 2002.

Vizenour, Gerald. *Bear Island: The War at Sugar Point.* Minneapolis: University of Minnesota Press, 2006.

von Clausewitz, Carl, Michael Howard, and Peter Paret, ed. and trans. *On War.* Princeton, NJ: Princeton University Press, 1976.

Wade, Gary W. *Rapid Deployment Logistics: Lebanon, 1958.* Ft. Leavenworth. KS: Combat Studies Institute, 1984.

War Department. General Orders Number 100. Washington, DC: Adjutant General's Office, 1863.

———. "Ordenance Report Number 15: Report of the Board of Officers Appointed in Pursuance of the Act of Congress June 6, 1872 for the Purpose of Selecting a Breach-System Muskets and Carbines for the Military Service, Together with Their Report on the Subject of Trowel-Bayonets." Washington, DC: War Department, 1873.

———. *The War of the Rebellion: A Compilation of the Official Records of the Union and Confederate Armies,* 130 vols. Washington, DC: Government Printing Office, 1880–1901.

———. General Orders No. 12, February 6, 1902.

———. Bureau of Insular Affairs. *What Has Been Done in the Philippines: A Record of Practical Accomplishments Under Civil Government.* Washington, DC: Government Printing Office, 1904.

———. "1st Aero Squadron, Signal Corps, War Diary: Period from March 12 to April 23, 1916."

———. *Annual Report of the Secretary of War to the President, 1917, 1.* Washington, DC: U.S. Government Printing Office, 1917.

———. *Annual Reports of the War Department for the Fiscal Year Ended June 30, 1901.* Washington, DC: U.S. Government Printing Office, 1901.

———. Headquarters, Department of the Pacific and Eighth Army Corps, and Office of the U.S. Military Governor in the Philippine Islands. *General Orders and Circulars, 1898-1900.* Manila, PI: Headquarters, Department of the Pacific, 1900.

———. *Official Army Register for 1900.* Washington, DC: Adjutant General's Office, 1900.

———. Headquarters, Department of Northern Luzon, General Orders No. 13, March 12, 1901.

———. Headquarters, Division of the Philippines, General Orders No. 63, August 8, 1902.

Van Fossen, Marion. Briefing: FCS Industry Day Brief, 11 January 2000. Warren, Michigan: Program Manager, Future Combat System, 2000.

Warner, Judith. *U.S. Border Security: A Reference Handbook.* Santa Barbara: ABC-CLIO, LLC, 2010.

Webb, William J., Charles Anderson, Dale Andrade, Mary Gillet, Glen Hawkins, Dave Hogan, Thomas Popa, Rebecca Raines and James Yarrison. *Department of the Army Historical Summary, Fiscal Years 1990-1991.* Washington, DC: Center of Military History, 1997.

Weidenbaum, Murray L. "The U.S. Aircraft Industry." Financial Analysts Journal 19 1963: 49-53. 1963.

Weigley, Russell F. *The American Way of War, A History of United States Military Strategy and Policy.* New York: MacMillan Publishing, 1973.

Welch, Richard E. Jr. *Response to Imperialism: The United States and the Philippine-American War, 1899-1902.* Chapel Hill: University of North Carolina Press, 1979.

Wellsome, Ellen. *The General and the Jaguar: Pershing's Hunt for Pancho Villa.* New York: Little, Brown & Co., 2003.

Westad, Odd A. *The Global Cold War.* Cambridge: Cambridge University Press, 2005.

Westmoreland, William C. *A Soldier Reports.* New York: Dell, 1976.

Westover, John, ed. *U.S. Army in Action Series: Combat Support in Korea.* Washington, DC: Center of Military History, 1955.

White, Herbert A. "The Pacification of Batangas." *The International Quarterly* 7, March 1, 1903, 431-44.

White, Paul K. "Airpower and a Decade of Containment." *Joint Force Quarterly* 35 (Winter 2000): 35-39.

Winert, Richard P. *A History of Army Aviation.* Fort Monroe, VA: Office of the Command Historian, U.S. Army Training and Doctrine Command, 1991.

Williams, James W. *A History of Army Aviation from Its Beginnings to the War on Terror.* New York: iUniverse, 2005.

Wilson, John B. *Maneuver and Firepower: The Evolution of Divisions and Separate Brigades.* Washington, DC: Department of the Army, 1998.

Wilson, Clay. "Improvised Explosive Devices (IEDs) in Iraq: Effects and Countermeasures, February 10, 2006."

———. "Improvised Explosive Devices (IEDs) in Iraq and Afghanistan: Effects and Countermeasures, September 5, 2006." Washington, DC: Congressional Research Service, 2006.

Winslow, Erving. "The Anti-Imperialist Position." *North American Review* (October 1, 1900): 460.

Woods, Kevin M., Michael Pease, Mark E. Stout, and Williamson Murray, Williamson. *Iraqi Perspectives Project: A View of Iraqi Freedom from Saddam's Senior Leadership.* Norfolk, VA: Joint Center for Operational Analysis, 2006.

Wright, Donald P. and Timothy R. Reese. *On Point II: Transition to the New Campaign, The United States Army in Operation Iraqi Freedom, May 2003-January 2005.* Ft. Leavenworth, KS: Combat Studies Institute Press, 2008.

Zeidel, Ronen. "Tribes in Iraq: A Negligible Factor in State Formation" in *Tribes and States in a Changing Middle East.* Oxford: Oxford Scholarship on Line, 2016.

Zeleznik, Frank J. and Bonnie J. McBride. *NASA Reference Publication 1094: Modeling the Internal Combustion Engine.* Cleveland, Ohio: Lewis Research Center, 1985.

Newspapers

Dallas Morning News, May 31, 2014.

Fort Campbell (KY) *Courier,* June 22, 1979.

Guardian (Manchester, UK), June 28, 1983

Korea Herald (Seoul, ROK), January 21, 2010.

Los Angeles Times, September 20, 2001; November 29, 2002.

Middletown (NY) *Times Herald-Record,* September 20, 2019.

New York Times, April 29, May 15, 29, July 12, 18, August 29, October 18, 20 1899; March 7, September 29, 1900; March 29, 1901; May 4, 30 1902; June 13, July 26, 1902; May 7, 16, 23, June 26, 29, 30, August 30, 1916; April 2, 23, 1919; November 13, 1921; January 7, 1963; July 7, 1970; July 11, 1991; March 6, June 2002; February 28, May 9, 2003; June 20, 2004; February 6, 2017.

Observer (London, UK), May 10, 1987.

San Francisco Chronicle. August 26, 2003.

United States Army-Navy Journal and Gazette, December 9, 1899.

Washington Post, February 8, 11, 17, April 20, 29, July 4, 6, 11, August 5, September 10, November 2, 28, 1899; February 6, April 20, May 19, 25, June 1, August 5, September 29, November 28, 1900; April 25, March 20, 29, June 11, December 16, 19, 1901; May 4, 6, 8, 21, 30, July 1, November 17, 1902; April 29, 1903; March 12, April 16, 1916; January 4, 1963; June 11, 1969; January 27, 29, 1992; May 17, 31, 2003; October 9, 2003; December 15, 2004; January 3, 2020.

Press Releases

Department of Defense Press Release 186-02, "DOD Identifies Soldiers Killed in Explosive Ordnance Accident," April 16, 2002.

Department of State Press Release No. 8, 1954. "Evolution Of Foreign Policy Text of Speech by John Foster Dulles, Secretary of State, Before the Council on Foreign Relations, New York, N. Y., January 12, 1954."

U.S. Army Europe and Africa, "Press Release: Ceremony to Establish V Corps Headquarters (Forward)," November 19, 2020.

Web Sources

1st Cavalry Division Association, "Vietnam War, 1965-1972." http://www.first-team.us/tableaux/chapt_08/ (accessed, 24 October 2018).

————. "256th Infantry Brigade History," https://1cda.org/history/256th -infantry-brigade-history/http://libproxy.temple.edu/login?url=https://www-proquest-com.libproxy.temple.edu/docview/443781947?accountid=14270 (accessed, 2 November 2020).

3rd Infantry Division (Mechanized) After Action Report, Operation Iraqi Freedom: Chapter 7. https://www.strategypage.com/articles/3IDAAR/chap7.asp (accessed, 17 September 2020).

4th Infantry Division, "4th Infantry Division History," https://www.carson.army.mil/units/4ID.html (accessed, 30 October 2020).

6th Cavalry Museum. "A Brief History." https://www.6thcavalrymuseum.org/regimental-history (accessed, 31 August 2020).

75th Rangers.com. "Discipline, Honor and Respect Must be Earned—They Cannot be Issued." https://www.75thrangers.com/#:~:text=The%20Black%20Beret%20has%20been%20worn%20exclusively%20by,Army%20were%20to%20remain%20silent%20on%20this%20issue (accessed, 16 December 2020).

173rd Airborne.com. "OPERATION NORTHERN DELAY," http://www.173rdairborne.com/manifest-iraq.htm (accessed, 22 September 2020).

Acquisitionnotes.com, "PPBE Process: Program Budget Decision." http://acqnotes.com/acqnote/acquisitions/program-budget-decision-pbd (accessed, 29 August 2019).

Airborne Ranger in The Sky. "MSG William "Chief" Carlson from Southern Pines, NC." http://arits.org/biographies/256-msg-william-chief-carlson (accessed, 16 July 2019).

Aldrich, Nancy W. "1st Aero Squadron." https://firstaerosquadron.com/articles/the-1st-aero-squadron-a-history (accessed, 20 January 2017).

Alexander, Bevin. "We Cannot Win the War in Iraq," https://www.bevinalexander.com/commentary/we-cannot-win-war-iraq.htm (accessed, 8 August 2020).

Army Aviation Association of America. "GEN Richard A. Cody." https://www.quad-a.org/Public/Awards/Awardees/CodyRA.aspx (accessed, 15 April 2019).

Association of 3rd Armored Division Veterans. "3AD Unit Pages." http://www.3ad.org/unit-web-sites/ (accessed, 22 October 2018).

Association of the United States Army. "Biography of Lt. Gen. Ricardo Sanchez, United States Army (Retired). https://lbausa.com/wp-content/uploads/Biography-of-Lt-General-Ricardo-Sanchez-US-Army-_Retired.pdf (accessed, 31 October 2020).

Asymmetric Warfare Group. "The Assessment and Selection Course," https://www.awg.army.mil/Recruiting/AWG-Recruiting/ (accessed, 30 July 2019).

Bay, Austin. "On Point: A Full Report on Operation Anaconda: America's First Battle of the 21st Century, A Complete After-Action Interview with Colonel Weircinski." https://www.strategypage.com/on_point/20020627.aspx (accessed, 29 September 2020).

British Broadcasting Corporation, "How Saddam Hussein Was Captured," http://news.bbc.co.uk/2/hi/middle_east/3317881.stm (accessed, 31 October 2020).

———————. "Top FARC Rebel Leader Alfonso Cano Killed in Colombia," November 5, 2011. https://www.bbc.com/news/world-15604456 (accessed, 2 February 2020).

BBC Monitoring Middle East. "Iraqi Tribes Seek Deal with Government to Delay Election," 5 January 2005. http://libproxy.temple.edu/login?url=https://www-proquest-com.libproxy.temple.edu/docview/458695511?Accountid=14270 (accessed, 4 November 2020).

BBC Monitoring South Asia. "Cold Weather Said to Hinder Taleban Operations in Afghanistan," February 20, 2005. http://libproxy.temple.edu/login?url=https://www-proquest-com.libproxy.temple.edu/docview/459924429?accountid=1427 (accessed, 2 October 2020).

Berryman, Clifford K. Cartoon:"Put 'Em Back in the Box: How Long is this to Keep Up?" https://www.loc.gov/pictures/item/2016678298/ (accessed, 29 August 2020).

Bush, George W. "Bush Ultimatum to Saddam: Text," March 17, 2003. http://news.bbc.co.uk/2/hi/americas/2859269.stm (accessed, 19 September 2020).

———. "Remarks by President Bush announcing the end of major combat operations in Iraq." https://www.cbsnews.com/news/text-of-bush-speech-01-05-2003/ (accessed, 3 November 2020).

———. "President Bush Discusses Freedom in Iraq and the Middle East." https://georgewbush-whitehouse.archives.gov/news/releases/2003/11/20031106-2.html (accessed April 9, 2020).

Byman, Danie, Kenneth M. Pollack and Gideon G. Rose. "Beef Up the Taliban's Enemy." https://www.rand.org/blog/2001/09/beef-up-the-tali bans-enemy.html (accessed, 10 September 2020).

California State Military Museum. "40th Armored Division Order of Battle, 1968. 40th Armored Division Order of Battle, 1968 (militarymuseum.org) (accessed, 28 October 2018).

———. "Californians and the Military: Major General Henry Wager Halleck." http://www.militarymuseum.org/Halleck.html (accessed, 19 March 2020).

Casey, Melanie. "Following in a Father's Footsteps: Keeping Military Service in the Family," February 16, 2011. northwestmilitary.com, http://www.northwestmilitary.com/news/focus/2011/02/northwest-military-ranger-airlifter-newspaper-JBLM-following-in-fathers-footsteps-military-service-i/ (accessed, 13 April 2020).

Censer, Marjorie. "DOD Formally Cancels FCS, ADM Details Four Follow-on Efforts," *Inside the Pentagon* 25 June 2009. https://signin-lexisnexis-com.libproxy.temple.edu/lnaccess/app/signin?back=https%3A%2F%2Fadvance.lexis.com%3A443%2Furlapi%2Flaapi%2Fsearch%3Fsource%3DMTA1NTMxMQ&aci=la (accessed, 22 January 2021).

Center of Military History. "From MAAG Indochina to MAAG Vietnam," https://history.army.mil/html/bookshelves/resmat/vietnam/part01/ch01/sec04.html (accessed, 25 April 2020).

———. "1999 Fletcher Conference, Wednesday, November 3, 1999 9:00 to 10:30 a.m., Panel 4, Serving the Nation in the 21st Century." https://www.history.army.mil/documents/Fletcher/Fletcher-99/F99-P4-T.htm (accessed, 29 October 2018).

———. "Army Chiefs of Staff," https://history.army.mil/faq/FAQ-CSA.html; (accessed, 17 January 2019).

———. "Civil War Campaigns." https://history.army.mil/html/reference/army_flag/cw.html (accessed, 21 August 2020).

———. "Lineage and Honors, Headquarters and Headquarters Battalion, 4th Infantry Division (Ivy Division). https://history.army.mil/html/forcestruc/lineages/branches/div/004id.htm.

————. "Lineage and Honors, 5th Special Forces Group, 1st Special Forces Regiment. https://history.army.mil/html/forcestruc/lineages/branches/sf/005.htm (accessed, 21 October 2018).

————. "Lineage and Honors, 7th Special Forces Group, 1st Special Forces Regiment, https://history.army.mil/html/forcestruc/lineages/branches/sf/007sfgp1sf.htm (accessed, 21 October 2018).

————. "Lineage and Honors, 10th Special Forces Group, 1st Special Forces Regiment. https://history.army.mil/html/forcestruc/lineages/branches/sf/010sfgp1sf.htm (accessed, 21 October 2018).

————. "Lineage and Honors, 19th Special Forces Group, 1st Special Forces Regiment. https://history.army.mil/html/forcestruc/lineages/branches/sf/019sfgp1sf.html (accessed, 21 October 2018).

————. "Listing of the Campaign Streamers Displayed on the U.S. Army Flag," https://history.army.mil/html/reference/campaigns.html (accessed, 13 March 2020).

————. "Index of Operational Code Names." https://www.history.army.mil/CODE.htm (accessed, 16 December 2020).

————. "Select Chronology of Army History for the Army's 240th Birthday," https://history.army.mil/html/faq/army_bday_chron.html (accessed, 31 October 2020).

————. "U.S. Advisory Support to the French." https://history.army.mil/html/bookshelves/resmat/vietnam/part01/ch01/sec04.html (accessed, 25 April 2020).

————. "United States Army Chiefs of Staff," https://history.army.mil/faq/FAQ-CSA.html (accessed, 18 December 2019).

Central Intelligence Agency. "The World Fact Book: Middle East: Iraq." https://www.cia.gov/library/publications/the-world-factbook/geos/iz.html (accessed, 23 July 2020).

Charlie Company Vietnam 1966-1972, Charlie Company, 1st Battalion, 22nd Infantry, 4th Infantry Division, "What Did Vietnam Soldiers Carry?" https://charliecompany.org/2013/01/19/what-did-vietnam-soldiers-carry/ (accessed, 21 January 2020).

Chiarelli, Peter. "The 100 Most Influential People: Martin Dempsey," *Time*, April 16, 2015, https://time.com/collection-post/3823200/martin-dempsey-2015-time-100/; (accessed, 14 April 2020).

Coalition Provisional Authority. Law of Administration for the State of Iraq for the Transitional Period, 8 March 2004, Article 59 (B). https://www.cesnur.org/2004/iraq_tal.html (accessed, 5 November 2020.

————. "Coalition Provisional Authority Order Number 2, May 23, 2003," https://permanent.fdlp.gov/lps45365/cpaord2.pdf (accessed, 29 October 2020).

Colombia Reports. "FARC," https://colombiareports.com/farc/ (accessed, 20 February 2020).

Combatting Terrorism Technical Support Office. "About IWTSD." https://cttso.gov/AboutCTTSO.html (accessed, 21 February 2020).

Defense Visual Information Distribution Service, "Smoother Transitions Planned for OIF 04-06," November 12, 2004, https://www.dvidshub.net/news/593/smoother-transitions-planned-oif-04-06#:~:text=For%20OIF%202%2C%20the%20rotation%20of%20forces%20took,spread%20over%20six%20or%20seven%20months%2C%20Carra%20said (accessed, 2 November 2020).

Department of the Army. "1st SFAB Hosts Activation Ceremony; Heraldry Announced." https://www.army.mil/article/200244/1st_sfab_hosts_activation_ceremony_ heraldry_announced (accessed, 16 December 2020).

———. 20th CBRNE Command, 20th CBRNE Command History. https://www.cbrne.army.mil/history.html (accessed, 14 June 2019).

———. 20th Support Command, "Mission." https://www.army.mil/20thcbrne#org-about (accessed, 14 April 2020).

———. Army Campaign Plan, 12 April 2004. https://digital.library.unt.edu/ark:/67531/metadc22340/m2/1/high_res_d/BRAC-2005_10368.pdf (accessed, 13 February 2020).

———. "Army Futures Command." www.army.mil/standto/archive/2018/03/28 (accessed, 10 December 2020).

———. "Army Modernization, January 16, 2018." https://www.army.mil/standto/archive/2018/01/16/?dmd (accessed, 10 December 2020).

———. "Fort Bragg, North Carolina." https://home.army.mil/bragg/index.php (accessed, 9 April 2020).

———. "Fort Carson, Colorado." https://www.carson.army.mil/allunits.html (accessed, 9 April 2020).

———. "Fort Hood, Texas." https://home.army.mil/hood/index.php/units-tenants (accessed, 9 April 2020).

———. "Fort Jackson, Victory Starts Here." https://home.army.mil/jackson/ (accessed, 6 September 2019).

———. "Fort Knox, Kentucky." https://home.army.mil/knox/index.php/about/history (accessed, 9 April 2020).

———. "Fort McCoy, Wisconsin," https://home.army.mil/mccoy/index.php/about/history (accessed, 9 April 2020).

———. "Fort George G. Meade History." https://home.army.mil/meade/index.php/about/history (accessed, 10 April 2020).

———. "History of the U.S. Army Beret," https://www.webarchive.org/web/20180704012922/ (accessed, 4 February 2021).

———. "Multi-Doman Operations." https://www.tradoc.army.mil/Publications-and-Resources/Article-Display/Article/1655556/multi-domain-operations (accessed, 16 December 2020).

———. "Permanent Order 092-02, dated 2 April 2014." https://history.army.mil/html/forcestruc/HRC/2014/092-02_20140402_HRCMD.pdf (accessed, 1 November 2019).

———. Rapid Equipping Force. "About Us." https://www.ref.army.mil/AboutUs (accessed, 18 April 2019).

———. "Security Force Assistance Command, 2nd Security Force Assistance Brigade activate at Fort Bragg." https://www.army.mil/article/214605/security_force_assistance_command_2nd_security_force_assistance_brigade_activate_at_fort_bragg (accessed, 16 December 2020);

———. How to Become an Army Ranger." https://www.goarmy.com/ranger/join-the-rangers/requirements-to-join.html (accessed, 30 July 2019).

———. Office of the Quartermaster General, "Armored Vest Fact Sheet," https://qmmuseum.lee.army.mil/korea/armored_vest.htm (accessed, 14 April 2020).

———. "Special Forces: Training—The Road to Becoming a Green Beret," https://www.goarmy.com/special-forces/training.html (accessed, 30 July 2019).

———. "U.S. Army Garrison, Fort A. P. Hill History." https://home.army.mil/aphill/index.php/about/history (accessed, 16 January 2021).

———. "U.S. Army Rangers." https://www.army.mil/rangers (accessed, 4 February 2021).

Department of Defense. "DoD Announce OEF/OIF Rotational Units." https://archive.defense.gov/Releases/Release.aspx?ReleaseID=8061 (accessed, 2 October 2020).

———. "Defense Department Briefing, November 29, 2005," https://search-proquest-com.libproxy.temple.edu/docview/190027264?accountid=14270&rfr_id=info%3Axri%2Fsid%3Aprimo (accessed, 27 October 2020).

———. "Combatant Commands." www.defense.gov//Our-Story/Combatant-Commands/ (accessed, 13 April 2021).

———. "Department of Defense Background Briefing on Enhancing Counter-ISIL Operations, October 30, 2015." https://www.defense.gov/Newsroom/Transcripts/Transcript/Article/626814/department-of-defense-background-briefing-on-enhancing-counter-isil-operations/ (accessed, 19 December 2020).

Department of Defense. Defense Casualty Analysis System. "U.S. Military Casualties - Operation Iraqi Freedom (OIF) Casualty Summary by Month and

Service." https://dcas.dmdc.osd.mil/dcas/pages/report_oif_month.xht ml (accessed, 21 September 2020).

———. "U.S. Military Casualties - Operation Iraqi Freedom (OIF) Casualty Summary by Month and Service." https://dcas.dmdc.osd.mil/dcas/pages/re port_oif_month.xhtml (accessed, 26 October 2020).

———. "U.S. Military Casualties – Operation Enduring Freedom (OEF) Casualty Summary by Month and Service." https://dcas.dmdc.osd.mil/ dcas/pages/report_oef_month.xhtml (accessed, 26 October 2020).

Department of Defense Historical Office. "Robert S. McNamara: John F. Kennedy/Lyndon Johnson Administration." https://history.defense.gov/Mul timedia/Biographies/Article-View/Article/571271/robert-s-mcnamara/ (accessed, 3 May 2020).

Department of Defense. Office of the Secretary of Defense, Director of Operational Test and Evaluation, "FY 2002 Annual Report." https://www.do te.mil/Portals/97/pub/reports/FY2002/other/2002DOTEAnnualRpt.pdf?ver =2019-11-07-180204-860 (accessed, 4 January 2020).

Detroit Historical Society, "Encyclopedia of Detroit: Ford Highland Park Plant," https://detroithistorical.org/learn/encyclopedia-of-detroit/ford-highland-p ark-plant (accessed, 8 April 2020).

Deutsche Welt, "Another year of 'Enduring Freedom' for German Troops." June 11, 2002, https://www.dw.com/en/another-year-of-enduring-freedom-for-g erman-troops/a-671240 (accessed, 30 September 2020).

Discover Southeast Arizona. "The Heliograph in the Apache Wars." http://ww w.discoverseaz.com/History/Heliograph.html (accessed, 22 April 2020).

Eighth United States Army. Korean Service Corps Battalion, "History." https:// ksc.korea.army.mil/ (accessed, 21 January 2020).

Enacademic.com. "Multi-National Division South." https://enacademic.com /dic.nsf/enwiki/8773315 (accessed, 30 October 2020).

Federal Information & News Dispatch. "Defense Department Outlines Links between Iraq, Iran: U.S. Military Intelligence Official Discusses Iran's Strategy in Neighboring Iraq," http://libproxy.temple.edu/login?url=https://www-proquest-com.libproxy.temple.edu/docview/189975438?accountid=14270 (accessed, 5 November 2020).

Filipinas Heritage Library. "The Moro to the Spanish Colonizers." https://ww w.filipinaslibrary.org.ph/articles/the-moro-to-the-spanish-colonizers/ (accessed, 13 December 2020).

Friedman, Herbert A. "Psychological Operations in Afghanistan, Operation Enduring Freedom, 2001." https://www.psywar.org/content/afghanistan02 (accessed, 6 October 2020).

Fulton, Williams S. "In Combat—Mobile Riverine Force." https://www.mrfa.org/wp-content/uploads/2016/06/General.Fulton.MRF_.pdf (accessed, 7 May 2020).

Garamone, Jim. "America Launches Strikes Against Al Qaeda, Taliban," American Forces Press Service, 7 October 2001, https://archive.defense.gov/news/newsarticle.aspx?id=44680 (accessed, 13 September 2020)

————. "Aircraft, Missiles Hammer Terrorists," American Forces Press Service, 7 October 2001, https://archive.defense.gov/news/newsarticle.aspx?id=44679 (accessed, 13 September 2020).

————. "U.S. Crews Assault Al Qaeda, Taliban Support Structures," Armed Forces Press Service, 9 October 2020. https://archive.defense.gov/news/newsarticle.aspx?id=44666 (accessed, 13 September 2020).

————. "Washington Provides Food, Medicine to Afghans." Armed Forces Press Service, 7 October 2001. https://archive.defense.gov/news/newsarticle.aspx?id=44677 (accessed, 13 September 2020).

————. "What Are Humanitarian Daily Rations?" Armed Forces Press Service, October 7, 2001. https://archive.defense.gov/news/newsarticle.aspx?id=44678 (accessed, 13 September 2020).

————. "U.S. Advisors Aid Northern Alliance, Build Cohesion," Armed Forces Press Service, 6 November 2001. https://archive.defense.gov/news/newsarticle.aspx?id=44481(accessed, 13 September 2020).

————. "Wolfowitz Shares Special Forces' Afghanistan Dispatches," Armed Forces Press Service, 15 November 2001. https://archive.defense.gov/news/newsarticle.aspx?id=44448 (accessed, 13 September 2020).

————. "Northern Alliance, U.S. Special Forces at Gates of Kabul." Armed Forces Press Service, 13 November 2001. https://archive.defense.gov/news/newsarticle.aspx?id=44461 (accessed, 13 September 2020).

————. "Taliban Surrendering Kandahar, U.S. Forces in Firefights." Armed Forces Press Service, 7 December 2001. https://archive.defense.gov/news/newsarticle.aspx?id=44387 (accessed, 13 September 2020).

————. "Anaconda Over, But Operations Continue in Afghanistan." Armed Forces Press Service, 18 March 2002, https://archive.defense.gov/news/newsarticle.aspx?id=44239 (accessed, 29 September 2020).

Garamone, Jim and Linda Kozaryn. "Still Plenty of Al Qaida in Afghanistan." Armed Forces Press Service, December 17, 2001. https://archive.defense.gov/news/newsarticle.aspx?id=44348 (accessed, 30 September 2020).

Gilmore, Gerry J. "Air Force Fliers Continue Food Drop Operations." Armed Forces Press Service, 12 October 2001. https://archive.defense.gov/news/newsarticle.aspx?id=44649 (accessed, 13 September 2020).

————. "Northern Alliance Progressing With U.S. Help, Pace Says." Armed Forces Press Service, November 7, 2001. https://archive.defense.gov/news/newsarticle.aspx?id=44478 (accessed, 13 September 2020).

————. "U.S. Troops Describe All-Day Shahikhot Battle." https://archive.defense.gov/news/newsarticle.aspx?id=44277 (accessed, 29 September 2020).

GlobalSecurity.org. "M114 155mm Howitzer." https://www.globalsecurity.org/military/systems/ground/m114.html (accessed, 6 May 2020).

————. "Operation Support Hope." https://www.globalsecurity.org/military/ops/support_hope.html; (accessed, 16 December 2020).

————. "Task Force Falcon/IFOR/SFOR Deployments." https://www.globalsecurity.org/military/ops/sfor-deploy.html; (accessed, 1 October 2018).

————. "US Army 1990 Divisions." https://www.globalsecurity.org/military/agency/army/division-90.html; (accessed, 26 September 2019).

Government of the Czech Republic, Ministry of Defence and Armed Forces. "Foreign Missions of the 601st Special Forces Group." http://www.601skss.cz/english/missions.html (accessed, 1 October 2020).

Government of France. "Afghanistan: Bilan Mitigé pour les Forces Spéciales Françaises," http://secretdefense.blogs.liberation.fr/2007/12/22/afghanistan-bil/ (accessed, 30 September 2020).

Government of the Netherlands, Ministry of Defense. "Dutch Contribution to Operation Enduring Freedom in Afghanistan," https://english.defensie.nl/topics/historical-missions/mission-overview/2001/operation-enduring-freedom/dutch-contribution (accessed, 30 September 2020).

Government of Romania, Ministry of National Defense "Historical Landmarks." https://english.mapn.ro/historical/index.php (accessed, 30 September 2020).

Grumman Aircraft Engineering Corporation." The Mohawk Surveillance System." https://dk-media.s3.amazonaws.com/AA/AA/moomoo/downloads/288676/1963-06_Mohawk_Surveilance.pdf (accessed, 28 May 2020).

Hall, John. "Combat Parachute into Bashur, Iraq—Fifteenth Anniversary." https://www.dvidshub.net/news/270671/combat-parachute-assault-into-bashur-iraq-15th-anniversary (accessed, 22 September 2020).

Hamm, Walter. "Veterans Remember Historic 2nd BCT Deployment," https://go-gale-com.libproxy.temple.edu/ps/i.do?p=AONE&u=temple_main&id=GALE|A368658350&v=2.1&it=r (accessed, 29 October 2020);

Haskell, Bob. "Guard Enhanced Brigade Prepares for Afghan Duty, Armed Forces Press Service, November 7, 2002, https://archive.defense.gov/news/newsarticle.aspx?id=27829 (accessed, 2 October 2020).

Hess, Pam. "IEDs Fewer but more Deadly in South Iraq," *UPI Security & Terrorism*, November 4, 2005, https://link.gale.com/apps/doc/A138322666/ITOF?u=temple_main&sid=ITOF&xid=39ea0721 (accessed 27 October 2020).

Howard, Christopher E. "This Vest May Save Your Life: U.S. Army Body Armor from World War II to the Present." https://arsof-history.org/articles/19oct_body_armor_page_1.html (accessed, 14 April 2020).

ICasualties.org, "Iraq Fatalities Total: 4092." http://www.icasualties.org/App/Fatalities?page=1&rows=0 (accessed, 21 September 2020).

Idc.com, "Radio Wave Propagation," http://www.idc-online.com/technical_references/pdfs/electronic_engineering/Radio_Waves_Propagation.pdf (accessed, 23 April 2020).

Incorporating Radio-Electronics.com. "Spark Gap Transmitter History." https://www.electronics-notes.com/articles/history/spark-gap-transmitters/history.php (accessed, 23 April 2020).

International Committee of the Red Cross. "Geneva Convention Relative to the Treatment of Prisoners of War, Article 4: Prisoners of War." https://www.icrc.org/en/doc/assets/files/publications/icrc-002-0173.pdf (accessed, 6 July 2020).

Jenkins, Brian M. "Is Al Qaida's Internet Strategy Working? Before the Committee on Homeland Security Subcommittee on Counterterrorism and Intelligence United States House of Representatives December 6, 2001." file://C:/Users/fogan/Downloads/ RAND_CT371.pdf (accessed, 19 October 2020).

Jewish Virtual Library, "Paul Wolfowitz (1943-)." https://www.jewishvirtuallibrary.org/paul-wolfowitz (accessed, 16 September 2020).

Kaplan, Fred. "Err War, The Army Buries Its Mistakes," https://slate.com/news-and-politics/2003/10/the-army-buries-its-mistakes.html (accessed, 30 October 2020).

Kozaryn, Linda. "Operation Anaconda is Winding Down," https://archive.defense.gov/news/newsarticle.aspx?id=44258 (accessed, 29 September 2020).

Lombardi, Ray. "Do you ever wonder where your role models end up? I do. Here are a few I tracked down." https://karlsendlessstories.blogspot.com/2008/11/do-you-ever-wonder-where-your-role.html (accessed, 13 April 2020).

M2 Presswire. "Briefing from Mosul, Iraq, on the Involvement of the 101st Airborne Division in the Conflict and in Post-War Stabilization Efforts," 14 May 2003. https://link.gale.com/apps/doc/A101702566/ITOF?u=temple_main&sid=ITOF&xid=b1c0 9456 (accessed, 30 October 2020).

———. "NATO: Poland Assumes Command of Multi-National Division in Iraq with NATO Support," 3 September 2003. http://libproxy.temple.edu/login?url=https://www-proquest-com.libproxy.temple.edu/docview/444657714?accountid=14270 (accessed, 30 October 2020).

———. "US DOD: Army Announces Mobilization of Two National Guard Brigades," 29 September 2003. http://libproxy.temple.edu/login?url=https://www-proquest-com.libproxy.temple.edu/docview/444647496?accountid=14270 (accessed, 31 October 2020).

———————. "Army Announces the Alert of Two National Guard Enhanced Separate Brigades" 28 July 2003. http://libproxy.temple.edu/login?url=https://www-proquest-com.libproxy.temple.edu/docview/445780550?accountid=14270 (accessed, 2 November 2020.

Marine Corps News. "155th BCT Hits the Ground Running," 19 February 2005. https://www.globalsecurity.org/military/library/news/2005/02/mil-050221-usmc01.htm (accessed, 2 November 2020).

McDonough, James. "Change Can Be Hard, Irrelevance is Easy." https://ivmf.syracuse.edu/2017/10/06/change-can-hard-irrelevance-easy/#:~:text=Retired%20U.S.%20Army%20General%20Eric%20Shinseki%20in%20December,change,%20you're%20going%20to%20dislike%20irrelevance%20even%20more" (accessed, 4 January 2020).

McGrath, Roger D. "The Patton You Didn't Know." *Chronicles*, February 2013, https://link-gale-com.libproxy.temple.edu/apps/doc/A328282473/ITOF?u=temple_main&sid=ITOF&xid=7c568f2d (accessed, 25 August 2020).

McKinley, William J. "Benevolent Assimilation Proclamation of December 21, 1898." http://www.msc.edu.ph/centennial/benevolent.html (accessed, 22 June 2020).

Mills, Nicolaus. "The General Who Understood Iraq from the Start." *Dissent*, April 8, 2008, https://www.dissentmagazine.org/online_articles/the-general-who-understood-iraq-from-the-start (accessed, 16 September 2020).

Mindanao (PH) State University. The Philippine-American War Documents, "Pedro Paterno's Proclamation of War. http://www.msc.edu.ph/centennial/pa990602.html (accessed, 10 June 2020).

Minnesota Historical Society. "The U.S.-Dakota War of 1862." http://www.mnhs.org/fortsnelling/learn/us-dakota-war (accessed, 16 March 2020).

Mitchell, Randy. "Afghan Food Drop Underscores Bush's Humanitarian Pledge." Armed Forces Press Service, October 9, 2001. https://archive.defense.gov/news/newsarticle.aspx?id=44669 (accessed, 13 September 2020).

Mobile River Task Force Association. "Mobile River Task Force 117." https://www.mrfa.org/us-navy/us-navy-mobile-riverine-force/task-force-117/ (accessed, 9 May 2020).

Multi-National Force and Observers. "About Us." https://www.mfo.org/about-us (accessed, 16 December 2020).

Naranjo, Roberto. "Partisan Warfare in the American Civil War." https://ehistory.osu.edu/articles/partisan-warfare-american-civil-war (accessed, 19 March 2020).

National Guard Bureau. "About the Army National Guard, History Since September 11, 2001." http://www.nationalguard.mil/About-the-Guard/Army-National-Guard/ (accessed, 9 January 2021.

National Public Radio. "Interview: Greg Mitchell Discusses Newspapers' Editorial Opinions About the War in Iraq." https://infoweb-newsbank-com.libproxy.temple.edu/apps/news/documentview?p=AWNB&t=pubname%3AMEN2M%21Morning%2BEdition%2B%255BNPR%255D%2B%2528USA%2529/year%3A2003%212003/mody%3A0131%21January%2B31&action=browse&format=text&docref=news/1577CFBD36BA1DD8 (accessed, 14 September 2020).

National Review. "Senior American Military Officials Gave a Briefing in Baghdad on Iranian Weapons that have Killed 170 Americans and Wounded 620 Since 2004: EFPs, or Explosively Formed Penetrators, that Release Armor-Cutting Molten Copper Balls," March 5, 2007. https://link.gale.com/apps/doc/A165364444/AONE?u=temple_main&sid=AONE&xid=18dd701d (accessed, 27 October 2020).

National Special Forces Green Beret Memorial. "Major General (Ret.) James A. Guest," https://specialforcesgreenberetmemorial.org/major-general-ret-james-guest/ (accessed, 28 May 2019).

North Atlantic Treaty Organization (NATO), Peace Support Operations in Bosnia and Herzegovina. https://www.nato.int/cps/en/natohq/topics_52 122.htm?selectedLocale=en (accessed, 3 December 2018).

————. NATO Kosovo Force, Enduring Commitment: About Us. https://jfcnaples.nato.int/kfor/about-us/ (accessed, 3 December 2018).

Obama, Barack. "Address to the Nation by the President," December 6, 2015. https://www.obamawhitehouse.archives.gov/the-press-office/2015/12/06/address-nation-president, (accessed, 16 December 2020).

O'Hanlon, "Shinseki vs. Wolfowitz," March 4, 2003. https://www.brookings.edu/opinions/shinseki-vs-wolfowitz/ (accessed, 14 September 2020).

OneTubeRadio.com, "Army Signal Corps Field Buzzer, 1916." http://onetuberadio.com/2014/09/21/army-signal-corps-field-buzzer-1916/ (accessed, 23 April 2020).

Pease, Lute. Cartoon: "Still Coming," https://www.loc.gov/pictures/item/2016683947/ (accessed, 29 August 2020).

Perito, Robert. The U.S. Experience with Provincial Reconstruction Teams in Afghanistan: Lessons Identified." https://www.usip.org/publications/2005/10/us-experience-provincial-reconstruction-teams-afghanistan-lessons-identified (accessed, 2 October 2020).

Political Transcript Wire. "Major General Peter W. Chiarelli Holds A Defense Department News Briefing Via Teleconference From Iraq," 5 January 2005. http://libproxy.temple.edu/login?url=https://www-proquest-com.libproxy.temple.edu/docview/467188216?accountid=14270 (accessed, 27 October 2020).

———. "U.S. Representative Peter Hoekstra (R-MI) Holds a Hearing on Al Qaida's use of the Internet for Strategic Communications." http://libproxy.temple.edu/login?url=https://www-proquest-com.libproxy.temple.edu/wire-feeds/u-s-representative-peter-hoekstra-r-mi-holds/docview/467170325/se-2?accountid=14270 (accessed, 13 April 2021).

PR Newswire. "Army Issues Notice of Crusader Contract Cancellation; United Defense Awarded Contract to Develop New Cannon Artillery System for the U.S. Army," 9 August 2002. http://libproxy.temple.edu/login?url=https://www-proquest-com.libproxy.temple.edu/wire-feeds/army-issues-notice-crusader-contract-cancellation/docview/450197373/se-2?accountid=14270 (accessed, 17 March 2021).

———. "Northrop Grumman Press Statement on Cancellation Announcement of RAH-66 Comanche Helicopter Program," 23 February 2004. https://link.gale.com/apps/doc/A131536326/AONE?u=temple_main&sid=AONE&xid=b2146bca (accessed, 22 January 2021).

Republic of Lithuania, Ministry of National Defense, "Lithuanian Special Force Troops Returned Home Upon Completing Mission in Afghanistan," http://kam.lt/en/news_1098/news_archives/news_archive_2004/news_archive_2004_-_12/lithuanian_special_force_troops_returned_home_upon_completing_mission_in_afghanistan.html?pbck=0 (accessed, 1 October 2020).

Rhem, Kathleen. "Two Soldiers Killed, Special Forces Assault Taliban Sites." Armed Forces Press Service, October 20, 2001. https://archive.defense.gov/news/newsarticle.aspx?id=44628 (accessed, 13 September 2020).

———. "Third U.S. Servicemember Succumbs to 'Friendly Fire' Wounds," Armed Forces Press Service, December 5, 2001. https://archive.defense.gov/news/newsarticle.aspx?id=44393 (accessed, 13 September 2020).

Roosevelt, Ann. "Aerostat to Guard Against Shoulder Fired Missiles, Mortars," *C4I News*, 19 February 2004. http://libproxy.temple.edu/login?url=https://www-proquest-com.libproxy.temple.edu/docview/232542058?accountid=14270 (accessed, 2 November 2020).

Rogers, William A. Cartoon: "The Punitive Expedition." https://www.loc.gov/pictures/resource/cph.3c30777/ (accessed, 29 August 2020).

Rumsfeld Foundation. "Donald Rumsfeld: Founder and President." https://www.rumsfeldfoundation.org/about/staff/donald-rumsfeld (accessed, 16 September 2020).

Sample, Doug. "DoD Joint Task Force Making Progress Against IED Threat." https://archive.defense.gov/news/newsarticle.aspx?id=31400 (accessed, 12 June 2019).

Smith, Bolling W. and William C. Gaines, "Coast Artillery Organization: A Brief Overview." http://cdsg.org/wp-content/uploads/pdfs/FORTS/CACunits/CACorg2011.pdf (accessed 20 October 2023).

SOFREP. "Polish SOF Reported in Iraq… Again." https://sofrep.com/news/polish-sof-iraq/ (accessed, 30 September 2020).

Special Dictionary. "Kurdish Proverbs." http://www.special-dictionary.com/proverbs/source/k/kurdish_proverb/#:~:text=%20154%20Kurdish%20Proverbs%20%201%20A%20camel,good%20companion%20shortens%20the%20longest%20road.%20More%20 (accessed, 22 July 2020).

Selective Service System. "The Vietnam Lotteries." https://www.sss.gov/About/History-And-Records/lotter1; (accessed, 13 June 2018).

Terrorism Research and Analysis Consortium, "Jaysh al Mahdi." https://www.trackingterrorism.org/group/jaysh-al-mahdi (accessed, 29 October 2020).

———. "Badr Corps." https://www.trackingterrorism.org/group/badr-corps (accessed, 29 October 2020).

Texas A&M University Corps of Cadets. "Blake Connors, Colonel, United States Army, Professor of Military Science." https://corps.tamu.edu/staff/pms/ (accessed, 2 January 2020).

United Nations. "Security Council Resolution 1267 (1999) of 15 October 1999" https://undocs.org/S/RES/1267(1999) (accessed, 20 July 2020).

———. "Security Council Resolution 688 (1991), of 5 April 1991." https://undocs.org/en/S/RES/688(1991) (accessed, 21 July 2020).

———. "Security Council Resolution (1994) 949, of 15 October 1994." http://unscr.com/en/resolutions/doc/949 (accessed, 21 July 2020).

United States Air Force. Beale Air Force Base, "First Reconnaissance Squadron." https://www.beale.af.mil/Library/Fact-Sheets/Display/Article/279940/1st-reconnaissance-squadron/ (accessed, 27 March 2020).

———. "Major General Benjamin Delahauf Foulois." https://www.af.mil/About-Us/Biographies/Display/Article/107091/major-general-benjamin-delahauf-foulois/ (accessed, 10 August 2020).

———. "2003-Operation Iraqi Freedom," https://www.afhistory.af.mil/FAQs/Fact-Sheets/Article/458942/operation-iraqi-freedom/ (accessed, 29 September 2020).

United States Army Aviation Center of Excellence. "1st Aviation Brigade." http://www.rucker.army.mil/usaace/1ab/ (accessed, 24 October 2018).

U.S. Army Europe and Africa, "Mission and History." https://www.europeafrica.army.mil/Mission-History/ (accessed, 10 December 2020).

———. "ATLANTIC RESOLVE." https://www.europeafrica.army.mil/AtlanticResolve/ (accessed, 10 December 2020).

———. "Press Release: Ceremony to Establish V Corps Headquarters (Forward)," 15 November 2020. https://www.europeafricat.army.mil/ArticleViewPressRelease/Article/2420593/press-release-ceremony-to-establish-v-corps-headquarters-forward/ (accessed, 10 December 2020).

United States Army Forces Command. "About FORSCOM," https://www.forscom.army.mil/Pages/About (accessed, 13 April 2020).

United States Army, General Officer Management Office. "Lieutenant General Daniel P. Bolger." https://www.gomo.army.mil/ext/portal/Officer/Resumes.aspx?Ltr=L&Type=Retired; (accessed, 21 December 2019); Bolger to Author, September 9, 2019.

———. "Lieutenant General James L. Lovelace." https://www.gomo.army.mil/ext/portal/Officer/Resumes.aspx?Ltr=L&Type=Retired (accessed, 21 December 2019).

———. "General David D. McKiernan," https://www.gomo.army.mil/ext/portal/Officer/Resumes.aspx?Ltr=L&Type=Retired (accessed, 8 December 2020).

———. "General James D. Thurman." https://www.gomo.army.mil/ext/portal/Officer/Resumes.aspx?Ltr=L&Type=Retired, (accessed, 21 December 2019).

———. "General William S. Wallace." https://www.gomo.army.mil/ext/portal/Officer/Resumes.aspx?Ltr=L&Type=Retired (accessed, 8 December 2020).

United States Army Materiel Command. "The Army's Materiel Integrator," https://www.amc.army.mil/ (accessed, 13 April 2020).

United States Army Pacific, "Major General Steven Silvasy, Jr." https://www.usarpac.army.mil/history2/cg_silvasy.asp (accessed, 28 May 2019).

United States Army Training and Doctrine Command. "About Us." https://www.tradoc.army.mil/About-Us/ (accessed, 13 April 2020).

United States Army Transportation Museum. "War on Wheels: The Mexican Punitive Expedition-1916." https://web.archive.org/web/20041017154729/http://www.transchool.eustis.army.mil/Museum/MexWar.htm (accessed, 23 April 2020).

United States Central Command. "CENTCOM Coalition Nations." https://www.centcom.mil/AREA-OF-RESPONSIBILITY/CENTCOM-COALITION/ (accessed, 5 August 2020).

———. FOIA request 07-0154, "2004-2007 Significant Activity Reports." https://www3.centcom.mil/FOIALibrary/Search (accessed, 27 October 2020).

———. U.S. Central Command News Release 04-02-08, February 5, 2004. https://www.globalsecurity.org/military/library/news/2004/02/mil-040205-centcom01.htm (accessed, 3 November 2020).

———. "Combined Joint Task Force-Operation Inherent Resolve (CJTF-OIR)." www.centcom.mil/OPERATIONS-AND-EXERCISES/OPERATION-INHERENT-RESOLVE/ (accessed, 11 April 2021).

United States Naval War College. *Maritime Advanced Warfighting School.* https://usnwc.edu/college-of-naval-command-and-staff/Additional-Academic-Opportunities/Maritime-Advanced-Warfighting-School (accessed, 22 June 2018).

United States Navy. Naval History and Heritage Command. "Battle of Manila Bay, 1 May 1898." https://www.history.navy.mil/research/library/online-reading-room/title-list-alphabetically/b/battle-of-manila-bay-1-may-1898.html (accessed, 10 June 2020).

United States Special Operations Command. "Introduction to U.S. Special Operations Command, 2006." https://www.socom.mil/FactBook/2006%20Fact%20Sheet.pdf (accessed, 16 December 2020).

———."Special Operations Aviation History." https://www.soc.mil/USASOAC/USASOAChomepage.html (accessed, 21 April 2020).

———. "Special Operations Forces Medal of Honor Recipients." https://www.socom.mil/hall-of-heroes/medal-of-honor (accessed, 26 October 2018).

———. "US Special Operations Command," in *USSOCOM Fact Book: Introduction to US Special Operations Command* (Tampa: US Special Operations Command, 2006). https://www.socom.mil/FactBook/2006%20Fact%20Sheet.pdf (accessed, 26 July 2018).

———. "Special Operations Command Factbook, 2010," https://www.socom.mil/FactBook/2010%20Fact%20Book.pdf (accessed, 16 December 2020).

United States State Department. United States Embassy Vietnam (South), *Major Religious and Armed Groups 1954,* http://libproxy.temple.edu/login?url=https://search-proquest-com.libproxy.temple.edu/docview/1679150229?accountid=14270 (accessed, 30 April 2020);

———. "U.S. Relations with Kuwait." https://www.state.gov/r/pa/ei/bgn/35876.htm; (accessed, 29 January 2018); *Washington Post,* January 29, 1992.

———. "Highlights in the History of U.S. Relations with Russia, 1780-June 2006," https://www.state.gov/p/eur/ci/rs/200years/c30273.htm#fall_communism (accessed, 29 January 2018).

———. "Humanitarian Aid to the Afghan People." https://web.archive.org/web/20070714002014/https://usinfo.state.gov/is/Archive_Index/Humanitarian_Aid_to_the_Afghan_People-4072c45bad525.html (accessed, 12 September 2020).

University of Quebec at Montreal. "Military Assistance Advisory Group, Indochina." http://indochine.uqam.ca/en/historical-dictionary/890-military-assistance-advisory-group-maag-indochina.html (accessed, 25 April 2020).

Vijh, Surekha. "Monuments of Destruction." *Insight on the News*, April 23, 2001, 24, https://link-gale-com.libproxy.temple.edu/apps/doc/A74011909/AONE?u=temple_main&sid=AONE&xid=1a7efe4b (accessed, 18 July 2020).

Washington National Guard, "History of the 81st Brigade Combat Team." https://mil.wa.gov/about-the-81st (accessed, 2 November 2020).

Watts, Clinton. "Beyond Iraq and Afghanistan: What Foreign Fighter Data Reveals About the Future of Terrorism." https://smallwarsjournal.com/blog/journal/docs-temp/49-watts.pdf (accessed, 16 March 2020).

West Point Center for Oral History, "Be Ready To Adjust Your Assumptions: Reflections On Adaptive Leadership From The 57th Superintendent, Franklin "Buster" Hagenbeck. https://www.westpointcoh.org/interviews/ be-ready-to-adjust-your-assumptions-reflections-on-adaptive-leadership-from-the-57th-superintendent (accessed, 29 September 2020).

White, R. W. "History of the 45th Infantry Brigade." https://signaleer.wordpress.com/2011/08/03/the-history-of-the-45th-infantry-brigade/ (accessed, 2 October 2020).

WrightBrothers.org. "Just the Facts: 1912-1913 Wright Model C." http://www.wright-brothers.org/Information_Desk/Just_the_Facts/Airplanes/Modelc.html (accessed, 27 March 2020).

Personal Papers and Archive Collections

Arsenault, Peter. This includes correspondence related to the battle of Objective BAKERSFIELD.

Gen. Casey, George W., Jr. National Defense University Library, Ft. Lesley J. McNair, Washington, DC. This collection includes Gen. Casey's papers from his service as the commanding general of Multi-National Forces Iraq and his tenure as the Chief of Staff, United States Army.

Brig. Gen. Hughes, Christopher. Gen. Hughes' collection includes papers pertaining to his service with the Joint Staff's Combatting Terrorism Technical Support Office.

Johnson, Lyndon B. LBJ Presidential Library, Austin, TX. The archive holds declassified National Security Council memorandums and transcripts of meetings, calls and interviews with selected officials pertaining to the war in Southeast Asia.

Brig. Gen. Schenk, Donald F. U.S. Army Tank, Automotive and Armaments Command, Warren, MI. This collection contains Gen. Schenk's papers from his time as the Program Manager, Future Combat System.

U.S. Army Training and Doctrine Command Archive, Ft. Eustis, VA. The archives holdings include all referenced material pertaining to Operation STEADFAST.

U.S. Army Asymmetric Warfare Group. This collection contained documents pertaining to the unit's formation and years of service. With the Asymmetric Warfare Group's deactivation, these files were dispersed to different holders with some going to the U.S. Army Training and Doctrine Command at Ft. Eustis, VA.

U.S. Army Training and Doctrine Command Archive, Ft. Eustis, VA. The archives holdings include all referenced material pertaining to Operation STEADFAST.

Further Reading

Bell, William G. and Carl Cocke. *Department of the Army Historical Summary, Fiscal Year 1971.* Washington, DC: Center of Military History, 1977.

Birtle, Andrew J. *U.S. Army Counterinsurgency and Contingency Operations Doctrine, 1942-1976.* Washington, DC: Center of Military History, 2006.

Bolkcom, Christopher. *Congressional Research Service--Army Aviation: The RAH-66 Comanche Helicopter Issue.* Washington, DC: Congressional Research Service, 2003.

Bradley, John H. and Dice, Jack W. *The West Point Military History Series: The Second World War—Asia and the Pacific.* Wayne, NY: Avery Publishing Group, 1989.

Briscoe, Charles H., Richard L. Kiper, James A. Schroeder, and Kalev I. Sepp. *Weapon of Choice: ARSOF in Afghanistan.* Ft. Leavenworth, KS: Combat Studies Institute, 2003.

Carter, Donald A. *Forging the Shield: The U.S. Army in Europe, 1951-1962.* Washington, DC: Center of Military History, 2015.

Clark, J.P. *Preparing for War: The Emergence of the Modern U.S. Army, 1815-1917.* Cambridge: Harvard University Press, 2017.

Cloud, David and Greg Jaffe. *The Fourth Star: Four Generals and the Epic Struggle for the Future of the United States Army.* New York: Crown, 2009.

Cocke, Karl E. *Department of the Army Historical Summary, Fiscal Year 1977.* Washington, DC: Center of Military History, 1979.

Cole, Richard H. *Operation Urgent Fury: The Planning and Execution of Joint Operations in Grenada, 12 October-2 November 1983.* Washington, DC: Joint History Office, Office of the Joint Chiefs of Staff, 1997.

Commonwealth of Australia. *Land Warfare Doctrine 1: The Fundamentals of Land power.* Canberra: Commonwealth of Australia, 2014.

Davis, Robert T. *The Challenge of Adaptation: The U.S. Army in the Aftermath of Conflict, 1953-1980.* Ft. Leavenworth, KS: Combat Studies Institute, 2008.

Department of the Army. *Army Doctrinal Reference Publication 3-0, Operations.* Washington, DC: Headquarters, Department of the Army, 2016.

————. *Field Manual 25-1, Training.* Washington, DC: Headquarters, Department of the Army, 1985.

————. *Field Manual 100-6, Information Operations.* Washington, DC: Headquarters, Department of the Army, August 1996.

Downie, Richard D. *Learning from Conflict: The U.S. Military in Vietnam, El Salvador and the Drug War.* Westport, CT: Praeger, 1998.

Douhet, Giulio. *Command of the Air,* edited by Joseph P. Harahan and Richard H. Kohn. Tuscaloosa: University of Alabama Press, 2009.

Feickert, Andrew. *CRS Report for Congress, U.S. Army's Modular Redesign: Issues for Congress (Updated December 11, 2006).* Washington, DC: Congressional Research Service, 2006.

Fuller, J. F. C. *The Foundations of the Science of War.* London, UK: Hutchinson & Co, LTD: 1926.

Land Warfare Development Centre. *Army Doctrine Publication 71940, Land Operations.* Warminster: Land Warfare Development Center, 2017.

Meilinger, Phillip S., ed. *The Paths of Heaven: The Evolution of Airpower Theory.* Maxwell AFB: The Air University Press, 1997.

Palmer, Robert R. *Reorganization of Ground Troops for Combat, Study No. 8.* Washington, DC: Historical Section, Army Ground Forces, 1946.

Panella, Rae T. *Department of the Army Historical Summary, FY 1975.* Washington, DC: Center of Military History, 1978.

Posen, Barry. *The Sources of Military Doctrine: France, Britain and Germany Between the World Wars.* Ithaca, NY: Ithaca University Press: 1984.

Rosen, Stephen P. *Winning the Next War: Innovation and the Modern Military.* Ithaca: Cornell University Press, 1991.

Senge, Peter M. *The Fifth Discipline.* New York: Doubleday, 1994.

Gen. Sullivan, Gordon R. and Col. James M. Dubik. *Envisioning Future Warfare.* Ft. Leavenworth, KS: United States Army Combat Studies Institute, 1995.

Gen. Taylor, Maxwell. *The Uncertain Trumpet.* New York: Harper and Brothers, 1960.

United States Government. *Pub. L. No. 99-433 (1986): Goldwater-Nichols Department of Defense Reorganization Act of 1986.*

————. *Pub. L. No. 109-148 (2005): An Act Making Appropriations for the Department of Defense for the Fiscal Year Ending September 30, 2006, and for other purposes.*

Winton, Harold R. and David R. Mets. *The Challenge of Change: Military Institutions and New Realities, 1918-1941.* Lincoln: University of Nebraska Press, 2000.

Wright, Donald P. *a Different Kind of War: The United States Army in Operation ENDURING FREEDOM, October 2003-September 2005. Ft. Leavenworth, KS: Combat Studies Institute.*

Index

1

2

3

4

5

D

E

F

G

H

I

T

U

V

W

X

Y

Z

www.ingramcontent.com/pod-product-compliance
Lightning Source LLC
LaVergne TN
LVHW010722210926
848485LV00007B/1423/J